During the early 1890s, the 10,000-acre Bragg Ranch was purchased south of Dodge City, Kansas. This cattle drive, circa 1895, included Thomas Bragg (second from left), who owned the ranch in addition to a shoe and leather factory in St. Joseph, Missouri, with his father, James Bragg. Thomas Bragg was the great-grandfather of Dr. James Coffman, professor emeritus of the College of Veterinary Medicine and provost emeritus at Kansas State University. For more on the Bragg family, please see chapter 5.

150 YEARS
OF KANSAS BEEF

150 Years of Kansas Beef logo designed by Mal Hoover, Kansas State University College of Veterinary Medicine

150 Years of
Kansas Beef

Justin Kastner and Blair Tenhouse, Editors

THE
DONNING COMPANY
PUBLISHERS

The Donning Company Publishers
184 Business Park Drive, Suite 206
Virginia Beach, VA 23462

Steve Mull, General Manager
Barbara Buchanan, Office Manager
Pamela Koch and Jan Martin, Editors
Chad Harper Casey, Graphic Designer
Priscilla Odango, Imaging Artist
Cindy Smith, Project Research Coordinator
Tonya Washam, Marketing Specialist
Pamela Engelhard, Marketing Advisor

Neil Hendricks, Project Director

Library of Congress Cataloging-in-Publication Data

150 years of Kansas beef / Justin Kastner and Blair Tenhouse, editors.
p. cm.
Includes bibliographical references.
ISBN 978-1-57864-685-2 (hard cover : alk. paper)
1. Cattle trade—Kansas—History. 2. Beef industry—Kansas—History. 3. Beef cattle—Kansas—History. I. Kastner, Justin. II. Tenhouse, Blair.
HD9433.U5K262 2011
338.1'762009781—dc22

2011012116

Printed in the United States of America at Walsworth Publishing Company

Contents

Acknowledgments

by Justin Kastner and Blair Tenhouse

The genesis of the *150 Years of Kansas Beef* project can be traced back to 2008, when Dr. Dan Thomson, director of the Beef Cattle Institute (BCI) at Kansas State University (KSU), provided initial financial support for a celebratory history project devoted to the Kansas beef industry. As scholars within the *Frontier* program—a historical studies unit in the College of Veterinary Medicine (CVM)—we were honored and humbled to assume responsibility for this important undertaking. Scores of individuals, families, and institutions have contributed to this project. While we acknowledge many of them here, their names, photos, and voices may be found elsewhere in this book and in the audio podcast series available on the BCI website (www.beefcattleinstitute.org).

As the dedication to Jack Vanier signals, we are grateful to the Vanier family for providing the funds necessary for this project. Mrs. Donna Vanier's financial contribution was a surprise gift that honors not only Jack but also Kansas agriculture. Significantly, *150 Years of Kansas Beef* is but one of many KSU projects that have been supported by the Vanier family, and we thank them for their generosity. We also thank Chris Stephens, former CVM development officer and the author of the dedication to Jack Vanier, as well as several other CVM Development Office personnel who have facilitated this project—Director Chris Gruber as well as Joe Montgomery, Diana Sarfani, and Cheri Ubel.

In addition to Dr. Thomson and the BCI, we express gratitude to Dr. M. M. Chengappa, head of the Department of Diagnostic Medicine/Pathobiology (DMP), and Dr. Ralph Richardson, dean of the college, for providing institutional support and encouragement. Several members of their staff provided critical administrative assistance, and we specifically thank Brandy Nowakowski and Wendy Michaels in the DMP main office as well as Lyndse Greenwood, Amanda McDiffett, Yvonne Bachura, and the rest of the capable staff of the CVM business office. Others in the CVM, including graphic designer Mal Hoover, who created the *150 Years of Kansas Beef* logo, have made valuable contributions.

This project—undertaken on behalf of the BCI by the *Frontier* program for the historical studies of border security, food security, and trade policy—involved several research assistants and students. We thank the following BCI and *Frontier* staff and students for providing historical data collection, podcast recording, website management, and other programmatic support during the last three years: Wrenn Pacheco (project coordinator, BCI), Steve Toburen (web developer and student program coordinator, *Frontier*), Gabe Webb (graduate research assistant, *Frontier*), Tiffany Lee, Tiffany Ebert, Brandon Greenwood, Evan Nave, Reagan Domer, and podcast-host extraordinaire Kelsey Rezac. Lest we leave anyone out, we thank the remaining students and staff in the *Frontier* and BCI communities for their support.

There is a biographical summary of the authors and photographers at the back of this book. As editors, we are grateful to each of these contributors for their diligence in helping author *150 Years of Kansas Beef*. In addition to the authors listed, we note the especially valued contributions of individuals and families who willingly shared ranch histories, family photographs, and other paraphernalia showcased in this book. We thank the following for their time and generosity:

Barbara Bragg, H. Hurst Coffman, Dr. Jim Coffman and family; Susan Russell; Lois Schlickau and family; Andy Schuler Jr.; Bill and Kathy Hogue; Galen and Lori Fink; Todd Johnson; George Crenshaw; Dr. Sandy (Steele) Killingsworth, Aaron Killingsworth, and the rest of the Steele and Killingsworth families; Mark Gardiner and family; Tom, Carolyn, Matt, and Amy Perrier and family; and Dr. Bill Stuart. A number of other ranching families were interviewed (and are cited) in chapter 3. Several others contributed advice and encouragement, including Dr. Abbey Nutsch, Kansas State University; Dr. Jim Hoy, director of the Center for Great Plains Studies at Emporia State University; and Gwen and Josh Hoy of the Flying W Ranch in Cedar Point, Kansas.

Many libraries and historical studies institutions supported our work. We thank Lisa Keys and Mary Madden of the Kansas State Historical Society (KSHS) for their timely assistance on different aspects of this Kansas sesquicentennial project, and we express gratitude to several other KSHS staff members who helped us: Barry Worley, Teresa Coble, Susan Forbes, Lin Fredericksen, Benjamin Epps, Robert Garcia, Donna Fulkerson, Hayley Greeson, Lori Roberts, and Sara Keckeisen. We also thank University Archivist Anthony Crawford and the staff of the Department of Special Collections at K-State Libraries; Sarah Biles, Joan Wells, and others at the Fort Worth Stockyards and the North Fort Worth Historical Society; Tonia Overstreet, Celecia Stoup, Bill Williams, and other friends and staff (including Ann and Carol) of the Kiowa County (Oklahoma) Historical Society; Lucy and Paul Baier, owner of the George Grant Villa in Victoria, Kansas; and many others. For extraordinary assistance provided in support of chapter 1, the project team thanks Richard and Marsland Moncrief, Kim Smith, Douglas Harman, and Steve Myers; in innumerable ways, they generously shared with Dr. Ashley Null their insights into the history of the Texas cattle trails to Kansas.

Finally, we thank members of our own families—Doug Tenhouse; Susie, Ian, and Sally Kastner; Jeff and Natasha Bryant; Curtis and Rebecca Kastner; and Gary and Donna Tenhouse—for their encouragement and support.

Justin Kastner and Blair Tenhouse, Editors
January 29, 2011

Dedication

Jack Vanier, Brookville, Kansas. (Source: Vanier Family)

For many Kansans, it is difficult to imagine showcasing the rich history of this great state without acknowledging the social and economic impacts of the beef cattle industry—an industry that has played a central role in the state's progress. You might say the same for one of Kansas's citizens who happens to be an internationally acclaimed cattleman and businessman. Jack Vanier has spent his life diligently working for the advancement of the state of Kansas and its vast agricultural-based industries. His primary focus has always leaned toward beef cattle production—an economic factor that touches the lives of every Kansan.

Jack Vanier's ranching dynasty began on the world-renowned CK Ranch, which was founded in 1933 by his late father, John J. Vanier. The CK Ranch is located in the Smoky Hills region of Saline and Ellsworth Counties, a part of central Kansas that is known to be one of the best cow-calf grazing areas in the country. The CK Ranch consists of approximately 15,000 acres of continuous native pastures and 1,000 acres of tillable, dry-land crop and alfalfa ground. Crops include grain sorghum, oats, and triticale—all for supplementing the cow herd. The CK Ranch was first stocked with steers, and registered Herefords arrived on the ranch in 1936. During the 1950s, the ranch had almost 2,000 head of registered cows, and approximately 1,200 calves were registered each year. For many years, the CK Ranch was the largest breeder of registered Hereford

Mr. and Mrs. John Vanier accepted the championship trophy in 1962 from Jack Sprengle and the entire staff of the American Hereford Association. The trophy was presented that year as a memorial to the late Ray Sprengle, who was a popular AHA fieldman. At the time of the presentation, the temperature was a bitter cold -24°F. (Source: Vanier Family)

cattle in the United States. The Vaniers have exported cattle to Japan, France, South America, South Africa, and Asia. Jack is internationally recognized in the livestock industry for leadership in marketing and for promoting the family's commercial bulls to cattlemen from coast to coast, primarily through the exhibition of carload bulls at the National Western Stock Show in Denver, Colorado, each January. The CK brand has been identified as a Brand of Significant Influence in the livestock industry.

Jack Vanier was born August 6, 1928, in Salina, Kansas. He attended Salina High School and St. John's Military School, which named a residence hall after him in 2007. He completed his undergraduate degree at Kansas State University in 1951. Upon graduation, Jack returned to assist his father with running the CK Ranch and the family's vast ranching and agricultural business entities. This soon opened the door to numerous leadership and philanthropic roles. Jack was active in the family's Gooch Milling Company until it was sold to Archer Daniels Midland (ADM) in 1970. He served on the ADM board from 1978 to 2004.

Many agricultural and business institutions have benefited from Jack Vanier's leadership. He has served as president of the Kansas Livestock Association, president of the American Hereford Association, and member of the board of directors of the National Cattlemen's Beef Association. He has consistently served the state of Kansas through many other civic, state, and national organizations. Beyond his ranching and business interests, Jack is an avid sportsman and a steward of the land; fishing and hunting with friends and colleagues are two of his favorite pastimes.

This painting depicts the CK Ranch in Brookville, Kansas. (Source: Vanier Family)

Jack is married to Donna L. Vanier, and they are the parents of John Vanier, Marty Vanier, and Mary Vanier. As active philanthropists, Jack and Donna have supported many community and Kansas State University projects through the Vanier Family Foundation. The Vaniers have served as honorary chairs for the K-State Changing Lives Campaign. The Vaniers' support of academic programs includes the Jack Vanier Chair of Innovation and Entrepreneurship at Kansas State University, Kansas State University at Salina, the Marianna Kistler Beach Museum of Art, the Terry C. Johnson Center for Basic Cancer Research, and the colleges of Human Ecology, Agriculture, and Veterinary Medicine. Their substantial support for K-State athletics is enshrined in the Vanier Football Complex, and the Vaniers have contributed to the health and wealth of Kansas through such community enterprises as the Colbert Hills Golf Course, the Salina Regional Health Center, and the Donna L. Vanier Children's Center.

The Denver carload show was held in the stockyard alley. This load of CK cattle is shown being considered by the judges *(lower right)*. Standing by the bulls are George Fritz, Les Townsend, and Claude Lecklider. (Source: Vanier Family)

Throughout his life, Jack Vanier has inspired many young people. His unpretentious demeanor is testimony to both a strong character and a genuine interest in the success of those around him. He has a personal sense of being the recipient of many blessings throughout his life and has sought to help others reach their full potential. It is indeed a privilege to honor this great man, who has built a strong foundation for future generations of cattlemen. With pride, we dedicate *150 Years of Kansas Beef* in honor of the hard work and entrepreneurial spirit of Jack Vanier.

Christopher M. Stephens
Formerly, Development Officer, College of Veterinary Medicine, Kansas State University
Currently, Director of Operations, EE Ranches, Inc., Dallas, Texas

Chapter 1
Cowboys, Cowtowns, and Cattle Trails

By Ashley Null

When I asked why my great-grandfather left the comfortably large family farming establishment in western Illinois and ended up homesteading in Logan County, Kansas, the answer was rather succinct, "to better himself." Kansas—the land of opportunity! That sense of optimism was the common possession of the many people who flocked to settle the state after the Civil War. They were drawn to Kansas by their own dreams as well as the concerted advertising campaigns of commercial interests.

Land developers boasted that Kansas was "the glory of the West."[1] Here "comparatively little labor" was required to richly "reward the husbandman for his trouble," and the cattleman routinely experienced "seventy-five to one hundred per cent. per annum" returns.[2] With possibilities as unbounded as the prairie horizon, those looking to better themselves were assured that here was a place "where they could prosperously grow up with the country."[3]

Previous: Flying W Ranch, Cedar Point, Kansas. (Source: Justin Kastner)

Above: (Source: Blair Tenhouse)

Hundreds of thousands of people from all over America and even Europe were convinced. They hitched their hopes for reaching the stars to working hard in Kansas. As a result, the population of Kansas swelled over fourteen times between 1860 and 1890.[4] Of course, as the ads suggested, at the very heart of this rapid development was the cattle industry. At first, Kansas was seen as a convenient middle meeting place between Texan supply and Eastern demand. Then, little by little, the state became home to its own vibrant beef production. The story of how the Kansas cattle industry grew—and helped to establish the state's economic prosperity in the process—is the story of this opening chapter.

Texas Longhorns

Texas longhorns are certainly distinctive.[5] They feature no defining coloring—usually motley and multi-hued. Their lean and lanky frame leaves them looking angular (no matter how well-fed), sway-backed, and bearing mammoth trophy-seized horns—now found mounted on countless walls and cars—that reach up to seven feet from tip to tip. Just standing still, these cattle seem to embody the very essence of that rugged self-confidence and natural pre-eminence which Texans pride themselves on having. It's little wonder then that Pig Bellmont, an English Bulldog, was replaced in 1916 by Bevo the Longhorn as the official mascot for the University of Texas football team. Yet their native state's obviously deep affection for these cattle is not merely because they typify legendary Texan swagger. The defining characteristics of the breed were also crucial to the longhorn's determinative role in the development of both Texas and Kansas history.

Descended from cattle brought over from Spain where the annual summer drought can last from two to five months,

(Source: Corinne Patterson)

longhorns were tough. They could fiercely defend themselves in the wild, for the tips of those famous horns had sharp points capable of penetrating deeply into any challenger's flesh, going in and ripping it thoroughly into pieces as they twisted their heads to pull them out. Possessing long powerful legs and resilient hooves superior to other cattle, they were able to travel long distances and be little worse for wear. They could also forage well enough on their own out on the open range, since they needed only limited water and vegetation. They could even withstand, year after year, the onslaught of ticks that infected lesser breeds of cattle with a disease to which the longhorns had, long ago, developed immunity. In short, survival was in their very DNA.

Because of Castilian hardiness, the Spanish explorer General Alonso de Leon left a bull and a cow at every river as his 1689 expedition crossed north of the Rio Grande. By 1715, a French trader, Louis Juchereau de St. Denis, observed that "the whole country" was covered with cattle by the "thousands."[6] After Mexican independence in 1821, more Anglo settlers emigrated to Texas, bringing with them domestic livestock descended from British breeds. Since range land was unfenced, interbreeding between the Spanish and English cattle was inevitable, and the result was the Texas longhorn. Because of Indian raids and the upheavals associated with the Texan War of Independence in 1836, many originally domestic animals ended up roaming freely as well. Consequently, on the eve of the U.S. Civil War, three kinds of longhorn cattle were foraging on the Texas range: (1) the domestic cattle bearing the brands of their owners; (2) *mestenas* (Spanish for "grazers"), unbranded cattle recently descended from domestic stock and, thus, still tame enough to be easily rounded up; and

(3) *cimarrones* (Spanish for "wild ones"), feral cattle which fiercely resisted being herded.

By 1860 Texans had begun calling unbranded grazing cattle *mavericks*, after Samuel A. Maverick, a signer of the Texas Declaration of Independence and rancher south of San Antonio. In 1847 he reluctantly entered the cattle business when he accepted a herd of four hundred head as payment for a $1,200 debt. However, the foreman he appointed to oversee the operation was too lazy to brand new calves and then let many of them stray as well. When Maverick decided to get out of the cattle business in 1856, he sold his herd to A. Toutant Beauregard, brother of the future Confederate general. Beauregard's cowboys proved far more efficient. They searched throughout South Texas, and whenever they found tame, unbranded cattle on the range, they claimed the animals for their boss as once being "Maverick's."[7]

During the Civil War, the northern troops cut Texas off from the traditional markets for its livestock.[8] Moreover, most of those responsible for the annual round-ups went off to fight. With no one to castrate young bulls and brand calves, the number of cattle on the open range greatly increased, the vast majority of them being mavericks. As the defeated veterans gradually straggled back home, they found little hope for a new beginning. The state's economy was in shambles and their herds dispersed. Of course, it was true that there were literally millions of mostly unbranded cattle waiting to be claimed on the open range. But what of it? Such an overabundance of mavericks meant that cattle in Texas would bring only $3 (maybe $4) a head—hardly the foundation for a future fortune. Those looking to better themselves had to think outside of the box—in fact, outside of Texas itself. The Civil War had severely strained the supply of beef in the North. At the meatpacking centers of St. Louis and Chicago, good cattle were selling for as much as $30 or even $40 a head. At that price, a man could certainly rebuild his family's war-devastated finances and, perhaps, even become fabulously prosperous. The only challenge was getting Texas cattle to eastern buyers.

First Attempts

What a challenge it was![9] No railroads yet knit secessionist Texas with Yankee markets in the Midwest. Because longhorns were tough, before the war adventuresome Texas cattlemen had simply driven them to St. Louis and even Chicago. They followed the Shawnee Trail from South Texas through Austin, Waco, and Dallas. They crossed the Red River at Rock Bluff, near Preston, where a natural formation provided a stone chute which funneled the cattle. Afterwards, the herds headed northeast, crossing the Arkansas River near Fort Gibson,

North Fork of the Red River in present-day southwest Oklahoma near the Great Western Trail. (Source: Justin Kastner)

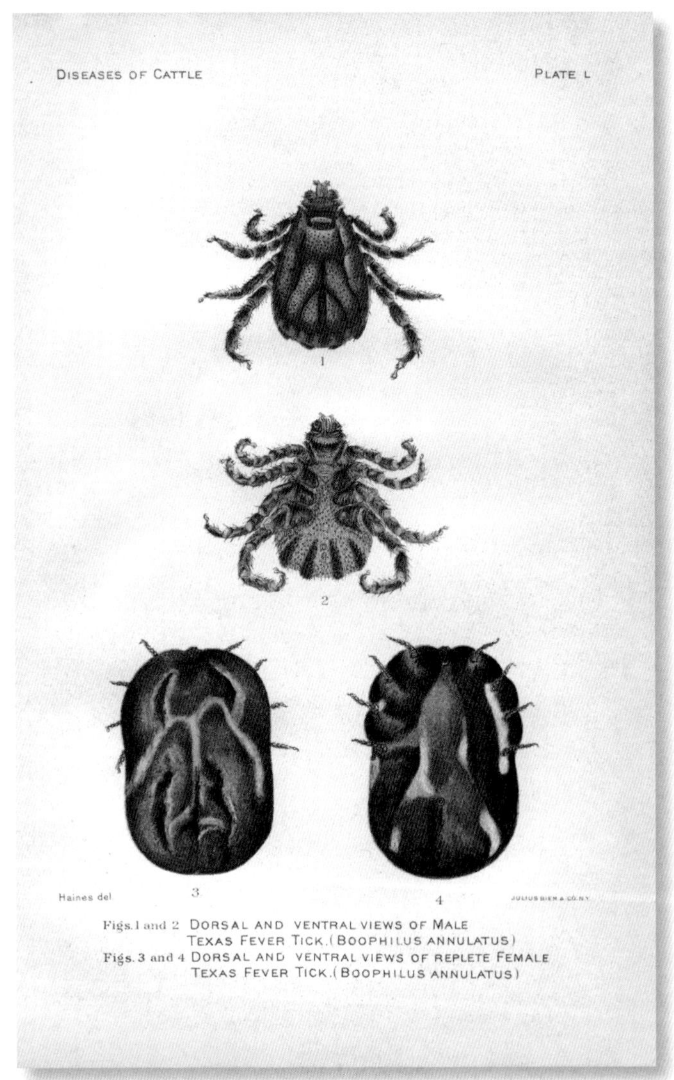

Male and female Texas Fever–transmitting ticks. (Source: U.S. Department of Agriculture, Bureau Animal Industry; *Special Report on Diseases of Cattle*, 1904)

and then followed the rivers toward the intersection of the Kansas, Missouri, and Indian Territory borders. Here the trail divided. The eastern fork headed across southern Missouri to St. Louis. The central branch ran through Sedalia, crossed the Mississippi at Quincy, and then headed for Chicago. Finally, the western option followed the rivers to Baxter Springs, Kansas, and then along the Kansas-Missouri border to Kansas City. With the high demand for stock by new emigrants wishing to settle, as well as the constant need to provide beef for western army outposts and Indian reservations, Kansas City was by 1855 a dominant market for cattle.

Yet even before the outbreak of Civil War abruptly halted all Texas droving on the Shawnee, serious resistance to the practice had already arisen. The longhorns' hardiness, which enabled them to be trailed a thousand miles and still not be much worse for wear, had a significant drawback. Since they themselves were immune to the tick-borne disease known as Spanish or Texas Fever, longhorn herds acted as carriers, depositing the infecting ticks all along their path. In settled country, area livestock would pick up the ticks, contract Texas Fever, and almost inevitably die.[10] By the mid-1850s, after a particularly virulent epidemic of Texas Fever in central and western Missouri, farmers and ranchers formed vigilante committees that threatened to kill any Texas cattle coming through their area and, if necessary, their drovers as well. When southern cattlemen responded in 1859 by trailing through Kansas territory to the new rail head at St. Joseph, Missouri, local livestock owners were just as hostile. Ultimately, by popular demand both states passed quarantine laws excluding Texas cattle from their settlements. However, since it had been observed that outbreaks did not occur in cold weather, Kansas and Missouri lifted their bans on December 1.[11]

After General Robert E. Lee's capitulation at Appomattox on April 9, 1865, most of the returning Texas ranchers spent that first summer back home rebuilding their herds. Working co-operatively together with their neighbors, they rounded up their own cattle, branded their calves as well as their fair share of mavericks, and made steers. When the trailing season opened the following April, at least 200,000 head were driven to Northern markets. Some went up the Shawnee Trail as far as Fort Gibson, and then sought to stay west of Kansas border settlements to avoid any confrontations over Texas Fever. They shipped out of St. Joseph to Chicago

Cattle Drive on the Great Plains was sketched by A. Cataigne and copied from *Scribner's Magazine*, 1892. (Source: Kansas State Historical Society, Topeka, Kansas)

or trailed further up to Iowa. Others tried to run the vigilante blockades, with varying success. Sixteen-year-old Texan Jim Daughtery was rewarded for his bravery in trying to lead a drive of his family's herd to Fort Scott; en route, his herd was stampeded and he himself was thoroughly whipped by Jayhawkers who chose to let him live only because of his youth. Writing back to Texas from Boggy Depot, Oklahoma, James Juvenal thought that his only realistic option was to trail his small herd to Baxter Springs, Kansas, although he feared that merely "a few hundred head would glut that market."[12] In the face of such fierce opposition and decidedly limited alternatives, perhaps as many as 100,000 head of cattle were forced to wait in Indian Territory until the onset of winter. Even then, a substantial number of owners chose to stay out on the unsettled ranges of the border areas until spring. As they waited far from home for a fresh chance to find a market in the new year, many owners must have realized that this was no way to run a business.

Joe McCoy's Better Idea

At the age of twenty-nine, Joseph G. McCoy was, by his own later admission, restless and speculative—more a natural-born promoter than long-term investor or manager. Although he had already worked hard and earned the right to become a partner in a successful cattle-feeding operation in Central Illinois with his two older brothers, he wanted to better himself further, to do business on a grand scale. When he learned through an employee recently returned from Texas of that state's enormous, pent-up supply of cattle, he resolved to break the bottle-neck preventing them from easily coming on the market and to make his own fortune in the process. "To him it became a

The Railroad King and the Illinoisan. This scene depicts the meeting between Joseph McCoy and the president of the Missouri Pacific Railroad in St. Louis. (Source: Joseph McCoy, *Historic Sketches of the Cattle Trade of the West and Southwest*)

waking thought, a sleeping dream."[13] At first he considered driving cattle up the Shawnee Trail to Fort Smith and then shipping them via the Arkansas and Mississippi Rivers to Cairo, Illinois, where they could then be sent to feedlots and fattened for eastern markets. But he soon abandoned the idea as impractical. He then decided to look west, not south. On the one hand, the latest Kansas quarantine law had exempted the southwest quarter of the state. On the other, the railroad had laid down tracks well-beyond the settled eastern portion of Kansas. Texans could drive unmolested through the western Kansas frontier, and the railroad could carry the cattle directly from there to the eastern packing plants in Chicago and St. Louis. All McCoy had to do was choose a good location along the railway and then secure the facilities necessary for buyers and sellers to meet there and conduct their business. While scouting for a suitable location in June 1867, McCoy had a chance, two-hour conversation with Colonel J.J. Myers in Junction City, Kansas. When the visiting Texas cattleman confirmed the soundness of the young man's plan, their handshake sealed McCoy's resolve to bring it to fruition.[14]

At first neither the railroad companies nor the rail towns seemed interested in McCoy's idea. Although the railroad was happy for the freight business, they were not willing to risk their own capital in providing the infrastructure necessary for the large-scale shipping of cattle. However, if the young man would invest his own money in constructing the holding pens and building the trade, they promised to lay down the rails for a siding and give him $5 for every car of cattle shipped.[15] Having obtained a contract, in June 1867 McCoy went to search out a desirable location for the new railhead. Junction City,

Col. J. J. Myers. (Source: Joseph McCoy, *Historic Sketches of the Cattle Trade of the West and Southwest*)

The tick that transmits Texas fever is shown on a steer's hide. (Source: U.S. Department of Agriculture, Bureau Animal Industry; *Special Report on Diseases of Cattle*, 1904)

the fourth city from the end of the line, had a small beef packing plant,[16] and their leading booster seemed open to expanding the city's cattle trade. However, the price demanded for the land for the future stockyards seemed exorbitant to McCoy.[17] Salina and Solomon, the last two towns, were more concerned with protecting their local livestock from Texas Fever than seizing the economic possibilities of becoming a cowtown. As an alternative, McCoy went to the third stop, a hamlet so small that the railroad had yet even to build a depot.

Although Abilene had been chosen as the county seat in 1861, six years later the town still consisted of little more than a dozen scattered log-cabins, including the requisite store and saloon. This water station along the rails was marked simply by a milepost with a hook for a mail bag and a board nailed above it to keep off the rain.[18] Sparsely populated, nested in a lush valley of the Smoky Hill River, surrounded by open range with excellent grass and plenty of cheap land, and located right next to the railroad, Abilene was the place, McCoy decided,[19] even though the hamlet still lay just within the quarantined zone. He bought the 480-acre parcel of land for $5 an acre, and the first Kansas cowtown was born.[20]

Abilene: The First Cowtown

All these arrangements were not concluded until July 1, 1867, mid-way into the trailing season; consequently, McCoy had to move quickly to get things underway. He immediately sent out an experienced surveying team to mark a trail from Abilene south down to the Oklahoma line. Then he sent a scout down that trail and then further

Abilene in 1867—Celebrating the Shipment of the First Train-Load of Cattle. (Source: Joseph McCoy, *Historic Sketches of the Cattle Trade of the West and Southwest*)

Loading the Stock at the Railroad was sketched by A. Castaigne and copied from *Scribner's Magazine*, June 1892. (Source: Kansas State Historical Society, Topeka, Kansas)

south into Indian Territory to find herds and direct them to the new route to Abilene.[21] While awaiting their arrival, he imported lumber from eastern Kansas via the railroad and began building accommodations for both beast and man. McCoy's projects included the Great Western Stockyard, an office with iron bars on its windows for protection, a large, twin-building livery stable, and an elegant three-story hotel to be called the Drovers Cottage where both cattle owners and cattle buyers could stay. The cowboys would be expected to overnight out on the range with the herds while the animals were fattening up on the rich local bluegrass before their sale.

In August, vigilantism reared its head again. Some of the few farmers living near Abilene formed a protective association to keep the cattle out of their area. They pledged to stampede any herds heading for Abilene. McCoy and some of the cattle owners who had arrived ahead of their herds to check out the new town agreed to meet with them. Realizing that personal economic incentive would prove the decisive factor, McCoy promised the local farmers guaranteed high prices for their goods and, by prior agreement, the drovers made contracts on the spot to pay up to twice what the settlers were asking for their produce. With the concrete prospect of bettering themselves, the farmers agreed on the spot to support the new venture. As their leader put it, "if I can make any money out of this cattle trade I am not afraid of 'Spanish fever;' but if I can't make any money out of this cattle trade then I am d----d 'fraid of 'Spanish fever.'"[22]

By August 31, the stockyard, stable, and office had been completed. The Great Western had a 10-ton Fairbank's scale and could accommodate 3,000 head. In two hours, 800 head could be loaded into forty cars. The first train of twenty cars pulled out of Abilene on September 5 in the presence of beef packers, cattle dealers, and railroad officials invited by McCoy to celebrate the new railhead. On September 24, Drovers Cottage was finished and furnished, although the official opening under an experienced hotelier had to wait to the following spring. The total cost was $15,000. By the end of that Fall, even with an aborted season, 35,000 head came to Abilene in 1867. Nearly 1,000 cars of cattle were shipped to Chicago. A few were driven up the Platte to supply government contracts for Indians. The rest of the herds either wintered in Kansas or were sold to the beef-packing plant in Junction City.[23]

The presence of so many Texas cattle in Kansas had not gone unnoticed. Many complained that the quarantine law was being broken, threatening an outbreak of Texas Fever. A September 27th editorial in the *Kansas Radical*, a Manhattan-based paper, objected for a different reason—namely, that McCoy's initiative threatened the emergence of a local cattle industry: "For every head of stock that is shipped from these stations one head belonging to some Kansas farmer is crowded out of the market . . . Kansans are . . . beef raisers. This matter is going to be one of importance before long."[24] To those who objected to Abilene's trail trade, Governor Crawford responded: "I regard the opening of that cattle trail into and across Western Kansas, of as much value to the State as is the Missouri river."[25]

The 1867 season had shown the structural soundness of McCoy's idea. Buyers and sellers could do business together unmolested in Kansas with the railroads successfully shipping the majority of the cattle exchanged to eastern markets. However, the financial soundness of the scheme was another matter entirely. In the end, the prices obtained for the cattle sent to Chicago were not good. Drovers lost heavily. So did McCoy. Not only had he invested a significant sum in building up the necessary infrastructure in Abilene, early on he bought 900 head himself, shipping them to New York at a loss of over $13,000.[26] Understandably, many predicted the demise of the whole enterprise. McCoy, however, was determined to prove the many nay-sayers wrong.

Shipping for the Eastern Markets appeared in the May 2, 1874, edition of *Harper's Weekly*. (Source: Kansas State Historical Society, Topeka, Kansas)

The 1868 and 1869 Seasons

McCoy realized the key to his vindication lay in publicity. Consequently, he organized a systematic advertising campaign which employed both the direct mailing of a circular to every name he could find connected with cattle in Texas as well as publishing the details in the state's many newspapers. He also sent two personal emissaries, Lt. Col. Sam Hitt and Charles F. Gross, "to spread in and through Texas by letters, circulars and public gatherings the information that a trail would be blazed on trees through the Indian Territory and by a furrow turned up by plow on open country from Red River to Abilene, Kansas, where all eastern buyers of cattle would meet them and buy or they could ship themselves."[27]

Finally, since the success of Abilene depended on the presence of both buyers and sellers, McCoy equally promoted Abilene as the place to obtain livestock with full-page advertisements in many northern newspapers. The ad in the March 27, 1868, *Missouri Republican* at St. Louis read:

> Cattle—the best grazing cattle in the United States can be had in any number, on and after the 15th of May, 1868, at Abilene, Kansas, weighing from 1,000 to 1,200 pounds, live weight. They reach Abilene from the Southwest and will take on from 50 to 150 pounds more than native cattle, and cost less than half as much. Until the 1st of May we may be conferred with personally

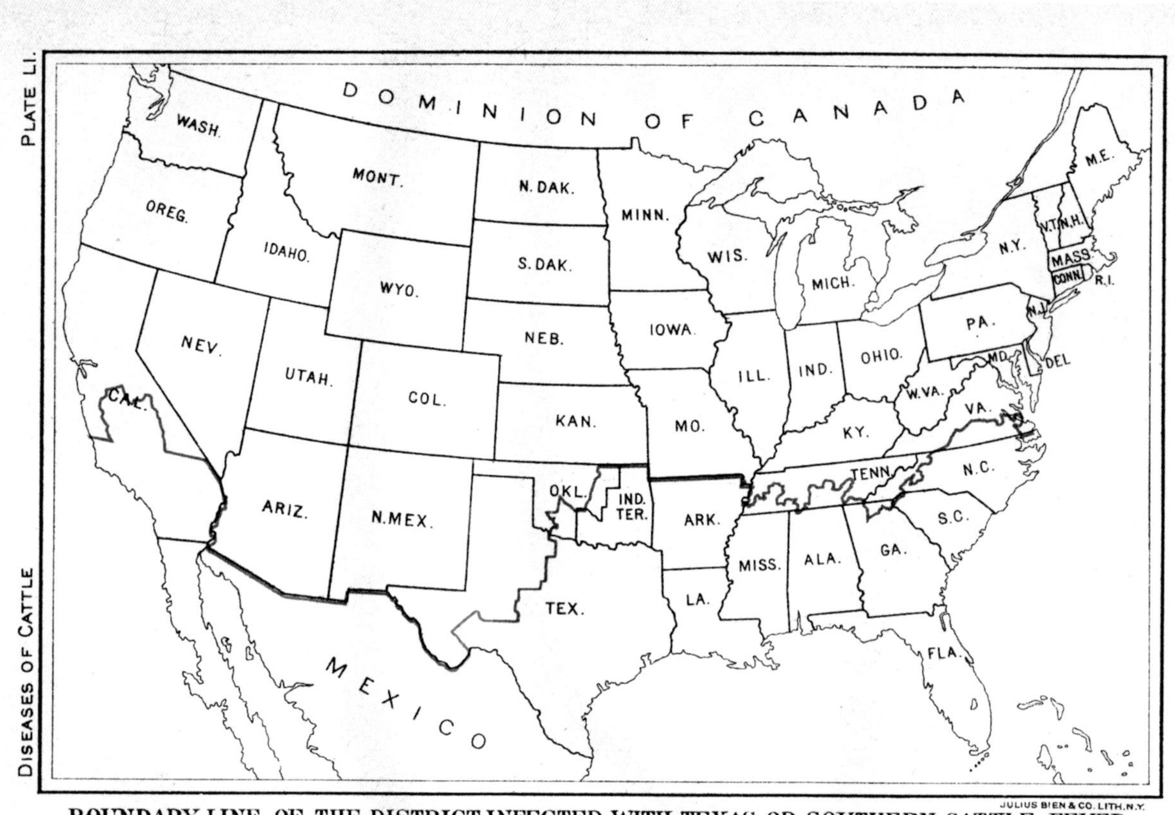

Boundary line of the district infected with Texas fever or Southern cattle fever. (Source: U.S. Department of Agriculture, Bureau of Animal Industry; *Special Report on Diseases of Cattle*, 1904)

at Springfield, Illinois, and thereafter at Abilene, Kansas, W. K. McCoy and Brothers.[28]

Joe McCoy estimated the total costs for all his advertising at $5,000. Leaving nothing to chance, he also sent a representative to the mouth of the Little Arkansas to direct the herds that came there onto Abilene rather than somewhere else.[29]

The initial results were encouraging. The June and July figures were good, both for number of cattle arriving in Abilene and for price per head there. However, fresh outbreaks of Texas Fever from longhorns in Illinois brought sales to a halt in August. In response, McCoy put together a buffalo roping and throwing show which traveled to both St. Louis and Chicago, its railcars advertising along the way the presence of 25,000 head of cattle in Abilene awaiting sale. According to Gross, who was then serving as his bookkeeper, the venture cost McCoy another $6,000.[30] Immediately afterwards, he organized a tour of Abilene and Western Kansas by Illinois cattle buyers. Both had the desired effect, and by the end of the season, 75,000 cattle had been trailed to Abilene where good prices were again given. The 1869 season was even more spectacular; 150,000 head were received at Abilene. McCoy's vision was now undisputedly established as the way forward for the cattle trade.

View of Col. O.W. Wheeler's herd en route to the Kansas Pacific Railway in 1867. The illustration was copied from *Historic Sketches of the Cattle Trade* by Joseph McCoy, published in 1874; the illustrator was Henry Worrall. (Creator: Baker-Co. Source: Kansas State Historical Society, Topeka, Kansas)

Nevertheless, local tensions remained high. Some Dickinson County livestock also died in 1868. According to Gross, "the farmers had but few cattle, little money and to see their small stock of cattle die off was to them almost unbearable. It was almost open gun war." Once again economic interests prevailed. McCoy arranged for compensation for Dickinson County farmers, and "a partial peace was obtained." Of the $4,500 paid out, McCoy's contribution was $3,300. The following year, settlers insisted on a $20,000 bond to protect them against damages.[31]

All these costs continued to add up. In the end, McCoy's vindication did not bring him financial prosperity. In fact, quite the opposite happened. Having spent so heavily to put Abilene on the map, when the Kansas Pacific failed to pay him, disputing what it owed him, McCoy was forced to sell his Abilene acreage, Drovers Cottage, and even the stockyard in 1869. By 1870 Joe McCoy was penniless. Although the Kansas Supreme Court sided with McCoy in 1871, by then it was too late. He had already been forced to sell the assets which could have allowed him to rebuild his fortune.[32]

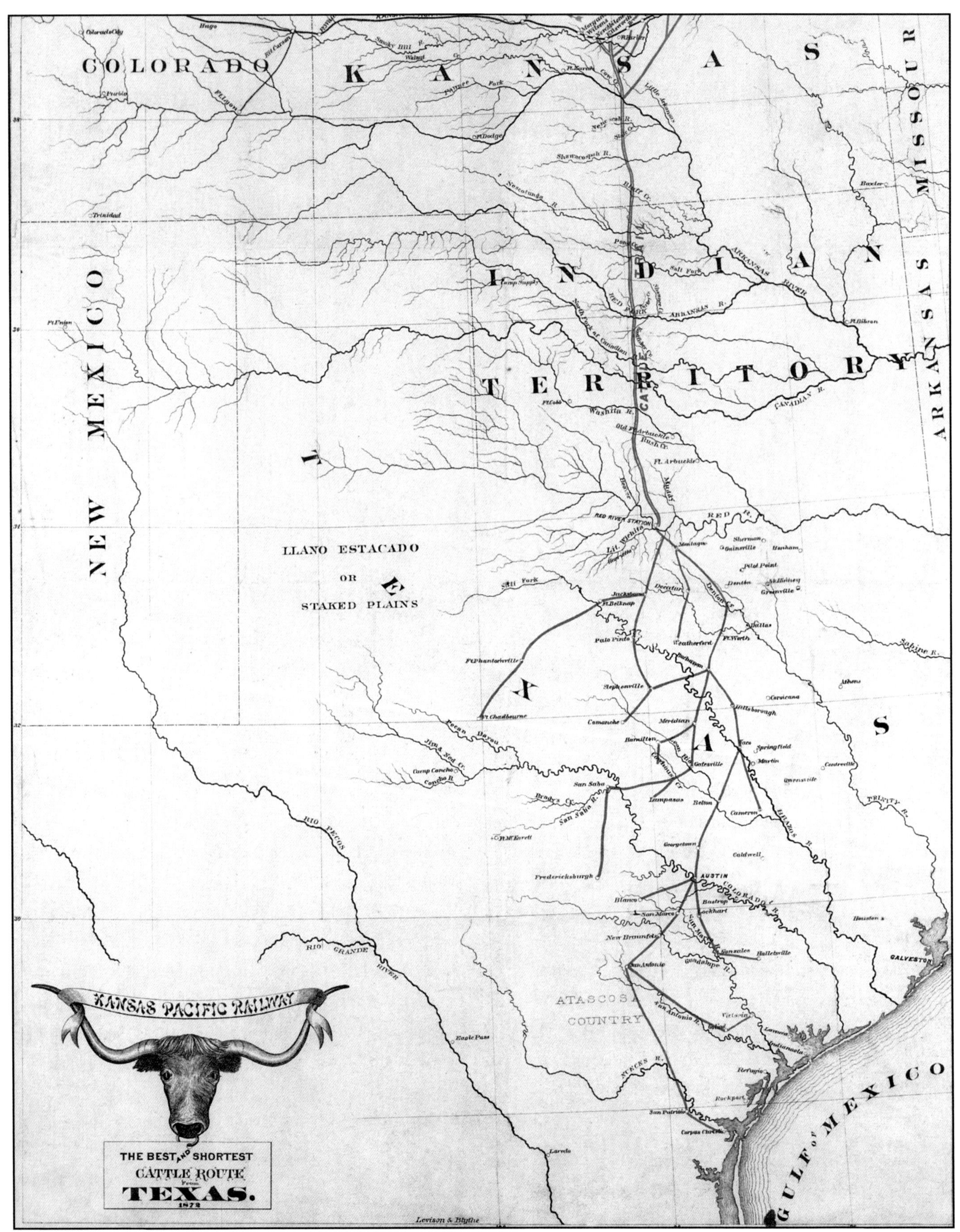
COLORADO
KANSAS
MISSOURI
NEW MEXICO
INDIAN TERRITORY
ARKANSAS
TEXAS
LLANO ESTACADO
OR
STAKED PLAINS
ATASCOSA
COUNTRY
GULF of MEXICO
RED RIVER STATION
AUSTIN
GALVESTON
ARKANSAS RIVER
CANADIAN R.
COLORADO R.
RIO PECOS
RIO GRANDE RIVER
Ft. Dodge
Trinidad
Ft. Gibson
Ft. Arbuckle
Sherman
Gainsville
Denton
Decatur
Dallas
Ft. Worth
Weatherford
Stephenville
Comanche
Meridian
Hillsborough
Waco
Gatesville
San Saba
Lampasas
Belton
Georgetown
Fredericksburgh
San Marcos
New Braunfels
Gonzales
Lockhart
Bastrop
San Antonio
Victoria
Indianola
Refugio
Corpus Christi
KANSAS PACIFIC RAILWAY
THE BEST AND SHORTEST
CATTLE ROUTE
From
TEXAS.
1872

Life on the Chisholm Trail

I woke up one morning on the Old Chisholm Trail
Rope in my hand and a cow by the tail.
Feet in the stirrups and seat in the saddle,
I hung and rattled with them Longhorn cattle.[33]

The route the drovers used early on from South Texas to Abilene went by many names, including the Abilene Cattle Trail (McCoy's name) and the Great Texas Cattle Trail (used on the Kansas Pacific Railway's map). By the end of the cattle-driving era, however, the trail was popularly known as the Chisholm Trail.[34] Jesse Chisholm (1805/6–1868) was a half-Scot, half-Cherokee pioneer trader, interpreter, and peace-maker. "Chisholm's trail" originally referred to the freight route he established between his trading post at Council Grove on the North Fork of the Canadian River (now part of Oklahoma City) and his trading post at the confluence of the Arkansas River with the Little Arkansas (the site of modern-day Wichita). Of course, Chisholm simply followed the shortest route between these two points that the natural topography of the area would permit. As such, the best route had long ago been trodden down by the buffalo during their annual migration between their summer feeding grounds up north and where they wintered down south. The Native Americans were very familiar with these paths, and scouts pointed them out to the army as the basis of their military supply routes. As an experienced guide himself, Chisholm was not so much the blazer of the route between the North Fork and the Little Arkansas as the inheritor of ancient wisdom who was the latest to put it into practice.[35] Eventually, the name for the path between his two trading posts was popularly applied to the whole route which the cattle followed from South Texas to Abilene, Kansas. Although not without continuing controversy, in western lore, *Chisholm's trail* became *the Chisholm Trail.*[36]

Previous: This map shows the Kansas Pacific Railway routes from Texas through Indian Territory (Oklahoma) to north-central Kansas. In Texas, there were thirty-five named stations or towns; the southernmost was Corpus Christi, and the northernmost was Red River Station. At the northern terminus in Kansas were the towns of Morgans, Wilson, New Abilene, Ellsworth, Brookville, Salina, Solomon, and Abilene. (Creator: Kansas Pacific Railway Company, 1872. Source: Kansas State Historical Society, Topeka, Kansas)

Above: (Source: Lewis Browder)

While there were feeder routes throughout Texas and variations by individual drovers, the main trail headed out from San Antonio and crossed the Brazos River at Waco, the Colorado River at the Montopolis ford by Austin, the West Fork of the Trinity River at Fort Worth, and the Red River at Red River Station, a former Texas Ranger outpost during the Civil War. Once in Indian Territory, twenty-seven miles north of the Red River the trail passed by Monument Rocks, a mesa with vast views up to fifteen miles in either direction. Next, the route crossed the Washita, the South Canadian, and then the North Canadian by Chisholm's trading post at Council Grove. From there, the cattle followed the deep ruts cut in the earth by his heavy freight wagons towards Kansas. The trail entered that state by Caldwell and continued north, crossing the Arkansas River at the Little Arkansas and then heading northeasterly over to Abilene.[37]

Cutting Out. The process of cutting out is one that requires skill and expert horsemanship. When a beef is selected to be cut out, he is adroitly and quietly maneuvered to the outskirts of the roundup, and when the opportune moment occurs, the cowboys dash at him to separate him from the herd. Once the beef realizes what is happening, he makes a desperate effort to regain his comrades, and this is where the skill of the cowboy is put to use. (Source: Joseph McCoy, *Historic Sketches of the Cattle Trade of the West and Southwest*)

Writing in 1892, Charles Harger described the trail as a "chocolate band amid the green prairies" varying between 200 and 400 yards wide, "beaten down into the bare earth" so as to be lower than the surrounding countryside, banked by drifted sand dunes as well as bleached animal skulls and skeletons—in short, wind-swept and desolate. Every so often, there appeared on the high ground neighboring the trail those barren circles which marked the "bedding grounds" where a great herd had spent the night. More infrequently, a slight mound in the verge indicated to the discerning eye the remains of a drover whose final fate was to have the loneliness of the trail sealed into his bones in perpetuity. Little wonder that the dying cowboy of the popular ballad asked to be buried beside his family back home and "not on the lone prairie."

This major regional artery, massive in both length and width, united the North and the South. For if the geography of Kansas before the Civil War made it the prognosticating bleeding battleground between the two protagonists, that same geography made it a meeting place and an eventual melting pot, bringing the former enemies together in a joint pursuit of greater economic prosperity for all.

The trail drive began on the open range in April when cattlemen and their hands rounded up the herd, cut out those to go north, roped and tied them, and applied any needed brands as well as earmarks.[38] Then, the local owners would need to decide who would take the cattle up the trail. As depicted in the famous scene in the movie *Red River*, in the early years the owners would often choose their best hands and drive the herds themselves. However, those who

On The Trail by Clyde Hostetter appeared in the May 2, 1874, edition of *Harper's Weekly*. (Source: Kansas State Historical Society, Topeka, Kansas)

went up the trail would be gone for many months. The drive itself could take two months or more. Then fattening the cattle on the lush grass around Abilene for one to two months before offering them for sale was also customary.[39] Finally, it could take up to a month for the hands to return to Texas by horseback, if the owner had been able to sell the herd and did not need to winter them up in Kansas. In order to keep their ranches in good working order during such a long period of time, many cattlemen soon chose to turn their herds over to contractors who would then join several smaller herds together to form a joint expedition under their command. Since the cattle usually came from multiple herds, a generic road brand was applied to all cattle going up the trail together. Usually, the owners gave the contractors letters empowering them to sell on their behalf once the drovers had brought the cattle to market. Some cattlemen, however, preferred to take the train to Abilene and find a buyer for their stock themselves.

Eventually, contractors became true middlemen. First, they would obtain an agreement from a buyer which specified the price, number, and kind of cattle desired (for example, young cows for breeding, or steers and dry cows for slaughter).

Swimming herds of cattle across swollen rivers was not listed as one of the pleasurable events during a drover's trip to the northern market. It was the scarcity of large rivers that constituted one of the most powerful arguments in favor of the Chisholm Trail. Nevertheless, the trail was not entirely free from this objection, especially during rainy seasons. (Source: Joseph McCoy, *Historic Sketches of the Cattle Trade of the West and Southwest*)

Then the drovers would contact Texas ranchers to take their animals on assignment. Finally, they would hire an outfit of between twelve to eighteen cowboys to make the actual drive. One of these was the trail boss who was normally paid $90 a month or more. He was older and experienced, often in his twenties. Second in command was the cook who ran the chuck wagon and received $75 a month. The remaining trail hands, known as "waddies," were just teenagers, often not older than eighteen. They received $30 a month. Many times the youngest cowboy (or an older man of limited physical abilities) was the wrangler who looked after the *remuda*, the herd of spare horses that accompanied the drove. Since a longhorn always challenged anything that approached it on the ground, cowboys always interacted with the cattle on horseback. As a result, each cowboy on the trail would have between four and six horses in the *remuda* from which he could choose a fresh mount when needed. Descended from range herds, mustangs were most often used, although because of their amazing speed over short distances, quarter horses could be trained to be outstanding cow ponies. To fulfill their many agreements, contractors would send up numerous such outfits during the trailing season.

A cowboy needed to be handy with a horse, a lariat, and a six-shooter. He was usually bowlegged, had long hair, and wore a red bandana about his neck. His hat was made of felt with a wide brim to keep the sun off his neck and the rain off his clothes. It was also big enough to be used as a signal to others on the trail. His shirt was a dull color of cotton or wool, with no collar (such things were for gamblers). His trousers were equally ordinary, canvas or leather. If he were Mexican or wished to look stylish, he wore a bright silk sash instead of a belt. His pistol belt hung loosely over either so that the six-shooter rode on his thigh. When going through country with a lot of brush, he wore chaps made of calf- or goat-skin for protection from scratches. His boots had two-inch heels to keep his feet from slipping through the stirrups. The obligatory spurs were large, usually nickel-plated, and designed as much to make noise jingling in town as to prick horseflesh on the trail. The more elaborate pairs were made of silver, and some even had tiny bells attached. Like any teenager, the cowboy was especially eager to set himself apart from the crowd by how he got around. His saddle was the object of special pride in this department, often more costly than his horse. As the lyrics from the "Old Chisholm Trail" put it,

> On a ten dollar horse and a forty dollar saddle,
> I was ridin', and a punchin' Texas cattle.

Of the estimated 35,000 cowboys during the trailing years, one-third are thought to have been either African-American or Mexican.

On the trail, the boss rode ahead of the herd to scout out good watering holes, green pasture, and suitable bedding grounds. He would then ride back toward the herd and use his hat to signal the direction ahead to his waddies. As a result, the trail boss rode "about three or four times as far as the herd."[40] The cook and chuck wagon also went ahead to the selected stopping points for midday and evening so he could have grub ready when the outfit arrived. Since the chuck wagon served as the moveable headquarters of the outfit, storing all their supplies, the wrangler and the *remuda* usually traveled right behind it. The size of trail herds varied, but rarely were more than 3,000 head sent up together. Any more cattle than that delayed getting the whole herd watered and complicated crossing the various streams. Even then, such large numbers required the longhorns to be strung out to prevent overheating, the narrow cattle queue often snaking over the trail like a backward tail for a half mile or more.

The cattle column came to a point at the front, where a lead steer set the pace. Many trail bosses found it much easier to get cattle to swim a river or go into a chute when following a lead steer. Charles Goodnight even put a bell on his famous "Old Blue," training the herd to move when they heard it clanking and to stay put at night when he muffled it. Goodnight used "Old Blue" on eight drives to Dodge City, and then gave him a well-earned retirement on his Texas Panhandle ranch. The animal became so accustomed to being a regular member of the trail outfit that he refused to associate

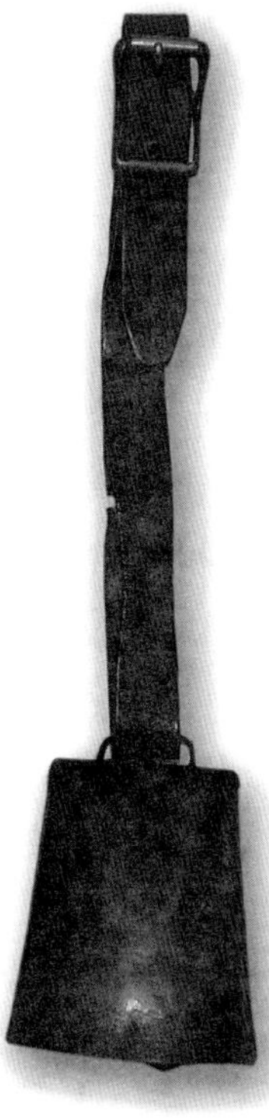

Blue's bell. Old Blue was the lead steer of Charles Goodnight's JA Ranch herd. It was said that Old Blue knew the road to Dodge City better than most cowboys, and the herds followed the sound of Old Blue's bell. Born in 1870, Old Blue was powerful, steady, and able to understand the slightest motion of the lead man. He proved himself to be worth a dozen cowhands. (Source: Cattle Raisers Museum, Fort Worth, Texas)

This photograph, which was taken with a flash sometime between 1891 and 1903, shows cowboys getting ready for guard duty while others are going to sleep next to the chuck wagon. (Creator: Francis Marion Steele. Source: Kansas State Historical Society, Topeka, Kansas)

with the herd at night. He grazed with the horses instead and mooched food from the cook as if he were a cowboy.

On either side of the lead steer, two cowboys rode "point," making sure that the herd stayed on course. As the cattle began to bunch up and the line became thicker, a cowboy on either side rode "swing" to prevent the longhorns from spreading out too widely or completely wandering off. A little further back, two more cowboys rode "flank" for the same purpose. At the crowded butt-end of the column, three or so cowboys rode "drag." Using short, rawhide whips, they had to make sure that the slowpokes kept up. Being at the tail end, these men got the worst of all the dust kicked up by the herd and blown about by the wind. Despite the popular television image, a cowboy with his bandana pulled up over his face did not normally mean he was about to rob a bank! The pointers were typically the most experienced men. Although they switched sides every day, they usually held their position at the front of the herd throughout the journey. In more democratic outfits, the rest of the hands rotated in sequence the remaining positions around the herd. However, it was not unheard of for "greenhorn" "tender-feet" to be assigned drag for the duration of the drive.[41]

A typical day began with both man and beast taking on food for the journey. The cook roused the men at daybreak for breakfast, while the last two from the previous night's guard duty got the cattle up for grazing. The first two to finish their grub relieved the guards so they could get their biscuits and thick, black coffee for the morning. After the cook had cleaned up and the wrangler brought out the horses, the hands mounted up and took their positions for the day. The trail boss signaled with his hat for the outfit to move out. As the cowboys passed the sign with their hats to each other down the line, the herd set to trailing. The cattle moved slowly, covering no more than between ten to twelve miles a day, the exact length of a day's journey usually determined by the location of the next good bedding ground. The herd stopped about 11:00 a.m. so the cattle could graze and the cowboys get grub. When the longhorns began to lay down, the trail boss knew it was time to move out again. After lunch, the two who were to be first on guard duty that night rode off with the chuck wagon. They would take their dinner ahead of the others, so that they could begin grazing the cattle as soon as the herd arrived. The drive ended when the boss waved his hat in a circle around his head to signal they had reached the bedding ground. The location would have a stream for watering the cattle, fresh grass for them to eat, and dry grass on which to bed. It also helped if the site was away from trees where wolves or Indians might be lurking, and high enough to catch a good wind. Sitting Indian-style, cross-legged on the grass, the other

R. K. Perry of Meade, Kansas, is shown in 1895 sitting on a hill overlooking his roundup herd of cattle at the Cimarron River, which was directly south of Englewood, Kansas. Visible in the background are the cowboys and the herd's 200 head of cattle. (Creator: Francis Marion Steele. Source: Kansas State Historical Society, Topeka, Kansas)

hands had their big meal for the day consisting of beef, beans, sourdough, and more hot coffee. At sunset, the cattle were herded into an ever tighter circle until they bedded down.

As the other cowboys stretched out around the campfire and pulled their blankets over their clothes, the two guards rode around the herd, riding in opposite directions. Telling time by the movement of the Big Dipper constellation, the pair would awaken the next two when their shift was done. This was done by speaking to the men, never by touching them, lest a startled cowboy in his grogginess reached instinctively for his six-shooter. Usually, there were three or four sets of guards a night, each watch lasting two hours or more.

The cowboys on duty would be on look-out for signs of intruders of the human kind—like Indians and other potential thieves. Between the Red River and the Cimarron, outfits had to deal with Comanche and Kiowa raiders; these Indians either simply demanded a cow (or two) as toll or tried to stampede the cowboys' herds and horses, or both—for coming in the day to beg allowed the Indians to scout out the possibilities for coming at night to steal. In addition, the cowboys riding guard also had to look out for intruders of the animal kind, since wolves and coyotes always welcomed the opportunity for a steak dinner.

Midnight Storm and Stampede. (Source: Joseph McCoy, *Historic Sketches of the Cattle Trade of the West and Southwest*)

Most important of all, however, the punchers on duty were on constant look out for signs of restlessness within the herd. Of all the dangers of droving, the stampede was the most dreaded of all. Anything could startle the longhorns and precipitate their wild, charging-off in one or more directions: the breaking of a twig, a crack of lightning, the lighting of a match, the howl of a wolf, the sneeze of a guard, or the surprise appearance of a flock of birds. To counter sudden fright at strange noises, the guards sung or hummed as the rode. Cowboy lullabies, ballads, and even revivalist hymns were used nightly to reassure the longhorns and deter their stampeding. Yet, once they were off—hooves clattering, horns clashing, heat radiating—the cowhands did not attempt to stop such an awesome force of nature. Mature trees had been known to be felled by a mass of rampaging longhorns. At first the cowboys simply tried to channel that fiery energy, directing the cattle away from cliffs and ravines. Ultimately, their goal was to gradually turn the flow, getting the lead steers to eventually circle the herd upon itself. That task could take a few hours, all night, or even days. As long as the stampede lasted, the punchers got no rest, as they rode for hours, usually at night, and often in blinding rain.

A Stampeded Herd was sketched by A. Castaigne and copied from *Scribner's Magazine*, June 1892. (Source: Kansas State Historical Society, Topeka, Kansas)

One contemporary New York reporter thought that the "swift gallop into the grassy darkness" required by a sudden stampede probably brought "secret delight" to "the drowsy guards," since it "enlivened" their otherwise dull daily routine:

> . . .every day so like every other that soon all reckoning of its place in the week is lost, each passing scene so much a mere bald repetition that the whole outlook in a short time becomes simply one vast, featureless, confusing impression, like the ocean. Indeed, being adrift on these great, vague and melancholy prairies is very similar to being out at sea. The drover of whom I have spoken told us that he never made the journey without a continual, torturing heartache and sense of exile . . . Always, after a few days of the march, and a fair start into what may be called the sorcery of the intervening desert, a habit of silence and reverie and depression comes upon the entire outfit . . . The songs and jokes die out, the storytellers cease to spin their coarse and knotted yarns, card-playing by the camp-fire is voted a bore, daily conversation dwindles to monosyllables. "Every man draws himself into his shell."[42]

Little wonder, then, that cowboys cut loose when they arrived at a cowtown.

Abilene in Its Cattle Glory Days

Despite Joe McCoy's personal misfortune, the Texas cattle trade to Abilene continued to double spectacularly each year: 150,000 head in 1869; 300,000 in 1870; and 600,000 in 1871—the peak of the whole trailing era. Naturally, such massive numbers of longhorns had an equally explosive effect on the number of cowboys in Abilene. During the longhorn season from June to October, 90 percent of the people in Abilene were connected with the cattle trade in some form. Fifty percent were Texans. Away from home, all responsibilities, and any accountability, with two months wages in hand and a determination to enjoy themselves freely after all the self-restraint on the trail, these teenage boys offered a golden opportunity for many a frontier merchant "to better himself."

When a cowboy came to town, the first thing he did was go to the barbershop for a bath, haircut, and shave. After spending far too much time in his trail duds, the next stop was to buy a completely new outfit, hat and boots included. In Abilene, he could go to Jacob Karatofsky's Great Western Store for clothes, hats, and gentlemen's accessories. A new Stetson alone could cost a cowboy more than two months pay. Custom-made boots by Tom C. McInerney's shop were especially popular. Even though he charged at least nearly two weeks pay for a pair, McInerney had to employ up to twenty men during the peak summer months to keep up with

(Source: Blair Tenhouse)

(Source: Blair Tenhouse)

demand for his red-topped leather boots with lone star and crescent moon tooling.

Once newly outfitted, these teenage boys sought three principal pleasures: alcohol, gambling, and female companionship. Starting in 1868, saloon keepers, gamblers, and prostitutes "came in swarms" to Abilene to oblige them. By 1871, Abilene had thirty-two places to buy liquor, including eleven saloons, sixty-four gambling tables, 130 known professional gamblers, and, in a specially fenced-off addition to the south, between ten and fifteen houses of ill-repute with maybe 100 "soiled doves."[43]

The two most famous saloons were the Alamo and the Bull's Head, both located on Texas Street, Abilene's main avenue. The Alamo was the most glamorous, with three double-glass doors, brass fittings throughout, an impressive mirror behind the bar which reflected all the colorful bottles of liquor, and walls hung with paintings of reclining nude women in imitation of the Venetian Renaissance masters. Gaming tables covered the entire floor space. Since the Alamo was open 'round the clock, an orchestra performed morning, afternoon, and all night.

The Bull's Head was the most notorious. Although named after the front of the animal, the two Texan owners painted a bull on the outside of the saloon which depicted very vividly his gender-identifying features at the other end. Immediately, for many of the settlers the artwork was "an insult to the good women of the town, a shame in plain sight of children." The proprietors refused to back down for two weeks. When it appeared that the issue would be solved by violence, a compromise of sorts ensued. The offended features were painted over, although their outline remained clearly visible.[44]

In the saloon, the cowboy found his three thrills. Here is one old-timer's account of how they mixed from him to the profit of others in Abilene:

> We ordered toddies like we seen older men do, and drank them down, for we were dry, very dry . . . The two toddies were making me feel different to what I had felt for months, and I thought it was about time for another, so I headed for a place across the street, where I could hear a fiddle. It was a saloon, gambling and dance hall. Here I saw an old longhaired fellow dealing monte. I went to the bar and called for a toddy, and as I was drinking it a girl came up and put her little hand under my chin, and looked me square in the face, and said, "Oh you pretty Texas boy, give me a drink." I asked her what she wanted and she said anything I took, so I called for two toddies. My, I was getting rich fast—a pretty girl and plenty of whiskey . . . My boss had given me four dollars spending money and I had my five-dollar bill, so I told the girl that she could make herself easy; that I was going to break the monte game, buy out the saloon, and keep her to run it for me when I went back to Texas for my other herd of cattle. Well, I went to the longhaired dealer and as he was making a new layout I put my five on the first card (a king) and about the third pull I won. I now had ten dollars and I thought I had better go and get another toddy before I played again. As I was getting rich so fast I put the two bills on a tray and won. Had now twenty dollars, so I moved my hat back as far as it would go and went to get a drink—another toddy, but my girl was gone . . . I went back and it seemed to me that I did not care whether I broke [the bank] or not. I soon lost all I had won and my old original five.[45]

Of course, a toddy not only made it easier to get a cowboy to open up his wallet; alcohol also removed whatever restraint the young man might have had about firing off his gun. Once well-liquored, a Texas cowboy "was extremely liable under the least provocation to use his navies [pistols], of which not less than one or two were

(Source: Blair Tenhouse)

always hanging to their belts. In fact if their fancy told them to shoot they did so, in the air, at anything they saw."[46]

C. F. Gross provided a sense of the skewed morality that gripped Abilene as a result.

> I have said nothing of the sordid witch who actually killed a man for money one winter, and we had to take up on Mud Creek to the mill and hang him. We were used to seeing men killed because someone disliked their looks, the color of their eyes, cut of their clothes, refused to take a drink or danced too much with one's girl. Such killings were all legitimate. But to kill for money—it was too much.[47]

In the Wild West, to kill on a whim was acceptable. To kill for an undesirable, but clearly understandable, motive such as greed was not. To contemporary Victorian morality, what the transient cattle folk thought wrong literally made more sense than what they thought right.

With the flood of settlers to Dickinson County from 1870 onwards, such an atmosphere in the town was no longer acceptable. That spring, the office of town marshal was created. The first several attempts to find an effective peace officer ended in abject failure, as was made clear by the glee with which cowboys, galloping by, shot up the posted ordinance which prohibited their wearing guns in town. Finally, it was decided to fight fire with fire, and employ someone "who could . . . do to others what was doing to us, and to do them first." They hired Tom Smith, a man whom they had passed over earlier because of his leading role in the mob violence at Bear River, Wyoming. Much to the marvel of the settlers and the cowboys alike, Smith quickly imposed order of the town. He was respected for his cool courage under pressure, but feared for the very quick use of his hands—not, however, as a gunman, but as a skilled boxer. Before his first two challengers could reach for their guns, Smith had laid them out on the floor with devastating blows to the head. There was no third challenger, nor any need for him to kill a man in town.[48] When Smith was tragically killed that autumn in a dispute with settlers at a dugout, Wild Bill Hickok was chosen to succeed him for the 1871 season. Already famous from a *Harper's Monthly* profile three years earlier,[49] Hickok effectively maintained order in Abilene during the peak year of Texas trailing. However, his style was very different from Smith. Relying on his quick draw, he killed his challengers before they could kill him. His violence was effective, although not without collateral damage. His policy of shooting first and asking questions later led to him killing his own friend as a tragic mistake in October.

The beef market could not immediately absorb all of the 600,000 head driven to Abilene in 1871. That winter,

(Source: Lewis Browder)

300,000 head were put out on the Kansas range. Then disaster struck.

> November came in with a blizzard, and with slight interruptions, kindly allowed by Nature for the purpose of affording us opportunities to skin dead cattle, the blizzard lasted until March, and the cold stormy weather for two months longer. There was no new grass until the middle of May. In all the Texas herds held in Kansas the losses were heavy. Hardly a herd lost less than 50 per cent, and 60, 70, and 80 per cent losses were common.[50]

With little reason or opportunity for the remaining cowboys to celebrate, Hickok was let go in December.[51]

Finally, the good citizens had had enough. They decided that the money the cattle trade brought in, when it was profitable, was not worth all that came with it. In February 1872, 80 percent of the residents of Abilene signed a notice asking "all who have contemplated driving cattle to Abilene the coming season to seek some other point of shipment, as the inhabitants of Dickinson County will no longer submit to the evils of the trade." The Texans respectfully did, and 80 percent of the business establishments in Abilene closed as a result.[52]

After Abilene

After four and one-half years, Abilene had abdicated as undisputed queen of the cowtowns. She had, however, blazed her own kind of trail, setting the standard for the towns that came after. On the one hand, southern cattle seller and northern cattle buyer would now plan on doing business together. On the other hand, teenage boys would seek to become men on the trail and expect to enjoy the benefits of their newfound adulthood at its end. Boomers wishing to put their town on the map overnight would work hard to accommodate both ends of the trade. Those who sold goods as well as those who sold pleasure knew a cowtown was a permissible place for them to make maximum profit. Yet if her success established a formula to be followed, her end equally foreshadowed a pattern to be repeated. As settlers moved in, the encroachment of civilization would time and again force the trade to move on further west.

In the spring of 1872, Ellsworth and Wichita fought for the new supremacy. As early as 1869, the city boomers of Ellsworth, located some sixty miles further west on the Kansas Pacific, had sought to save their struggling new township by diverting the cattle trade their way. In March of that year, Kansas Governor Harvey signed a law establishing a cattle trail from Fort Cobb, Indian Territory, to Ellsworth. With that in hand, one month later the leaders of Ellsworth petitioned Governor Harvey to enforce the quarantine law of 1867 which would force the Abilene trade their way. Then, copying McCoy's own successful tactics, they also built stockyards, flooded Texas with advertisements, and sent out agents to direct drovers toward their city. In 1871, the railroad purchased the stockyard outright and enlarged it to handle two hundred cars of cattle a day, reputedly the biggest in the state. That same year, 18 percent of all cattle shipped on the Kansas Pacific came from Ellsworth.[53] Among them were cattle belonging to W.F. Burk, whose wife Amanda had accompanied him on the trail in a buggy all the way to Ellsworth.[54]

(Source: Blair Tenhouse)

Naturally, then, the boomers of Ellsworth saw their city as Abilene's successor. When the latter published its announcement in February 1872, the Ellsworth *Reporter* sent its own circular throughout Texas proclaiming the end of Abilene as a cowtown and Ellsworth as its replacement. In March, the *Reporter* boasted that half of Abilene would relocate to Ellsworth in two months.[55] Two mainstream mainstays of Abilene did so. Moses B. George, the current owner of the Drover's Cottage, literally packed up the three-story portion of his hotel and sent it by flatcar to Ellsworth. On about May 1, Jacob Karatofsky also began selling supplies from his Great Western Store in Ellsworth. Of the pleasure merchants, at least seventy-five gamblers were active in Nauchville, the rough area one-half mile away down in the river bottoms. The following year, thirteen licenses would be issued for saloons and three for hotels to serve liquor. As for prostitutes, the city fathers felt "as long as mankind is depraved and Texas cattle herders exist, there will be a demand and necessity for prostitutes." Consequently, the city simply fined them to raise revenue, and over thirty names appear in the police records as a result.[56]

Perhaps one story best sums up the fragile hopes and hard practicality of Ellsworth during its cattle days. In February and March 1872, the *Reporter* did enjoy telling the sad tale of a man who burned down his estranged wife's place of business in "Nauch," when she grew bored of farm life and returned to supporting herself. In retaliation, she feigned reconciliation so as to set up an opportunity to shoot him dead. Afterwards, left unmolested by either her dead husband or the law, she carried on practicing her profession.[57] Although 60,000 head of cattle came up the trail to Ellsworth in 1872, joining the 40,000 head that had survived the harsh winter in the Smoky Hills, the Abilene *Reflector* consoled as follows: "Business is not as brisk as it used to be during the cattle season, but the citizens have the satisfaction of knowing that hell is more than sixty miles away."[58]

Wichita

Not to be outdone by the Kansas Pacific, by 1871 the Santa Fe Railroad was equally eager to get into the cattle shipping business. When Abilene could not handle all 600,000 head that year, they turned to a newly connected city on their system which lay right on the Chisholm Trail: Newton, Kansas. Joe McCoy came down from Abilene to supervise the building of shipping pens, and the town blossomed accordingly. That summer, Newton had all the accouterments of a cowtown: the expectation of shipping out 40,000 head for the season; twenty-seven saloons; eight gambling establishments; a bordello district called Hyde Park (sometimes referred to as Hide Park); and multiple deaths from gun battles. Nevertheless, on May 11, 1872, the Santa Fe had reached Wichita twenty-five miles south, and the cattle trade moved on.[59]

At the mouth of the Little Arkansas, the site of Jesse Chisholm's trading post, Wichita was ideally situated to take advantage of traffic coming up his trail. However, the city fathers left nothing to chance. In now time-honored fashion, they hired a well-known Texan to direct southern cattle sellers to Wichita, and Joe McCoy himself to promote the city to northern cattle buyers. The Santa Fe also hired three people: a young Texas cowman named Shanghai Pierce as well as McCoy's two old hands, Colonel Hitt and Charles Gross, to assist him. With their help, Wichita surpassed Ellsworth in 1872, shipping out 70,600 head of cattle. Of course, such success attracted the cowtown veterans. Jacob Karatofsky (Abilene and Ellsworth) became a resident, as did Rowdy Joe Lowe and his wife Kate (Ellsworth and Newton), the operators of a dance hall in Delano, the entertainment district west of Wichita across the river.[60]

Ellsworth, however, was not prepared to give up supremacy without a fight. To make Ellsworth more competitive with Wichita's better location, its town leaders commissioned a new short-cut which saved thirty-five

Wichita—Street Scenes—1874 by Clyde Hostetter appeared in the May 2, 1874, edition of *Harper's Weekly*. (Source: Kansas State Historical Society, Topeka, Kansas)

miles and three days. Peeling off the trunk of the Chisholm Trail at Pond Creek Ranch in Indian Territory, the new branch was more westerly, crossing the Arkansas at Ellinwood, thus avoiding both Wichita and the settlements springing up in central Kansas. The result was at least 150,000 head driven to Ellsworth, over two times as many longhorns as the previous year. Of course, Hell was still in session. On August 15, Ben Thompson, former co-owner of the Bull's Head in Abilene, got into a confrontation which led to his brother Billy accidently shooting Sheriff Chauncey B. Whitney. Thus, 1873 was the high-water mark for both Ellsworth's cattle business and its cattle violence.[61]

Wichita was able to hold its own against Ellsworth in the lawlessness department. Rowdy Joe had a spectacular shoot-out with another dance hall owner, leaving one man blind and another dead, and forcing Lowe and his wife to move on once again.[62] The city seems less successful in the cattle department, with evidence of only 65,831 head having been shipped out.[63] Sadly, Margaret Borland's cattle

Halting-Place on the Ninnescah River appeared in the May 2, 1874, edition of *Harper's Weekly*. (Source: Kansas State Historical Society, Topeka, Kansas)

were not among these. Although she had accomplished the singular feat of actually driving the cattle herself to Wichita as owner and trail boss, the financial paralysis following the 1873 panic had dried up the market.[64]

Economic conditions in general did not improve much in 1874. Trailing numbers overall were down by 50 percent. To make matters worse for Ellsworth, despite hiring Shanghai Pierce away from Wichita, the momentum had decidedly shifted to the more southern city. The Kansas Pacific shipped only 18,500 head for the whole system. The Santa Fe shipped 49,730 from Wichita alone. That year, Wyatt Earp moved down from Ellsworth and was hired as

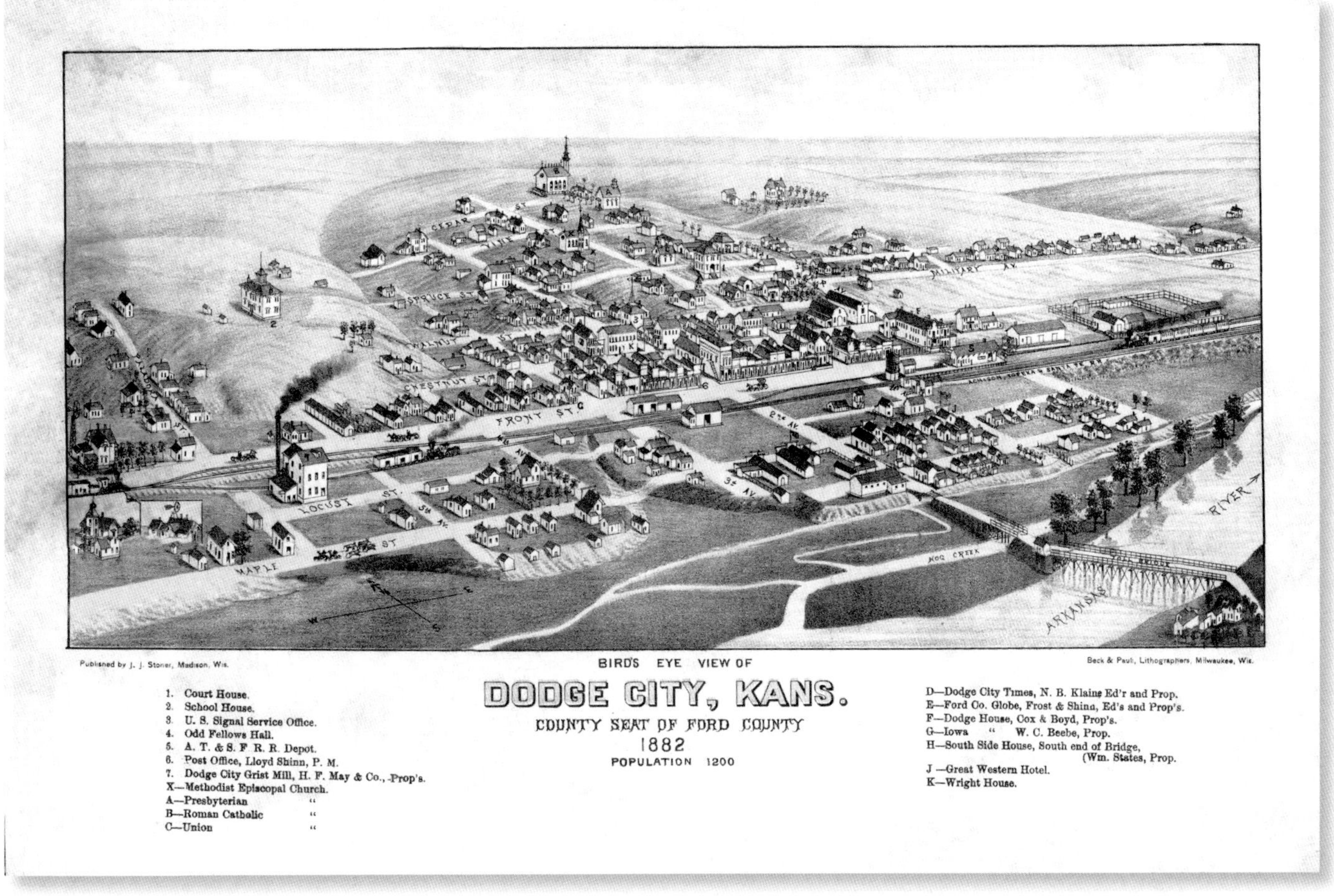

Dodge City, Kansas. This 1882 lithograph provides an annotated bird's-eye view of the streets and buildings in Dodge City, which was the county seat of Ford County. At that time, the population was 1,200. (Source: Kansas State Historical Society, Topeka, Kansas)

a policeman in Wichita.[65] At the same time, more settlers moved into Ellsworth County, increasingly cutting the railhead off from open range. By 1875 Ellsworth was no longer a cattle town.[66] It was Wichita's turn to reign once again as the top Kansas cowtown with 22,569 head shipped on the Santa Fe. In 1876, Wichita would still hold on to the title with 14,643 head shipped, but it was an increasingly diminished prize. Finally, in 1876 the Kansas Legislature passed a new quarantine line to the west of Wichita, ending its cattle days for good as well as the Old Chisholm Trail beyond the border of Kansas.[67]

Dodge City and the Great Western Trail

Situated on the Arkansas River in southwest Kansas, the Santa Fe railroad did not reach Dodge City until September 1872. Initially, the buffalo trade was the mainstay of the new town's economy. Robert M. Wright bought and shipped more than 200,000 hides during the first winter after the rails had arrived.[68] Three factors, however, worked together to soon change the town's financial basis.

First, in 1874 Captain John T. Lytle pioneered a new trail west of Fort Worth through Fort Griffin, Doan's Crossing on the Red River, Fort Supply in Indian

Red River. (Source: Justin Kastner)

Territory, and then to Dodge City and from there up to Ogallala, Nebraska (what became known as the Great Western Trail). As the declining shipping figures for both Wichita and Ellsworth show, in the 1870s the demand for cattle trailed to Kansas railheads increasingly came from the northern Indian agencies and the new ranches in the Northwest. Lytle discovered that his more westerly route saved him between twenty and thirty days. However, the ongoing threat of Indian attack that far west meant that, for the time being, the route was impractical for large-scale trailing.[69]

Second, in 1875 the leaders of the Santa Fe Railroad were told at their annual board meeting that the area about Wichita, like Abilene and Ellsworth before it, was becoming too crowded for droving. To avoid the constantly moving quarantine line for as long as possible, the board chose to develop facilities at Dodge City to carry on their cattle trade.[70] When the Kansas Legislature passed the 1876 Quarantine Law, the initial effect was to reroute traffic on the Old Chisholm trail to Dodge City via the Cimarron River.[71]

Third, when the U.S. Army removed the Indians to reservations by 1878, they dramatically increased both the supply of and demand for beef coming through the remaining Kansas cattle railhead. The Western Trail was now safe for walking longhorns through Dodge; consequently, the city had not one but two supply routes for its markets. Moreover, the U.S. government's actions had greatly expanded the markets themselves. On the one hand, since they could no longer visit their hunting grounds, the new inhabitants of the reservations would need to be supplied with meat. On the other hand, the settlers taking over their land would need livestock for their farms and ranches.

Thus, by 1878 Dodge City was poised to become the greatest cowtown of the trail driving era.[72] The figures for the longhorns coming through Dodge on the Western Trail alone are impressive enough: 201,159 in 1877; 265,000 in 1878; 287,000 in 1880; and 300,000 in 1884.[73] Little wonder then Robert M. Wright, in his role as the city's elder statesman, insisted that "there were more cattle driven to Dodge any and every year that Dodge held it, than to any other town, and for about ten years, Dodge was the greatest cattle market in the world."[74] While Abilene's last year was twice as good as Dodge's best year, even Abilene's impressive record does not come close to Dodge's consistently high average over a period twice as long.

Of course, such prolonged success at serving the cattle trade had to have a profound effect on the local culture. After all, "they only raise cattle and hell in Dodge."[75] Perhaps this is best observed not so much from another catalogue of the number of establishments catering to cowboy vices as the ability of leading figures in the pleasure industry to assume significant roles in respectability society. According to Bat Masterson, in Dodge City, "Gambling was not only the principal and best paying industry . . . at the time . . . it was also reckoned among . . . [the] most respectable."[76] Consequently, one contemporary observer noted that "At Dodge [Luke Short, the gambler-gunfighter-saloon-keeper] associates with the very best element, and leads

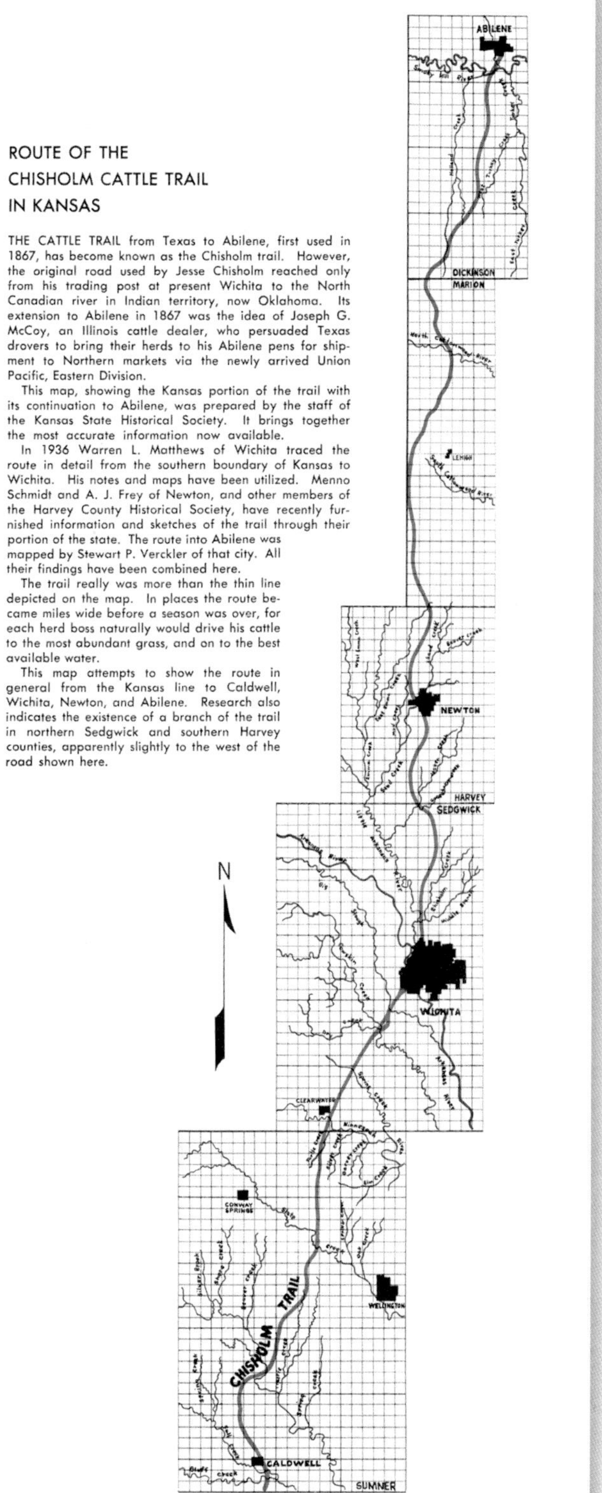

(Source: Kansas State Historical Society, Topeka, Kansas)

(Source: Lewis Browder)

was emerging—barbed-wire. Invented in 1874 and successfully promoted in 1876, this inexpensive means to fence claims on the open range became wildly popular. In 1877 thirteen million pounds were produced. The following year the figure was double: twenty-six million. In 1879, nearly double again: fifty million. In 1880, it was eighty million.[79] Due to fencing by the Cherokee Strip Live Stock Association, the Cimarron Cut-off to Dodge City was closed in 1883. Fencing pressures closed down Caldwell's cattle trade a year later, thus bringing an end to the Chisholm Trail as a sinew of American commerce.[80]

in almost every social event that is gotten up."[77] Indeed, Bat himself, although a professional gambler by trade, was not only an officer of the law, but also awarded a gold watch and a gold-topped cane as the prize for winning the most popular man in Dodge contest at the Fourth of July celebrations in 1885.[78] He, along with Wyatt Earp, were even deacons in the town's first church.

Although pre-eminent, Dodge City was not completely without Kansas cowtown competitors. The Old Chisholm Trail in Kansas had a brief reprieve when the Santa Fe extended its line down to Caldwell, the route's traditional entry point into the state, in June 1880. Being so near the Oklahoma border, it was not affected by the Quarantine Law. However, a new threat to the cattle trade

Perhaps the prospect of impending financial insecurity is what motivated John Henry Brown, the marshal of Caldwell, to turn to a life of crime. Although he had ridden with Billy the Kid, Brown had married a local girl and been given a gold-mounted Winchester in 1883. However, in early May 1884, he and three accomplices tried to rob the bank in Medicine Lodge. They succeeded only in killing two people and getting caught. Brown himself was killed trying to escape a lynch mob, much to the approval of the citizens of Caldwell.[81] Like Charles Gross, those good people could tolerate cowboy violence, but not killing just for money.

By 1884, fencing was even beginning to affect the Western Trail. Following in the footsteps of Joe McCoy, Dodge sought

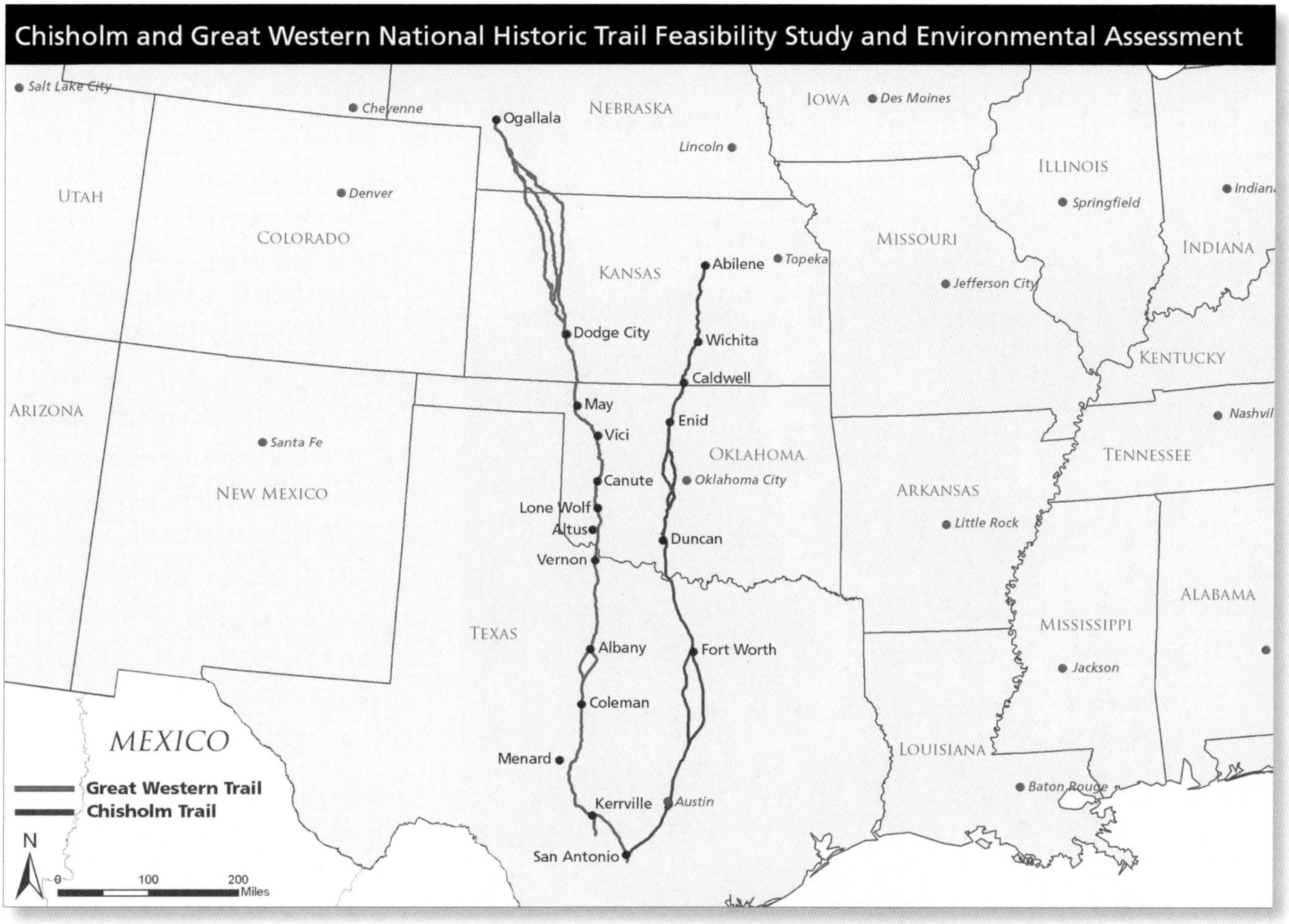

Chisholm and Great Western National Historic Trail Feasibility Study and Environmental Assessment

Salt Lake City
Cheyenne
Ogallala
NEBRASKA
IOWA
Des Moines
Lincoln
UTAH
Denver
ILLINOIS
Springfield
Indian
COLORADO
MISSOURI
INDIANA
KANSAS
Abilene
Topeka
Jefferson City
Dodge City
Wichita
KENTUCKY
Caldwell
ARIZONA
May
Enid
Nashvil
Vici
Santa Fe
OKLAHOMA
TENNESSEE
Canute
Oklahoma City
NEW MEXICO
ARKANSAS
Lone Wolf
Little Rock
Altus
Duncan
Vernon
ALABAMA
TEXAS
MISSISSIPPI
Albany
Fort Worth
Jackson
Coleman
MEXICO
Menard
LOUISIANA
Great Western Trail
Chisholm Trail
Baton Rouge
Kerrville
Austin
N
San Antonio
0 100 200 Miles

Above: Proposed national historic trail routes for the Great Western and Chisholm Trails. (Source: Frank Norris, U.S. National Park Service)

Right: (Source: Justin Kastner)

to bolster its image with an authentic Fourth of July bullfight. Several hundred cowboys turned out to watch five Mexican matadors face off against longhorn bulls. Yet, even such an extraordinary publicity stunt was not enough to stave off the inevitable. The cattle business in Kansas was changing.

End of the Road

As the proposed National Historical Trail feasibility and environmental assessment on the Chisholm and the Great Western Trails will undoubtedly remind us, the trail drives taught a nation about the benefits of bluestem grazing. As a result, many of those who brought cattle to Kansas decided to stay. W.E. Campbell heard the glowing reports of the state's pastures while working on the railroad in Wyoming. Consequently, in 1868 he settled nine miles south of Wichita

(Source: Blair Tenhouse)

and built his fortune pasturing Texas cattle on Kansas grass.[82] E. S. Root did the same in the Smoky Hills.[83] When J.C. Juvenal lost heavily in the 1873 panic, he sold his Indiana operations, stopped trailing, and kept his Russell, Kansas, cattle ranch as his sole means of support.[84] In 1877 Charles F. Colcord's father moved him and his cattle from Texas to near Coldwater, Kansas. There his father established a ranching operation which eventually included the introduction of high-grade livestock.[85] Such pioneers made up the emergent native Kansas cattle industry, and its growing opposition to transient cattlemen ultimately brought trailing to an end.[86]

As local cattle ranchers increasingly invested in upgrading the quality of their herds, the financial loss from continuing outbreaks of Texas Fever in the state mounted. In 1884, the worst-ever epidemic broke out in Kansas, costing cattlemen between $250,000 and $500,000. Not surprisingly, that same year the Kansas cattle barons demanded that Governor George W. Glick institute stricter quarantine regulations. Fulfilling the 1867 prediction of the *Kansas Radical*, one rancher spoke the mind of many when he queried, "[H]ow can it be expected that, with our countless thousands of high grade cattle, we should continue to permit the passage of hundreds of thousands of . . . Texas cattle annually in our tracts, through the heart of our grazing country, leaving germs of disease which have already cost stockmen hundreds of thousands of dollars from losses thereby incurred, without any compensation or direct benefit whatever[?] We are convinced that the time has now come when this should cease."[87]

In 1885 the U.S. Bureau of Animal Industry estimated that Kansas had 1.3 million head of cattle, 28.2 million acres of pasture land, and a cattle industry valued at $56.8 million.[88] Little wonder then that Glick acceded to the ranchers' demands. That year he announced a statewide quarantine during his address to the Second Annual Meeting of the Kansas State Short-Horn Breeders' Association. Three years later the new Kansas Sanitary Commission admitted the obvious—that the state-wide ban on trailing seriously hindered "the commerce in Southern cattle." Nevertheless, its members concluded that it was "absolutely necessary for the protection of our own interest."[89] To paraphrase Cotton Mather, the Texas cattle drives had birthed the Kansas ranching industry, but in the end the daughter ate the mother. From henceforth, Kansans seeking to better themselves through cattle would no longer look to open southern ranges to make their speculative fortune but rather to their own state's fenced-in ranches to provide a faithful livelihood and leave a family legacy.

Chapter 2
Interest and Investment from Abroad

By Justin Kastner

For more than forty years, Garden City, Kansas—the legendary southwest Kansas bastion of cattle feeding in the United States—has been home to the annual celebration that is fittingly named Beef Empire Days.[1] The festive event, which includes educational and recreational activities associated with Kansas beef, continues to be inspired by the presence of pioneering companies, such as Brookover Feed Yards (established in 1951), and many other names in the industry. Beef Empire Days reminds Kansans of both the innovations of southwest Kansas and the "empire status" that characterizes the entire Kansas beef industry, which has consistently attracted interest and investment during the past 150 years.

Almost immediately after joining the Union as the thirty-fourth state, Kansas was the target of noteworthy and strategic investment. Beginning in the 1860s, both cowtowns (like Abilene, mentioned in chapter 1) and populated eastern-market terminals (including Kansas City, featured in chapter 4) attracted cattle and capital from outside of the state. From the historic economical courtship between Texas drover Colonel J. J. Myers and Illinois businessman Joe McCoy[2] to the livestock-raising ventures of George Grant's colony in Victoria,[3] Kansas has been an object of investment, including investment from foreign investors. This chapter paints a picture of this important period in U.S. and Kansas history, which established a solid foundation for future generations of Kansas beef leaders.

The Development of the American Beef Export Trade

To properly understand the almost perfect timing of the Kansas beef industry's birth, it is useful to look back at international relations between the United States and its trading partners, particularly Great Britain, before Kansas statehood.

After Britain acquiesced to American sovereignty on September 3, 1783, the United States and Britain resumed what the Revolutionary War had temporarily disrupted—international trade between the two countries.[4] Later, during the 1830s and 1840s, British economic leaders and activist groups made the case for free trade in agricultural products. Despite their appeals, which stressed the importance of American-British trade relations, the British government did not immediately embrace free trade.[5] Gradually, however, free trade ideas were embraced, and in 1842, livestock imports were permitted into Britain for the first time.[6] The logic of free trade in agricultural products became more compelling when Britain experienced food crises during the mid-1840s. Substantial harvests were reaped in 1842, 1843, and 1844, but a poor yield in 1845 reminded British farmers and consumers of the elusiveness of plenty. During that year, Britain's food supply was beleaguered by bad weather and poor harvests; in Ireland, it was the beginning of a potato famine. The woes of 1845 appeared to prove what free trade advocates had previously claimed: agricultural imports

Previous: (Source: Lewis Browder)

(Source: Lewis Browder)

were needed to ensure a cheap, plentiful food supply for the working classes of an increasingly urban Britain. After the bad harvest of 1845, Prime Minister Sir Robert Peel agreed, and in 1846, Britain completely repealed a number of trade-restricting policies and eliminated all import tariffs on livestock imports.[7] This enabled grain, livestock, and meat to flow freely into Britain.

Despite these changes, American livestock, meat, and grain supplies did not reach British markets in significant amounts for a few more decades.[8] During the 1850s, the United States provided an insignificant amount of British grain imports, which remained low and were primarily sourced from France, Russia, and the Baltic states. American meat exports, which initially were salt-preserved pork and dried beef, also were insignificant and accounted for only 2 percent of the British meat supply. That the United States did not immediately take more advantage of open British markets may be explained in part by (a) the infancy of American and transatlantic transportation and (b) the American Civil War, which served to disrupt

(Source: Lewis Browder)

transatlantic trading.[9] Indeed, several controversial issues—from the American Civil War to American westward expansion—frustrated Anglo-American relations during the nineteenth century.[10] Nevertheless, during the second half of the nineteenth century, Britain's commitment to free trade grew, and agricultural trade with America did, too.

The spirit of free trade was alive and well not only in Britain but also in continental Europe. Following the example set by Britain, Belgium repealed its protectionist agricultural trade policies in 1850, and by the early 1860s,

France, Prussia, and Britain had all entered into a variety of free trade treaties. This commercial diplomacy spawned increased cross-border trading in agricultural products, including livestock and meat.[11] In Britain, where a growing population hungered for more livestock and meat products, producers experienced difficulty in meeting the demand. Britain's domestic meat production was increasing at only one-half of the general population rate of increase, and imports were needed.[12] From the middle of the nineteenth century onward, Britain increasingly met this need through imports, primarily European livestock but also American preserved meat (dried, pickled, and cured beef as well as cured pork products).[13]

By the 1860s, the profile of Britain-bound American exports had noticeably expanded from the staple goods (e.g., cotton and tobacco) that had characterized transatlantic agricultural trading since colonial days. Richard Cobden, one of Britain's most popular advocates of free trade, explained this to British textile manufacturers not long after Kansas became a state. During an October 1862 meeting of the Manchester Chamber of Commerce, he remarked, "You get an article even more important than your cotton from America—your food."[14] Cobden's remarks were made in the unique context of Britain's food shortages (Cobden's remarks came as Britain found itself heavily dependent upon U.S. wheat between 1860 and 1862),[15] but his comments would prove prophetic for the remainder of the nineteenth century.

During the second half of the nineteenth century, Britain was well aware of the importance of international food supplies, particularly those from the United States, and in May 1867, the food committee of Britain's Society of the Arts investigated the prospect of further increasing imports of American meat.[16] The committee interviewed a businessman and a professor who were acquainted with the transatlantic trade in preserved American meat. While one noted quality concerns, both interviewees agreed that American meat would help supply Britain's working classes with nutritious, inexpensive food.[17] During the 1870s, the Liverpool Provision Trade Association and similar groups promoted salt-preserved meat imports that were packed by Cincinnati- and Chicago-based firms such as Armour & Co.[18]

Salt-preserved meat, however, did not satisfy British consumers' preference for mild-tasting food,[19] and businesspeople in the transatlantic region contemplated providing American meat in other forms. This included livestock, which could be shipped alive across the Atlantic, imported into Britain, and slaughtered closer to meat markets. During the late 1860s and early 1870s, firms in Scotland and the United States conducted experiments in the transatlantic shipping of North American cattle.[20] Although the live-cattle trade was not immediately successful, by the mid-1870s, shipping entrepreneurs—most notably New York's Timothy Eastman—were regularly sending American cattle to Britain.[21] Initially, American cattle were allowed into Britain alive. These so-called "store cattle" were fattened by British farmers, slaughtered, and sold as fresh beef. Additionally, entrepreneurs tried shipping refrigerated, fresh beef. Unlike the live-cattle experiments, refrigerated beef shipments were immediately successful. In 1875, Eastman sent the first-ever transatlantic shipment of chilled beef to London's Smithfield Meat Market. A sample of the inaugural shipment was taken to Windsor Castle, and Queen Victoria declared it to be "very good." When American chilled beef reached the port of Liverpool, customers bought up supplies by mid-afternoon. For British consumers, American beef was deemed superior to continental European beef.[22] In 1877, an Englishman who had sampled American beef reported, "It was perfectly fresh, and so pleasant, both to the nose and eye . . . In eating, the grain seemed a little coarser than our English meat, but that was the only difference one could see. It was as sweet; the osmazome seemed more abundant—as odor, fine brown color and good gravy testify."[23]

While the aroma and the flavor of American beef may have been inviting, it was not the only reason why Britons

(Source: Lewis Browder)

were interested in American agriculture. Great Britain was equally, if not more, intrigued by North America's vast agricultural resources and what these resources might mean in economic terms—if transported to the eastern seaboard and across the Atlantic Ocean, American wheat and meat could be a bountiful source of inexpensive food for Britain. Nineteenth-century British politicians and investors actively pursued this possibility by infusing capital into American railroads and transatlantic shipping. Strategically, the British did this to cheapen domestic food prices.[24]

Investment from Abroad

During the second half of the nineteenth century, the United States was the most important recipient of British international investment.[25] British investors first sought to put their capital to work in U.S. railroad enterprises for two reasons: railroads were a necessary part of Britain's strategy for ensuring a cheap food supply, and railroad ventures allowed British investors to earn dividends. Both the British government and its people invested in American railroad securities; after 1850, sales of British railroad securities subsided "only to give place to great expectations from railroad enterprise in the United States."[26] By the 1870s, American railroad securities were being purveyed in Britain courtesy of Anglo-American banks and an increasingly sophisticated stock market culture.[27] Indeed, after the introduction of limited liability laws, the British people found it relatively easy to invest in different ventures related to the transatlantic food trade.[28] In addition to American railroads, transatlantic shipping received attention and money from British investors, and British ships carried most of the transatlantic commerce.[29] By the 1870s, British transportation investment was paying off, and American food supplies arrived into British ports in increasing amounts.

Just as British investment in transportation aimed to secure access to American food supplies, British capital also was devoted to a wide variety of American agricultural enterprises. These ventures ranged from wheat farming and milling to cattle ranching and meatpacking.[30] Journalists such as James MacDonald, who wrote for *The Scotsman* in 1877 and 1878, traveled to North America and reported favorably on agricultural investment opportunities.[31] In 1878, MacDonald's report was published as a handsome book titled *Food From the Far West: American Agriculture with Special Reference to the Beef Production and Importation of Dead Meat From America to Great Britain.* The book has two full chapters on Kansas alone, and topics indexed at the beginning of the two chapters are telling of what intrigued foreign investors during the 1870s. Chapter X features the following preview:

> THE POSITION AND EXTENT OF KANSAS.—ITS GOOD NAME.—SOIL.—CLIMATE.—RAINFALL.—SCARCITY OF WOOD.—THE LOVELINESS OF THE PRAIRIES.—GENERAL ATTRACTIONS AT A CASUAL GLANCE.—IMPRESSIONS MODIFIED AND CURIOSITY AROUSED BY MINUTE INQUIRY.—CROPS : YIELD AND VALUE.—THE DISTRICT INTERSECTED BY THE CENTRAL BRANCH OF THE UNION PACIFIC RAILWAY.—HALF-CULTIVATED HOMESTEADS DESERTED : HOW THIS HAPPENS.—THE DIFFICULTIES EARLY SETTLERS HAVE TO CONTEND WITH.—SUCCESS OF SETTLERS WITH SUFFICIENT CAPITAL.—FAILURE OF SETTLERS WITH LIMITED MEANS.—THE NECESSARY CAPITAL FOR A HOMESTEAD.—MIXED FARMING THE MOST PROFITABLE.—CATTLE IN NORTH-EAST KANSAS.—TEXAS CATTLE IN BAD FAVOUR.—THE VALLEYS OF THE REPUBLICAN AND SMOKYHILL RIVERS.—MR HENRY'S WHEAT-FIELDS.—GRASSHOPPERS.[32]

The next chapter of MacDonald's book features these additional headings:

> NUMBER AND VALUE OF LIVE STOCK IN KANSAS.—TEXAS CATTLE IN KANSAS.—THE CHARACTERISTICS OF KANSAS CATTLE.—CARELESSNESS IN BREEDING.—DEATH-RATE AMONG CATTLE.—THE SUPPLY OF PASTURE.—INTRODUCTION OF SHORTHORNS.—MR CRANE'S RANCH AT DURHAM PARK.—MR GEORGE GRANT'S VICTORIA COLONY.—ITS FORTUNES AND MISFORTUNES.—PRESENT PROSPECTS.—VALUE OF ITS LAND.—MR GRANT'S HERD OF CATTLE.—HOW BRED

The George Grant Villa is located southeast of Victoria, Kansas. Grant and his descendants owned and resided at this homestead from 1872 to 1980. (Source: Blair Tenhouse)

AND MANAGED.—POLLED CATTLE ON THE PRAIRIES.—MR GRANT'S FLOCK OF SHEEP.—THE HON. MR MAXWELL'S LAUDABLE ENTERPRISE.[33]

Indeed, George Grant's colony, which was established in Victoria, Kansas, was a major subject of MacDonald's report on American agriculture. As James Forsythe explained in a 1986 issue of *Kansas History,* Grant came to America because of investment reasons: "Grant hoped to capitalize on the growing demand for beef in England and the desire of wealthy young Englishmen to purchase vast landed estates and to entice Scottish and British capital to America for investment in his venture in Ellis County."[34]

At the conclusion of this chapter appears a scanned copy of a well-preserved investment prospectus that was provided to prospective settlers in London; this copy is worth reading and has been reproduced almost in its entirety. The prospectus notes, among other things, that Victoria is "only thirteen days' journey from London" and has "many advantages for stock farming over the distant colonies of Australia, New Zealand, and the long winters of Canada."[35] While both truth and myth may be found in accounts of Grant and his colony, what is agreed upon is that Grant's colony was the first site in the United States for Aberdeen Angus cattle (in September 2008, the Kansas Angus and American Angus Associations celebrated 135 years of Angus cattle in the United States by hosting a commemorative event in Victoria).[36]

In 1943, a limestone memorial statue was erected in the cemetery where George Grant lays in Victoria, Kansas. In 1973, it was topped with a statue of an Angus Bull. The memorial reads, "To the memory of George Grant, 1822–1878, Founder of the British Colony, Victoria, Kansas. The first Aberdeen Angus cattle imported into the United States were brought to this locality by Mr. Grant on May 17, 1873. His faith in the future of agriculture and stock raising in Western Kansas, evidenced by the purchase of nearly 100,000 acres from the Kansas Pacific Railroad, helped promote the early development of this region." The memorial was rededicated on September 27, 2008. (Source: Blair Tenhouse)

Interest and investment in the American beef industry was frequently tied to international trade. Newspaper reports noted the industriousness of American farmers and America's growing expertise in the handling and transporting of food products.[37] In 1878, an English businessmen involved in the transatlantic cattle trade declared in the pages of the *New York Times* that "every facility is offered on both sides of the Atlantic for the development of a large business."[38] A special Parliamentary commission later reported in 1880 that 33 percent profits

The original stone corral still stands at the George Grant Villa. (Source: Blair Tenhouse)

were possible in the American ranching industry, which prompted more British investment; encouraged by this news, Scottish businessmen from Edinburgh and Dundee proceeded to invest in cattle ranches that stretched from Texas to Saskatchewan.[39] One historian has summarized why Britain was in love with American ranching for both food consumption and investment reasons:

> By 1882, the beef-hungry and capital-laden British had been convinced, through their home press and the government, that the western cattle industry was the panacea for their problems. The investment public needed to invest its surplus capital because it lived off that capital. The English had the money to buy the beef, and the western plains appeared to be the prime producer of it. All that was really necessary was to place cattle, capital, and consumer into the right relationship, and the needs of all could be met.[40]

By the late 1880s, several British companies had established residency in cattle-raising states such as Colorado, Kansas, and Texas, and many of their legacies are still visible today.[41] An interesting example includes the business- and family-related activities of the family responsible for running the Manhattan, Kansas, office of the British Land and Mortgage Company, Ltd. Mrs. Stuart James Hogg, the wife of the company operator, kept a detailed diary of her experience moving from Britain to Kansas, and in 1951, excerpts of that diary were published in the *Kansas Historical Quarterly*.[42]

(Source: Blair Tenhouse)

A Foundation Established

Although foreign capital was generally welcome in agricultural America, "foreign control" was in many cases resented. Consequently, many British-owned agricultural enterprises began to retain American managers.[43] Nevertheless, foreign investment contributed to the westward expansion and the rapid commercialization of North American agriculture, including that in Kansas.[44] As America expanded westward, so did ranching. During the 1870s, Kentucky and Illinois were the states from which Timothy Eastman sourced his export cattle,[45] but by the mid-1880s, 95 percent of American cattle that were raised for the British export market were sourced west of Chicago.[46]

Nineteenth-century international interest and investment created a foundation for the Kansas beef industry. While primarily Great Britain has been referenced here, other countries like Germany also contributed; as chapter 5 reveals, many of the families (e.g., the Schlickau family) who emigrated to Kansas were German. The remaining chapters of this book showcase how Kansans have worked diligently to build a beef production empire.

ENGLISH ENTERPRISE

IN

AMERICA.

NOTES

ADDRESSED

TO INVESTORS AND SETTLERS

CONCERNING THE ESTATE OF

VICTORIA

(ELLIS COUNTY, KANSAS, U.S.),

THE PROPERTY OF

Mr. GEORGE GRANT,

(Late of the firm of GRANT & GASK, *now* GASK & GASK,
Oxford Street, London.

LONDON:
PRINTED BY T. PETTITT & CO., 22 & 23, FRITH STREET, SOHO.

1876.

NOTES

ON THE

VICTORIA ESTATE

(ELLIS COUNTY, KANSAS, U.S.),

THE PROPERTY OF

Mr. GEORGE GRANT,

(*Late of the firm of* GRANT & GASK, *now* GASK & GASK, *Oxford Street, London.*)

LONDON:
PRINTED BY T, PETTITT & CO., 22 & 23, FRITH STREET, SOHO,

1876.

LONDON,

April 10th, 1876.

MR. GEORGE GRANT informs his friends, and those who may desire to correspond with him, that he leaves England, per the steamship "Adriatic," White Star Line on the 13th April, and that for the next six months his address will be Victoria, Ellis County, Kansas, U.S., where he will be glad to see friends and intending colonists who will be welcome to his hospitality and every information concerning the Victoria Estate, and the country generally.

The first shipment of thorough-bred cattle, sheep, dogs, &c., purchased by Mr. Grant during the past two months, will be dispatched per steamship "City of Limerick," May 10th, from Liverpool to Philadelphia.

Amongst them are four heifers and one bull, **ROYAL GEORGE,** by Mr. BOOTH'S celebrated bull, ROYAL BENEDICT, and DAM CREAM THE FOURTH (England's Glory), the pick of Her Majesty's herd, from the Home Farm, Windsor.

These animals will be landed at Philadelphia and exhibited at the **GREAT CENTENNIAL SHOW** before being forwarded to Mr. Grant's Home Farm, Victoria, Kansas, U.S.

INTRODUCTION.

VICTORIA, the subject of this pamphlet, is the name of an estate belonging to Mr. George Grant, late of the firm of Grant and Gask (now Gask and Gask), Oxford Street, London, and situate in the western portion of the State of Kansas, which is itself the central State of the Union. In the following remarks it is proposed to afford intending settlers some concise information as to the property, which covers an extensive area of rich and well watered grazing land in Ellis County, of that State. Further information may be obtained from any of the gentlemen whose names are on page 29.

A 2

INFORMATION TO INVESTORS AND SETTLERS.

SINCE Mr. Grant's return to Europe in the end of December last, he has received numerous letters requesting detailed information about his property, the class of persons who should emigrate to it, and the amount of capital emigrants should possess in view of a fair start, with reasonable hopes of success, in the agricultural pursuits of the country.

Mr. Grant's knowledge and experience—gained by three years' residence in Kansas—of the stock which he has himself maintained and sustained there, and of the crops he has himself raised during the past year, enable him to give the desired information, but he prefers doing so, as far as practicable, on the evidence of others rather than on that of himself; and because, being the owner of the land, he might be thought by some to be prejudiced in its favour.

The following, published at New York, in the "American Agriculturist," of February, 1876, are facts supplied from the actual entries in the books of his farm at Victoria, and gives a correct statement:—

7

During the past three years Victoria Colony has been visited by a large number of gentlemen, many of them acknowledged skilful agriculturists, and well known in Europe and America; some of the opinions expressed by them, and by representatives of the Press, are annexed, and which, Mr. Grant thinks, will give all necessary information to such as may contemplate stock farming, &c., &c., &c., with the advantages of good sporting, good society, and a fine climate, with certain fortune to the industrious and enterprising; and as Victoria Colony is only thirteen days' journey from London, it must be evident it has many advantages for stock farming over the distant colonies of Australia, New Zealand, and the long winters of Canada.

OPINION OF BRITISH AGRICULTURISTS.

VICTORIA, KANSAS, *September* 10, 1873.

To Geo. Grant, Esq., Victoria, Kan.

DEAR SIR,—Having spent some time at Victoria, we wish to give some expression of the opinion we have formed, after the most careful investigation we are able to make, that your extensive property is possessed of extraordinary resources, and offers a very inviting field for agricultural enterprise. Having visited Kansas in August and September, when the land is at its driest, we imagine we have seen its least inviting aspect, and, at first sight the prairies seemed arid and parched. A very little experience and investigation has proved to us, however, that this is far from really the case. We find that the grass which looks withered and burnt up, is in fact a natural hay, which feeds as well as when green. Sheep, cattle and horses (like the native buffalo) eat it readily, and fatten quickly upon it. A special instance of this is given by your own thirty English rams, which we observe were in full condition, though fed upon the natural grass alone, and while it is well known that cattle thrive on the prairies, we are inclined to think that even more money might be made by the breeding of sheep, and we would suggest that the native breeds might be greatly improved by crossing with English rams, such as the Leicester, Cotswold, Shropshire or Oxford Downs. While flocks and herds can be maintained on the prairie in its natural condition, we are well assured that the land may be most profitably cultivated, and is capable of producing almost any kind of crop.

8

The soil consists of a rich dark brown sandy loam, extending downwards far beyond the reach of any plough. It is entirely free of stones and of every other obstruction; but limestone, suitable for every building purposes, crops up in every portion of the property. The stratification and general lie of the beds render them capable of being easily quarried, while the general softness of the stone, on just being exposed, makes it easy working for building purposes. Like many of the softer limestones of England, it hardens by exposure, and at the same time has a soft and attractive appearance to the eye, and is in every way eminently adapted for building purposes, as is evinced by the numerous structures of similar stone erected along the railway. Good water for household purposes is always found at from ten to forty feet below the surface. While some of us believe that the ordinary plough is amply sufficient for the cultivation of the soil of Victoria, we are unanimously of opinion that no spot of the world can offer a finer field for the unobstructed operations of steam-ploughing. Considering that one of the great lines of railway passes through Victoria, thus securing a ready market for every description of produce, and affording communication with every part of the United States, we cannot but believe that its rich and fertile soil offers an exceptionally favourable field for the arable as well as the pastoral farmer.

We have enquired very particularly as to the health of the inhabitants, and are convinced that this part of Kansas will contrast favourably in this all-important respect with any part of the world. In conclusion, we may state that so convinced are we that the settlement of Victoria will rapidly rise in market value, that we have ourselves made investments in it, which we confidently expect will prove profitable.

We remain,

Yours faithfully,

(Signed) R. SCOT SKIRVING.
A. BETHUNE.
CHARLES PRESCOTT.
ROBT. W. EDIS, F.S.A.
JOHN FERGUSON.
THOS. R. CLARK.

MEMORANDUM.

Mr. R. SCOT SKIRVING, Campton, East Lothian, President of the Royal Physical Society, and ex-President of the Scottish Chamber of Agriculture, is an extensive practical agriculturist, and familiar with the cultivation of land for all crops and with cattle raising and feeding. Mr. BETHUNE, of Blebo, a Deputy-Lieutenant of Fifeshire, is a director of the Scottish Chamber of Agriculture, and of the Scottish Steam Cultivation Company; besides having a thorough knowledge of cattle and crop raising, he has a practical knowledge of steam cultivation. Mr.

PRESCOTT is an English country gentleman. Mr. EDIS, F.S.A., 14, Fitzroy Square, London, is an architect in extensive practice. Mr. FERGUSON, Brae of Coynach, Mintlaw, Aberdeenshire, combines with agriculture a knowledge of civil engineering at home and abroad. Mr. CLARK is an American gentleman, who, after seeing Victoria, has invested in 1920 acres of Mr. Grant's land. Any of the above gentlemen may be communicated with on the subject.

STOCK FEEDING.

The stock these gentlemen saw grazing and thriving at Victoria, on the natural grasses of the plains, without any addition whatever of other food, included the short-horn and polled bulls, and Scotch and English rams, Mr. Grant had taken out in the April preceding from this country. In so far as stock feeding is concerned, there is thus direct evidence that cattle or sheep of the finest kinds thrive in this portion of the State. That such stocks thrive in the more eastern portions, is proved by the fact that at the Annual State Fair in 1872, there were exhibited 95 entries of shorthorns, Jerseys, Devons, and grades of a quality able to take a position in the front ranks of any State fair; 12 thorough-bred stallions, 12 thorough-bred mares, 50 horses of all-work, 50 mares of all-work, 17 draught-horses, 16 draught-mares, 30 matched horses and mares, 38 geldings and mares for harness, 20 saddle-horses, 30 asses and mules. The exhibition in sheep consisted of 22 entries of long wools, and 20 entries of fine wools. The swine shown comprised 74 entries of Pollard China, Chester whites, and other large breeds, and 67 entries of Berkshire, Suffolk, Essex, and other small breeds. The poultry showed 94 entries of Cochin China, Game, Dorkings, Pollards, Spanish, Brahmas, Bolton Greys, Dominiques, Bantams, Hamburgs, turkeys, ducks, geese, Guinea fowls, pea-fowls, pigeons, Howdans, Leghorns, and black Incas. In the implement section there were shown ploughs, thrashing, mowing, and other machines, steam engines and machinery, worked metal goods of all kinds, vehicles, and furniture manufactures. In farm produce were exhibited winter and spring wheats, rye, oats, Indian corn, buckwheat, Timothy seed, clover seed, blue grass seed, with crossed varieties. In vegetables were potatoes (late and early), sweet potatoes, onions, mangolds, parsnips, celery, cabbage, tomatoes, egg-plants, white beans, lime beans, castor beans, pumpkins, squashes, water melons, musk melons, carrots, tobacco, pea-nuts. There were also entries of flour, starch, butter, cheese, bread, cakes, and pickles; also trees and shrubs, with apples, pears, plums, peaches, quinces, nectarines, grapes, and a great variety of flowers and vegetables familiar to all in Britain.

In the middle and eastern divisions of the State there are thus raised most of our British crops and stocks, and all of them are brought to maturity under the most favourable conditions for utilization. The whole middle division of the State needs only to be occupied and cultivated to realise like products, in the opinion of the friends who visited Mr. Grant in July and August last.

OPINIONS OF THE PRESS.

VICTORIA COLONY, KAN.

RICH NEW YORKERS MAKING THEIR CHILDREN ABSOLUTELY INDEPENDENT.

Mr. J. G. Gunther, of the great Broadway fur importing firm, recently said:—

Each of my younger sons, Frederick W. and John J. Gunther, is the owner of a square mile of land in Victoria Colony, Ellis County, Kan. The Victoria River flows near my sons' farms, and increases their natural fertility. The virgin soil, upon which buffaloes have herded for many years, is in the most productive quarter of the colony, sixteen feet thick, and the least fertile three feet, and black as ink. Owing to its heaviness, due to its richness, the earth has to be tilled with ploughs drawn by mules. Fertilizers will not be necessary there for fifty years. My boys raise "sod" corn, "alfalfa," coarse grass for the stock, and millet principally. Last year grass-hoppers devoured their crops, but this year my boys have secured far more corn, alfalfa, and millet than even the old settlers, the corn harvest, amounting to thirty bushels per acre, and the alfalfa being mowed three times. My sons have 1200 sheep, recently purchased at a stock sale in Pueblo. Great corrals enclose these sheep, and a herder, aided by two Scotch sheep-dogs that I imported, takes care of them. Twelve rams, of the merino variety, blend the characteristics of the Scotch and the Kansas sheep, and the result of the alliance is a superior animal. The short-horned cattle of England flourish in Victoria.

Mr. George Grant, owner of almost the entire county, lately bought Flodden, the half-brother of Thormanby, who won the Derby two years ago, and 500,000 dollars in one year, for 16,000 dollars, and the gentry about Mr. Grant are buying all the retired racing mares that they can find hoping to obtain a fine class of horses through Flodden. My boys' neighbours, for the most part, are the younger sons of English noblemen. Being unable to put their sons into the army, the navy, or the church, these noblemen have established them in Kansas, at the expense of about 15,000 dollars per man, choosing to see them growing old as tillers of the soil rather than as fixtures in banks or insurance houses, or tradesmen's shops.

Mr. Grant sells land to deserving young men for eight or ten dollars an acre, permitting them to pay for it when they can, charging them, interest at the rate of 6 per cent. per annum. My boys are unmarried, and have only two servants, a man and his wife. The man attends to the horses, and does the heavy work about the house, and his wife is cook, chambermaid, and dame of the poultry yard. Frederick and John work in the field when they have reason to, and when they have not they mount their horses, collect their dogs, and join their neighbours in a hare or antelope hunt.

St. Louis being only a day's ride from them, my boys have a large and accessible market for all their wares. In ten years they will be independent. My outlay in establishing them was only 16,000 dollars, and, I assure you, that I deem it well spent money. If my boys had stayed here they would have been in some bank or insurance office, fixed for life, with some 1200, 1500, or 2000 dollars, unable to support a family, and exposed to the danger of forming bad habits. They are happy, and would not return to New York.

THE VICTORIA COLONY,

Stock-raising in Kansas—The Grant Estate—Covered Corrals for Sheep—Profits of Farming on the Prairies.

[From an occasional Correspondent of the *Tribune*.]

Victoria, Ellis County, Kan., Aug. 23.—Seated on the piazza of George Grant's house, with the cooling breezes constantly blowing across the plains, it is difficult to believe that the thermometer stands at 90°, or that the summer weather in other parts of the country is un-

B

comfortably hot and oppressive. The nights here are always cool and refreshing. Three years ago Victoria was unknown, but with the advent of the Kansas Pacific Railway a new order of things has taken the place of the old savage life which had existed for centuries. All along the line of the road towns and villages, with here and there a comparatively large city, have grown up, making central points from which civilization is extending for many miles into the interior. Settlers are becoming more numerous, and are taking advantage of the provisions of the Homestead Law, which gives them a sufficient number of acres on which to build a home and to secure a comfortable existence. Law and order prevail, and property is as safe and well protected as in the more densely populated States of the East. These plains of Middle Kansas are pre-eminently adapted by nature to the nourishment and protection of cattle, sheep, and other live stock. They have been aptly termed the stock grazing ground of the nation. Millions of cattle, sheep, and horses can be fed, watered, and cared for on these plains, summer and winter, at the lowest possible cost. Nature has bountifully provided nutritious grasses for sheep and cattle, which with the exception of a very few days in the the depth of winter, is accessible the year round. This new ground, opened up for the enrichment of the nation and the development of its resources, is also remarkably fertile in the production of grain. Rye, wheat, barley, millet, corn, and alfalfa clover, yield immense crops to the acre, and besides furnishing winter feed for the horses, cattle, and sheep, return a large money income to the farmer.

Mr George Grant, of London, England, soon after the Kansas Pacific Railroad began to run its through trains, was induced to visit this part of the United States, with a view of investment, and after spending several weeks in travelling over Kansas, saw so many evidences of the natural advantages of these plains, that he decided to invest a large sum of money in founding a colony, which should test practically and intelligently the special adaptation of these prairies to stock-raising, and bring into the market a part of the country, hitherto regarded as barren and unfruitful. Mr. Grant subsequently entered into a contract with the Kansas Pacific Railway for the purchase of a large tract of land. This immense estate he divided into sections or stock farms one mile square, which are well watered and suitable for raising stock and grain. Many of the sections, from the peculiar formation of the land, contain natural corrals or shelters for sheep and cattle, and need only a short line of fence to afford a perfect winter protection for the stock. The colony is also advantageously situated, a portion of the estate lying contiguous to and surrounding Fort Hayes, one of the principal permanent military stations of the West. Denver, with its wealth and comforts of civilized life, is within an easy journey westward by rail, when one leaves Victoria in the early morning, is reached at supper time without fatigue or discomfort. Equally distant from Victoria, eastward, is Kansas City, and a nights' ride beyond terminates at St. Louis. New York is only sixty-two hours from Victoria, and the traveller may breakfast at the Langham Hotel, London, in less than 14 days after leaving Mr. Grant's hospitable mansion.

The Victoria estate differs from other settlements in this country in several important particulars. Mr. Grant has subdivided his estate into large farms, and sells only to settlers of assured character and position, who will devote themselves to raising improved breeds of sheep, cattle, and horses, and cultivate the land according to the progressive farming ideas of the present day. The unsurpassed advantages of the climate, soil, grass, water, and shelter of this district are peculiarly favourable to the growth and improvement of live stock, and Mr. Grant believes that a breed of sheep, cattle, and horses will be developed on these plains fully equal to the best breeds of Europe. Foot rot and kindred diseases are unknown here, owing to the high elevation and dryness of climate. Mr. Grant began his new enterprise by importing English long-woolled rams, selected from the first flocks of England, chiefly Oxford Downs, Cotswolds, Lincolns, Leicesters, and Shropshires. He then purchased Colorado graded ewes for breeding, and the experiment of crossing the merinos with the long wools resulted most favourably. Besides the general improvement in form, size and weight, they clipped from seven to ten pounds of wool, being nearly double that of the native stock. The English rams cost from 150 dollars to 250 dollars each, but this outlay has been amply repaid in the marked improvement shown in the merinos crossed with this best imported blood. The sheep are divided into flocks of about 1,000 each, under the charge of separate shepherds, and remain out in the open prairie more than three-fourths of the year. Mr. Grant has built extensive covered corrals of the Kansas stone, found here in great quantities, for winter shelter, and in stormy weather feeds crushed corn. The cost is comparatively small, and out of a flock of 7000 ewes, there was a loss of only 4 per cent. during the winter. This Mr. Grant attributes to the perfect shelter and corn feed. Experiments made by the head shepherd with certain flocks, by feeding hay instead of corn, produced a greater death-rate, and corn was substituted with most beneficial results. Mr. Grant sums up his experience on this point in the opinion that proper shelter, with a feed of crushed corn, will carry a flock of sheep safely through the winter with no appreciable loss. Mr. Grant has increased his flock this season to 7000 ewes, and he is arranging to have on his estate 100,000 of improved breeds within five years. An estimate has been made of the probable results of sheep-farming at Victoria, basing the figures on past experience, and it shows that, beginning with a flock of 1000 breeding ewes and twenty rams, the annual increase and profits in ten years, at the lowest estimate, will amount at least to 200,000 dollars. As to cattle, Mr. Grant uses only first-class imported short-horns, of the Booth and Bates strain, some of which are the finest animals ever seen in this country. He has also

B 2

four bulls, known in Europe as the pure Angus Poll, but almost unknown here, and which, in the London market, bring a higher price as beef than any other cattle. In the ultimate success of this breed in crossing native stock, Mr. Grant has the fullest confidence. These noble-looking animals, full, round, and deep-chested, with short legs, are exceedingly robust in appearance, and may be described as "all beef." At their present age, three years, they weigh on an average 2200 pounds each. The 200 young calves sired by them are fine specimens of the improved breed, and closely resemble their sires. In all, Mr. Grant has 500 native cows of the best class and approved colours, from which, and his frequent purchases, he will possess, in a few years, one of the finest herds in the United States. These cattle are under the charge of herders, and are quartered for the summer on what is known as the Smoky Hill Ranch, which is supplied with running water from the Smoky Hill River, and sheltered by cottonwood and elm trees. This ranch is about fifteen miles from Victoria depôt. The winter ranches are on the Victoria River, and the sheep and cattle occupy a frontage of clear running water extending fully ten miles in length. In horses, Mr. Grant has about thirty very fine brood mares, most of them of high pedigree, by sires such as Bonnie Scotland, and other thorough-breds. These have been served by his imported stallion Flodden, sired by the celebrated English stallion Thormanby, the winner of the English Derby in 1860. The pedigree of Flodden dates back to the reign of Charles II. He is a gold bay, with black points, 16½ hands high, and shows his lineage in every movement. The crops of Middle Kansas, this season, are unusually fine.

(*From the* Colarado Farmer, *December* 2, 1875.)

Observing notices in the papers about Victoria Colony, and having a desire to look over that place and examine the stock kept upon it, we took the morning train at Denver, and by two o'clock next morning landed at Victoria, and were soon in the comfortable stone mansion of Mr. George Grant, taking our morning nap. After breakfast a stroll about the immediate premises occupied the forenoon. It is our purpose to write more particularly about the stock, the facilities for handling it, and its condition. A full description of the blood of this stock, the course pursued in breeding, together with the whole minutia of its management in this colony, would far transcend the limits of our space, we will therefore attempt only a general outline.

THE HOME CORRALS

Consists of about fifteen acres, which includes the hay-stack yards, stable yards, one cattle yard, sorting and dipping pens for sheep, &c. Here is stabling for about thirty head of horses, and, at this time, it is full, and adjoining is the blacksmith shop. One of these yards is fitted with shelter for the imported sheep, of which there are about two hundred ewes and seventy rams, selected from the different varieties, and combing wool and mutton breeds. Among others were Lincolns, from the flock of Mr. Kirkham, Grimsby, Lincolnshire, England, whose name is a sufficient guarantee for their superior quality and purity of blood. These sheep are excellent animals of their kind, and Mr. Grant expects to set at rest, at least in his mind, whether they are the sheep for this country and for crossing with the Mexican or common sheep of the country. In looking over the flocks of grade sheep, we were very much pleased with the Lincoln and Shropshire Down crosses —the former on account of the improvement of fleeces, and the latter for the improvement in carcass as well as fleece.

THE STABLE OF IMPORTED CATTLE

Are especially worthy of notice. In this royal row of cattle is the two-year-old short-horn bull Count De Brunnow. His is the highest pedigree of the Booth strain, and he is worthy of it. (See pedigree 13,725, American Herd Book.) Mr. Grant remarked that this animal was named after his old friend, the late Count De Brunnow, who was for more than thirty years Russian Ambassador at the English Court. Among other Durhams, are two heifer calves from this bull that show progress in breeding—first, Lady Mary, dam Fairy Vernlane (American Herd Book, vol. xiii, page 584). Also Ida, dam Niobe VIII., imported (American Herd Book, vol. xiii, page 839). The fine appearance of this calf will attract the attention of the most casual observer, while in fancy points of excellence she will bear the inspection of the most critical. She bids fair to grace the royal ring and bear off many a blue ribbon. In this stable are found

IMPORTED BLACK POLLED ANGUS BULLS.

This blood being but little known in this country are perhaps not appreciated. As evidence of the fine quality of their beef it is only necessary to state that Mr. Grant showed us a letter from a friend, Mr. W. Grant, Drumdelgie, Scotland, which says that the beef of these cattle is now selling in the London market by the carcass for 22 dols. per hundred. They will, when full grown, weigh 1600 to 2200 pounds. The grades from this stock out in the herds exhibit strongly the characteristic of the sires. In driving through a herd of eight hundred head, it was singularly remarkable to see red, dun, and even light mixed coloured cows, with calves jet black, hornless and otherwise so perfectly resembling the sire. This is a most favourable recommendation of their blood. They were imported by Mr. Grant for use upon the Victoria stock farm, and bred by the late George Brown, Esq., Westertown, Fochabers, Morayshire, Scotland.

Near the cattle stable is the yard of imported combing wool and mutton sheep above mentioned, among which are thorough-breds of Cotswolds, Lincolns, Leicesters, South and Oxford, and Shropshire Downs. For lustre combining wool, Mr. Grant prefers the Lincolns, and for mutton and medium wool the Shropshire Downs. Not only in the thorough-breds does this choice seem to be a wise one, but they are markedly preferable in the grades bred from a cross of them on his large flocks of common coarse wool sheep. In the afternoon we took a short drive out seven miles to the

BIG SHEEP AND CATTLE CORRAL.

This sheep corral encloses five acres, and its roofing will shelter eight thousand sheep. It is located on the south side of the Victoria stream, and is enclosed by an excellent stone wall, which takes in three acres in the bottom and rising over a high bluff, includes two acres overlooking the bottom. This lofty eminence is used for summer corrals. In the lower part in one corner is a stone house for shepherds. The sheep sheds extend around the wall and through the middle, dividing the corral in two. The whole length of the shedding is 1000 feet, and 22 feet wide, with a comb roof made of 12-foot pine lumber, with one eve resting on the stone wall, and the other coming within four feet of the ground, making at once a complete roof, wind-break, and shelter. About three miles farther up the stream, on the site of old Fort Harker, is the

GREAT WINTER CATTLE CORRALS.

Here are about thirty acres, nearly surrounded by the water of the stream, with plenty of timber on its banks. The old fort has been under the command of many well known officers, such as Generals Custer, Sherman, Sheridan and others, while many weary soldiers have bivouaced on its grounds. It is now to be made the comfortable winter quarters of hundreds of cattle. The shelter is made by a high stone wall laid in live mortar, the top of which reaches to the comb of the shed roofs, which are sixteen feet wide, and extends the whole length of the walls on both sides. The combined length of these walls (there being two parallel with each other) is twelve hundred and forty-eight feet. It matters not from what direction the storm may come, the cattle find protection on the opposite side. It will shelter from one thousand to one thousand five hundred of yearlings and calves. The extensive and liberal provision for stock in these corrals is not excelled even in Kansas, or elsewhere in the older settlements.

From this place we returned to Victoria, and next morning started out on a

FORTY MILE DRIVE.

Mr. Grant had a number of visitors, who accompanied him to-day, among whom were two ladies—Miss Alice Matthews, of Chicago, and Miss C. W. Davis, of Greenfield, Massachusetts. At the end of five miles we passed the residence of Mr. Maxwell. He was absent, but his improvements showed his handiwork. These are most substantial —a large mansion, stables, and corrals of stone, &c. At least 25,000 dols. has been expended here. Next we come to the stone mansion of the Gunther Brothers, another 25,000 dol. investment. They are young

17

gentlemen from New York City. This year they raised twenty-five bushels per acre of sod corn, besides good crops of Hungarian, oats, and alfalfa; also melons, pumpkins, and potatoes in abundance. They sowed fifteen pounds of alfalfa to the acre, and it grew up and was cut with the oat crop for feed. It bids fair to do well. They have a fine flock of one thousand six hundred improved sheep, upon which they are going to use Merino rams, selected from the celebrated flock of Samuel Archer, Kansas City, Mo. From here we drove, in a south-west direction, several miles, to Mr. Grant's summer cattle ranch, on the Smoky Hill River. Here were eight hundred head of cows and heifers, and calves, in very fine condition, full and comfortable, basking in the sunshine of that lovely Indian summer day. On the high bluffs of this river at this point, our eyes were treated to the grandest view of rolling prairie country which we ever beheld. Even the ladies could scarcely refrain from waving their handkerchiefs at the sight, and they thought it would be a fine place to live if it were only the suburbs of a large city. Mr. Grant informed us that he camped here the first night when out with his survey, or prospecting the country, with a view to purchasing. Leaving here, and travelling eight miles homeward, we came to the handsome residence of Mr. Shield, a New Yorker, who also has a stone mansion, barn, corrals, and other improvements, well arranged for sheep. At this place Mr. Grant showed us his picked lot of fifteen hundred yearling ewes. For sheep of their grade, considering the number of their flock, the size and condition of the animals, we never saw them excelled. As the sun was now on the wane, we began to hasten homeward, passing another flock of sheep, and still another, until the ladies began to consider the sheep business rather monotonous. Having had a drive of forty-four miles, we reached Victoria as the shades of evening gathered, where the evening repast awaiting us, was not only a festive board, but a substantial luxury.

This day of drives among the flocks and herds, ranches and ranchmen—lords of this little creation of their own—was a day of pleasure and profit. The next day a shorter drive, of some twenty miles, took us by several flocks of sheep, corrals, and the bottom lands and hay fields, where are some seven hundred tons of Hungarian and millet, carefully stacked. The buffalo, or gramma grass, passed over in these two days, and which is the summer and winter feed, cannot be excelled either in abundance or for quality. It is now cured, and is so thick upon the ground, that in walking over it we could only compare it to a velvet carpet.

Mr. Grant has about ten thousand sheep, all fat—all his stock are fat, nor could they be otherwise with such grazing. He prefers to secure for winter from thirty to fifty tons of millet, or hay, to each one thousand sheep. Last year the feeding and keeping of his sheep cost him 76½ cents per head. In this estimate were the imported long wools, which cost 11 dols. per head, mainly because the flock was so small. Last winter he experimented with two flocks—feeding one lot each day two

18

pounds of hay per head, and the other flock half pound of ground corn per head each day. Those on the corn did much the best, and came out in the spring in the best condition. He hopes this will be remembered by our flock masters the coming winter, and the plan generally tested.

His arrangements for dipping sheep are complete. There are two boilers of one thousand gallons each, made of wood, lined with zinc, and having iron bottoms. They are set together in a stone furnace, arranged with faucets to draw off the liquid into the dipping vat, which is eighteen feet long, three feet wide, and six feet deep, with bottom sloping up into the dripping floors, each of which hold one hundred sheep. The temperature of the liquid in the vat is kept at 100 degrees for long or loose wools, and 120 degrees for Merinos. The liquid is made by steeping twenty lbs. of plug tobacco, and five lbs. of sulphur, to one hundred gallons of water. No arsenic or other poisons are used, and it is claimed that none is needed. The sheep are made to approach the vat through a long shute, and when started will jump in of their own accord. In this way five men will put through three thousand sheep per day in good shape. The blood of stock kept, and the liberal provisions here for keeping it, gives assurance that the business will be a great success.

We did not intend to allude to anything but the stock; yet to show the inclination of the settlers, we noticed that preparations are being made to put the public school system in operation as soon as possible. Measures are taken, and trustees appointed for the purpose of building a 15,000 dol. church, to be erected on two acres of land, inside the town limits of Victoria, donated by Mr. Grant, who gives, in addition, 1,000 dols. towards the good work.

Our visit to Victoria, commenced among strangers, was socially pleasant and profitable, and ended, we trust, among friends, and will surely be remembered as a bright spot in the pathway of life.

U. S. PATENT OFFICE, WASHINGTON, D. C., Jan. 16, 1876.
(To the Editor of the U. S. Economist).

DEAR SIR,—Believing it will interest you, if not your readers, I send you a late account of the very extensive operations in land owning and stock breeding of Mr. George Grant, in Western Kansas. Mr. Grant is a Scotch gentlemen of intelligence, wealth and enterprise —formerly a London merchant—who, coming to America less than three years ago, first broke ground on what is now one of the largest estates in a single hand on the globe. It includes the site of old Fort Harker, that well-known frontier post, which has been commanded at different times by such distinguished officers as Generals Sherman, Sheridan, and Custer. It is drained by the Smoky Hill and Victoria Rivers, and penetrated by the Kansas Pacific Railway.

Mr. Grant's experiments in sheep-breeding (his specially) seem to have been eminently successful thus far. Eighteen months ago he wrote to me, stating his determination to cross the English long-wooled breeds imported by him, upon our American merino and grade merino, as well as upon the common Mexican ewes. He then said that, although many American breeders are opposed to such a cross, he was of opinion that it would prove a success, not only in improving the wool, but the mutton as well, and that he hoped to be able in less than a twelvemonth from that time to report the good results which he anticipated from the proposed step. The results announced in the account which I send you, more than bear out the early anticipations of the sagacious breeder, and fully vindicate his judgment upon this point.

Mr. Grant sailed from New York, for Europe, on the 8th January, in the *Celtic*, and expects to return about May next, bringing with him blooded cattle, sheep, and horses. These, with selections from his Kansas rancho, he intends to exhibit in the live stock department of the Centennial Exhibition at Philadelphia. He is doing a grand, good work for himself and the country. Yours, H. G. OTIS.

STOCK FARMING IN KANSAS.

GEORGE GRANT'S VICTORIA COLONY—SOME FACTS CONCERNING ONE OF THE LARGEST LAND AND STOCK ENTERPRISES IN THIS COUNTRY.

(Letter in the *New York Times* of January 3, 1876.)

Probably one of the largest and most interesting experiments at stock farming on a mammoth scale in this country, is that which was instituted some three years ago, in the then uninhabited region of Kansas, by Mr. George Grant, of London, England. Mr. Grant's estate is known as the Victoria Colony, and is situated in Ellis County, Kansas, on the Kansas Pacific Railroad, 250 miles west of Kansas City. In 1873 Mr. Grant, who was in this country, was induced to visit that region, with a view of investing in real estate, and, after travelling over various portions of Kansas, finally settled at the point named, and purchased the greater part of Ellis County. His first purchase gave him an area of many square miles, to which he has since added several other purchases, so that he now owns one of the largest tracts of land in this country. The first arrival of settlers to the new colony was in May, 1873, and at that time, with the exception of the railroad depôt, and the reception-house, built by the railroad company, under contract with Mr. Grant, for the use of the colony, there was not a house within ten miles. At the present time, so rapid has been the growth and improvement in the wealth and material prosperity of the colony, that there is not an acre of land for sale within ten miles of Victoria, on the south, and there are, along the line of the Victoria River, a number of fine limestone residences. Victoria Colony differs widely, and in many important respects, from any other colony in the United States. Mr. Grant does not sell to any one party less than a section, or 640 acres, of land, while some of the residents, or those about to become so, have purchased as many as eight sections, or over 5000 acres, and the most of them own two sections, or 1280 acres. The settlers are, therefore, far above the condition of ordinary farmers, being made up of the sons of noblemen, tradesmen, artisans, and well-to-do farmers. The first purchaser and settler in the Colony was Hon. Walter Maxwell, a son of Lord Herries, an English nobleman. This gentleman owns 2000 acres of land, upon which he has built a fine stone residence. There are also two sons of Mr. J. G. Gunther, the brother of ex-Mayor Gunther, well-known residents, and business men of this city. These two gentlemen own three sections of land, and have also built a fine stone residence, with outhouses, and have nearly 2999 fine breeding sheep. They have been there only a little more than one year, and so highly pleased is Mr. J. G. Gunther with the experiment, that he pronounces it the best investment of money he ever made. The young men are delighted with their prospects for the future, and their life on the stock farms. At the opening of the coming spring, there will be a large increase in the population, from England as well as from this city. Ample arrangements are being made for the building of a church, and the establishment of schools, which will be amplified as they are required. The church, though nominally Episcopal, will be free for the use of all denominations. The Trustees are Right Rev. Thomas H. Vail, Bishop of Kansas; Mr. John D. Perry, of St. Louis; Mr. Thomas R. Clark, of New York; and Mr. George Grant, of Victoria.

In the matter of stock farming, Mr. Grant's experiment has been upon a very large scale, and eminently successful. Although his crops for the first year were nearly destroyed by grasshoppers, that did not materially interfere with the feeding of cattle upon the buffalo-grass of the plains, which is affirmed to be the most nutritious grass that grows —even more so than the famous blue-grass of Kentucky. Mr. Grant's specialty is sheep. He began about eighteen months ago with a flock of 3555 breeding ewes, and sixty long-wooled English rams of the highest pedigree, and from the first flocks in England, consisting of Oxford Downs, Leicesters, Lincolns, Cotswolds, and Southdowns. Taking the past year as an average of the next ten years, the profits of this flock, allowing 15 per cent. less than shown in the last year's increase, will amount to at least two million and a half of dollars, on the plan of reinvesting yearly two-thirds of the proceeds of the sales of wool, mutton, wethers, &c., in the purchase of fresh breeding ewes.

VICTORIA COLONY, ELLIS CO., KAN.

(From the WESTERN RURAL, *Chicago, July* 17, 1875.*)*

EDITORS WESTERN RURAL.—I have just completed a trip of over 200 miles, west of Topeka, up the valleys of the Kansas and Smoky Hill Rivers, on the line of the Kansas Pacific Railway. The crops are all exceedingly promising. Immediately after leaving Topeka, the signs of the locusts cease and the stand of corn is good; many fields are ready to be laid by. The wheat and rye are already ripening, and the reports made by the owners is very cheering. The wheat field of Mr. Henry, near Abilene, is adjunct to the railroad, and is about four miles in length, the area 1300 acres. Salina County holds good her prestige of being one of the best, if not the best, wheat county in the State. Last year her wheat crop was over 300,000 bushels, and it is estimated that it will be double that amount this season.

Two years ago I wrote you a short account of the Victoria Colony, in Ellis County, and I have now had the opportunity of visiting it again and seeing the growth since that time. As this enterprise is about 300 miles west of the Missouri River, and in a region that but a few years since was supposed to be altogether unsuited to general farming and fit only for stock growing, it may be of interest to speak of results that to-day have come under my observation.

Geo. Grant, Esq., the proprietor, has commenced his rye harvest of 225 acres. The grain is so high that it is quite difficult to cut it, expecially with the harvester that carries the binders. The average height is over five feet. It is well filled and must yield a large crop. So well satisfied is he with this kind of grain that he proposes to put in from 600 to 1000 acres this fall. Rye at present is worth $1 50 per bushel in Denver, and $1 10 in Kansas City. It is sure and of great value for the fall and winter pasturages it affords. Several farmers informed me that they kept their cows fat on it and made nice yellow butter all last winter. Mr. Grant has also 100 acres of millet and 80 of oats. He hopes to double these areas next season.

In a ride with him of about forty miles on his lands, he pointed out the beautiful mansion of Walter Maxwell, Esq., son of Lord Herries, who has 2000 acres; of Charlie Gunther, nephew of ex-Mayor Gunther, of New York City, both of whom were at work in the harvest, and of Messrs. Walter and Charles Shields, of New York. These have all made a fine start, and their houses already remind one of suburban residences in the older parts of the country.

Building rock of fine colour and quality is found on the banks of the streams, the Victoria, Big Tinbury and Smoky Hill. The well water is good and abundant. At present the buffalo grass predominates, but blue stem, wild oats and other grasses are making their appearance.

The whole colony, purchased, is smooth and gently undulating, except a few bluffs on the creeks, which, with the woodland on them, render many parts very picturesque and beautiful.

The four Angus Polled black yearling cattle, imported two years ago, have now become a marvel to behold, without horns, short-headed, heavy-necked, short-legged, and compact, they are fat as butter, black as jet, and glossy as silk. They weigh from 2000 to 2400 pounds each. Their calves and yearlings by choice Cherokee cattle, are as fine a lot as I ever beheld. Indeed, in a herd of over 500 young cows I did not see one poor, scrawny calf, the result of crossing with these Aberdeen cattle and his imported short-horns. These Cherokee cattle and their calves are all fat enough for the butcher, and were kept on the range all the winter.

The experiment of crossing native Merinos and Colorado graded ewes with Cotswolds, Oxford Downs, Leicesters, and other imported long-wooled sheep, has been very successful. The lambs are large, and strongly resemble their sires. The great size of the sheep, and high price of the long wool, renders the good judgment of Mr. Grant manifest. The sheep are perfectly healthy; no scab, no foot-rot. He marches his 10,000 sheep twice a year, in June and September, through a bathing trough, four feet deep and fifteen feet long, filled with a solution of twenty pounds of tobacco and five pounds of sulphur, to 100 gallons of water, at a temperature of 120 degrees. The sheep are completely submerged, and in their efforts to get out become perfectly soaked. They ascend to a pen, from which all the dripping flows back into the trough; 2000 can be thus soaked per day. This process prevents scab, kills vermin, and renders the sheep healthy.

The ewes are kept in flocks of from 1000 to 1200. The lambs aggregate full 100 per cent. of the ewes. The ewes having twins and triplets, are collected in a separate flock, as they need special care. These sheep ranged out all winter, except ten or twelve days, when they were sheltered, and fed a pint of corn meal a day. The fleeces of native sheep average four and a fourth pounds, of improved sheep seven pounds.

In the diary of N. I. Irons, Esq., Hornsey, England, of his travels in America, he says :—

But the most important and promising enterprise of Western sheep farming, is that of Mr. George Grant, an English gentleman, who has purchased a vast estate in the middle of Kansas, eminently adapted for grazing purposes.

The proprietor is desirous of developing this large tract by creating pastural farms of not less than a square mile, or 640 acres; the average price of which, for well-watered lands, runs about 25s. an acre; the taxes on this quantity of land is from £4 to £7 per annum per square mile.

A fifth of the purchase-money is required on selection, and like payments, with 6 per cent. on the balance. The advantages of farming on the estate are—that already a necleus of first-class families, representing wealth, position and culture, have already settled on these

lands, and will attract persons of similar tastes and means. The owner having entered into the spirit of colonization with original ideas, his object is to surround himself only with men of sufficient capital to engage in sheep farming and sheep raising, to encourage which he has imported animals of the highest breeds, of which the settlers can avail themselves.

Seven thousand sheep and 500 cows, and bulls of the short horn and Angus Polled breed, is the example of what Mr. Grant is doing to develop this wealth-promising enterprise.

It is well known that cattle raising with the semi-wild breeds of Texas average profits of 33 per cent.

The experience of Mr. Grant's enterprise, by the judicious crossing with his imported animals and introducing new blood, has raised the per centage much higher, and the results of his sheep farming operations may safely be put down at 80 per cent.

PROFITABLE INVESTMENT FOR CAPITAL.

All sorts of live stock, wool, &c., &c., command a ready market, at good prices, as do crops for feeding stock; and these are the branches of industry Mr. Grant recommends settlers to cultivate, as he himself cultivates them.

FACILITIES FOR BUILDING.

There is abundance of fine stone and lime readily accessible for building purposes on the property. A house, sufficient for a family, may be erected of wood at from £40 to £120; and in stone from £100 to £500, fully one-third less than in England as to cost of materials. The stone is similar to Bath freestone, and when first quarried, can be faced by a hatchet, but becomes hard after exposure to the air.

It will be seen from the enclosed circular that a church is in progress, and Mr. Grant fully expects that it will be finished before the autumn, and in connection with the church there will be ample school accommodation.

WHO SHOULD EMIGRATE.

In answer to other queries naturally arising:

The Classes of Persons who are recommended to emigrate to Victoria are gentlemen and farmers, with moderate capital, to enter upon stock-farming, which is by far the most profitable industry in the United States.

The AMOUNT OF CAPITAL required to purchase and stock a square mile of land may be approximately estimated at from one to two thousand pounds, according to situation. Smaller parcels of land will, of course, necessitate less stock and less capital.

PROVISIONS generally, and FARMING IMPLEMENTS, command about European prices, with the exception of beef, mutton, game, and fruit, which are about 50 per cent. cheaper.

There are three completed RAILWAY STATIONS upon or adjacent to the property; that of Victoria, on the Kansas and Pacific line, which traverses the .estate, places its inhabitants in communication with markets for the sale of produce and the purchase of supplies.

The WATER SUPPLY, as before stated, is ample.

THE GAME

On the property is abundant, consisting of antelope, wild turkey, prairie grouse, &c., and quail in season. Buffalo in hundreds, black-tailed deer, &c., may be found west of Victoria, near Wallace, Greeley, &c., in a day's journey by railway. Also Beaver, speckled trout, &c., in the streams and lakes near the Smoky Hill River, and ducks, widgeon, teal, and other wild fowl.

ROUTE FOR TRAVELLERS.

The following route for travellers is recommended by Mr. Grant. Per White Star Line from Liverpool to New York, or by American Line, United States Mail Steamships, between Liverpool and Philadelphia. Either of the companies will give travellers to Victoria, Kansas, any consistent advantage and information. From New York or Philadelphia to St. Louis; thence by the St. Louis and Kansas City Northern Railway (P. B. Groat, Esq., Passenger Agent) to Kansas City; thence per Kansas Pacific Railway Company (Beverley R. Keim, Esq., Passenger Agent) to Victoria.

Either of the above gentlemen will give personal information, or otherwise, and friendly advice, to travellers unacquainted with the country.

TEMPERATURE AND CLIMATE.

A gentle wind constantly traverses the State, favouring the works of IRRIGATION, Grist, &c., by means of windmills.

An idea of the CATTLE TRADE to be looked for, may be gathered from Mr. Hutchinson's excellent treatise, published at Topeka, the capital city of Kansas, in 1871, which has been already referred to :—

From 50 to 75 per cent. per annum is a common estimate of profit on stock growing in Central Kansas. According to the United States census, there were in Kansas, in 1870, no less than 968,000 head of cattle.

The soil is of a very productive character, deep and rich. There is no forest land. Owing to the absence of wood on the lands, excepting on the banks of the rivers, the soil can be prepared for the first field crops at one-fourth of the expense that is necessary in Canada and in the Northern States. Ordinarily, the first year's crop of "sod corn" will more than pay for the breaking up of the ground. Indeed, the records of the Agricultural Bureau at Washington show that Kansas has grown a larger amount of wheat per acre than any other Western State. Hemp, flax, and cotton are staple productions; sorghum and tobacco are profitably cultivated. Melons, gourds, and sweet potatoes grow luxuriantly.

The following passages and letter are extracted from Mr. Hutchinson's book, before referred to, and will, no doubt, prove acceptable to inquiring readers:—

Many flowing sentences and well-rounded periods have been framed in the endeavour to describe the climate of Kansas. It has been called "Arcadia," but more frequently travellers who have been around the globe, and enraptured citizens who write to their friends in the East, call it an "Italian clime." In truth, it is neither Arcadia or Italy—at least it is not one unbroken round of golden days and halcyon nights, but it is quite certain that there is no region in the United States, east of the Rocky Mountains, where there are more bright, sunshiny days than we have in Kansas. The winters are more mild than in the same latitude east of us, and the thermometer rarely sinks below zero. During midsummer the heat at noonday sometimes ranges for several days from 80 to 100 degrees, but the air is so dry and pure that one scarcely realises the range of the mercury, while the nights are invariably cool and refreshing. Men work on buildings and other exposed situations with safety, at a temperature which would be fatal in the Eastern States.

The soil is so fruitful that farmers never feel obliged to expose themselves to severe weather, summer or winter. Especially is our climate held in high esteem by those who escape to it from the extreme northern frigidity, or from the torrid heats of southern latitudes.

I think the higher and drier portions of Western Kansas are in some respects superior as a winter stock range. The less rain falls upon the grass, the richer it will be. This is not a theoretical opinion. Stock that ranged on grass during the hard winter of 1860-61, which succeeded the famous "dry season," came out in the spring in better order than usual. The grass was short but very nutritious—having been cured on the ground. The time is not distant when the western portion of the State, one hundred by two hundred miles in extent, will be selected as the choice pasture land of the continent. Its altitude of fifteen hundred to twenty-five hundred feet above the ocean level, makes the climate all that could be desired. It is plentifully watered for stock purposes by springs and running streams, whose water is palatable to the herds and flocks, and upon the banks are small timber growths and high bluffs for shelter. There are also stone quarries, from which houses may be cheaply constructed.

I think it desirable that everywhere in Kansas a little hay should be put up, as a safeguard against light snows, accompanied by wind, which may render grazing difficult for a few days at the time. Such snows occur every two or three years in all the region we have been considering, but are much less severe in Kansas than farther north. Whenever they occur great herders expect to lose more or less stock. It is one of the chances they take, and the aggregate results for a series of years prove that, with all risks, the business is still very profitable. But in every part of Kansas there is grass in abundance to make hay. The wide bottoms afford from one to three tons per acre, even at the western limits of the State, and on ground as smooth as a floor, it is little trouble to put up hay with machinery. Perhaps half the year it would stand untouched, while stock was fattening on the natural buffalo grass. But it is better to provide against contingencies, and if not used it will keep over in good condition, if well stacked. The estimated amount that ought to be put up per head in the buffalo grass region is from four hundred to six huudred pounds. Among scores of experienced stock-men, with whom I compared notes on this subject, none set it higher than the latter figures.

John S. Chisum, one of the most noted stock dealers and breeders of Texas, a man who handles cattle by the ten thousand head, said, "For Kansas, from four to five hundred pounds." Major H. Shanklin, of Lawrence, who has wintered cattle in the Arkansas Valley several seasons, said, "Five hundred pounds, and it may rot down unfed every other year." Rev. L. Sternberg, who lives at Fort Harker, on the

Kansas Pacific Railway, said, "Five or six hundred pounds, and probably not half that amount would be fed out." Nor is this precaution desirable for Kansas alone. In Colorado, prudent persons provide a little hay for their stock, and think it pays them a profit to do so, and with the rearing of improved breeds this will be an acknowledged necessity. Large herders, with thousands of cattle, do not consider the loss of a few score head of cheap Texas stock as a matter of importance. But when each bullock comes to be worth fifty or seventy-five dollars the case will be different.

The foregoing was written in December, 1870. It is now April, 1871, and we have passed through a very severe winter, snow having lain on the ground longer than ever before known. During this winter many thousand head of cattle have fed on buffalo grass and winter grasses, without any hay or grain whatever. The result has been surprising to all. Among Texas cattle, or stock bred from them (and there is little other stock in the buffalo grass region), there has been less loss than in the more eastern or southern portion of the State, where they were fed on hay, or hay and corn. The cattle thus wintered will soon fatten upon the fresh grass. It is natural for this stock to get its own living on the range, and they do not do well on corn the first year they are brought from Texas. Next to their native range a field of standing corn stalks, after the ears have been plucked, seems the best suited to their wants. Sheep have also done well in Western Kansas this winter, on grass alone. I am convinced that herders, with several hundred or thousands of cattle, will do better to seek some of the many *canyons*, or sheltering bluffs, or timber patches, to be found in the buffalo grass region, with plenty of water, and graze stock all the season, than to cut hay for them. The loss in the former case will not equal the additional expense in the latter case. Small stock raisers and farmers will undoubtedly do well to put up a little hay.

In short, Western Kansas and its buffalo grass offers the BEST ADVANTAGES IN THIS STATE, OR IN ANY STATE FOR STOCK RAISING. I do not advise people to rush into that region—that is, to the remote, high prairies where nothing grows but buffalo grass—who are destitute of means; but with a little capital to invest in stock, a living is certain and easy. There is certainly no need for any to suffer for meat in that region, for buffalo meat is toothsome and nutritious, and to be had for the killing, while the peculiar waxy fat furnishes to the hardy frontiersman a sweet and healthful substitute for bread.

The geological formation called carboniferous, occupies the entire eastern portion of the State—having a general width from east to west of about 180 miles. Its western limit crosses the Kansas River through Davis and Riley Counties, and the area is about 17,000 square miles. There are outcroppings of Bituminous coal throughout the entire extent of this vast surface—an area more than twice the size of the State of Massachusetts. . . . This coal is of a superior quality. There is considerable lustre to its broken edges, and it does not crumble to dust by handling and shipping as does much of the coal in other western States. It is retailed in our towns and cities from 25 to 30 cents a bushel (of 80 lbs.).

Yet a book about Kansas would be incomplete unless it contained something concerning the people who live in Kansas, especially as our friends "within the bounds of civilization" have decidedly erroneous opinions concerning us. Those who take the trouble to examine what is shown in these pages about the institutions of Kansas—its Churches, Schools, Newspapers, Railroads, Cities, and Public Buildings, must conclude that although distant from the homes of our fathers, we have not lapsed into barbarianism. In fact we all had fathers and mothers who lived in "the East," or some other place, and it is but a few years since we left those dear old homesteads to make homes for ourselves on these lovely prairies. The light has gone out from many of the places where we once lived and loved, but all the way from the Mississippi to the Atlantic, and even beyond its billows, there are fires burning on hearth stones at which we find a welcome and a chair. These people at the East who suppose that Kansans are necessarily uncouth and ignorant, will do well to reflect for a moment as to the character of those who have left their own vicinity for distant Western States. They know too well that the best blood and brain of the Eastern, Middle, and Southern States is seeking for itself a new domain in the boundless West. Do all these people forget their cunning because they "come to Kansas?" Let their institutions and their works answer for them.

No other State ever had among its early settlers so many well-educated men and women as Kansas. There were seventy college graduates among the four hundred voters at the first election held in Lawrence.

It was formerly one of the staple objections against coming West, that there were so many foreigners here, but now there are nearly as many in the East as in the West in proportion to population, especially in New England, which is fast becoming old Ireland. There are comparatively few foreigners in Kansas, but we heartily wish there were more of the same sort. Here there is "room and verge enough for all," whether they pronounce our "Shibboleth" or not.

TERMS OF SALE.—The price of land in the Victoria Estate, necessarily varying with local conditions, *i.e.*, its position with respect to rivers, woods, and communications, may be briefly said to range from ten shillings to forty shillings per acre, payment to be made either in cash at time of purchase, or in yearly instalments extending over five years.

For further information apply to
ROBERT W. EDIS, ESQ., F.S.A.,
Architect to the Estate,
14 *Fitzroy Square*,
LONDON, W.;

To ROBERT TOMPKINS, ESQ.,
Carlisle House,
Reading,
BERKSHIRE;

To WILLIAM STEELE, ESQ., S.S.C.,
7, *George Street*,
EDINBURGH,

OR TO ANY OF THE FOLLOWING GENTLEMEN IN NEW YORK, WHO HAVE VISITED THE ESTATE, OR WHO HAVE SONS OR RELATIVES SETTLED UPON IT:—

JOHN C. GUNTHER, Esq., 504, *Broadway, New York.*
GEO. MOLLER, Esq , 96, *Wall Street.*
PETER MOLLER, Jr., Esq., *Albemarle Hotel.*
THOS. R. CLARK, Esq., 690, *Broadway.*
Messrs. KOBBE AND FOWLER, 52, *Wall Street.*

OR TO ANY OF THE GENTLEMEN IN EUROPE, WHOSE NAMES ARE IN THE LETTER TO BE FOUND AT PAGE 8.

P.S.—Mr. Grant, or his Superintendent, will at all times be happy to receive all who may desire to visit the estate, as ample accommodations are provided for their entertainment.

APPENDIX.

In addition to the stock purchased in the United King dom by Mr. Grant last year, and imported into Victoria, he acquired from Senator Cochrane, of Hillhurst, Lower Canada, two of that celebrated breeder's highest class short-horn bulls, one of which, "Count de Brunnow" (now eighteen months' old), is said by competent judges to be one of the most promising animals on the American continent. He also purchased the famed blooded stock of Charles Clinch, Esq., of Wakefield, Kansas, and Witney, Oxon (England), whose certificate is subjoined :—

WAKEFIELD, KANSAS, *October 1st*, 1873.

I hereby certify that I have this day sold to George Grant, Esq., of Victoria, Kansas, my whole stock of imported blooded sheep, sixty in number—twelve rams and forty-eight ewes, selected from the flocks of Mr. Lane, Broadfield; Mr. John Gillet, of Minster; Mr. Richard Lord, of Stanton Harcourt; and Mr. John Walesby Kirkham, of Cadeby Hall, England. Also, three short-horn bulls and two heifers, Bate's breed, with pedigrees furnished; also, one stallion, "Lord Clyde," bred by and bought from Mr. John Hutt, of Waterston. "Lord Clyde" is by "Nugget of Gold," a prize horse at the Royal Agricultural Show of England, held at Oxford, in England, in 1870.

(Signed) CHARLES CLINCH.

1st February, 1874. VICTORIA, ELLIS COUNTY, KANSAS, U.S.

The St. Louis Times states: "One of the bulls in Mr. George Grant's herd of twelve, Count de Brunnow, whose pedigree is annexed, is, at his present age (about three years) the finest animal of the Booth strain on the Continent of America."

PEDIGREE OF THE SHORT-HORN BULL,

"COUNT DE BRUNNOW,"

(No. 23725 American Herd Book).

THE PROPERTY OF GEO. GRANT,

"Victoria Stock Farm," Victoria, Ellis Co., Kansas.

COUNT DE BRUNNOW—Roan, calved Nov. 15th, 1872, bred by the Hon. M. H. Cochrane, Hillhurst, Compton, Ontario, Canada.

SIRE "ROYAL COMMANDER" (29857)
DAM BRIGHT LADY by Lord Blithe (22126)
gr d Bright Countess by Breast Plate (19337)
g gr d Bright Princess by British Prince (14197)
g g gr d Bright Dawn by Vanguard (10994)
g g g gr d Bright Phœbus by Crown Prince (10087)
g g g g gr d Blanche 2nd by Zadig (8796)
g g g g g gr d Blanche by Auld Robin Gray (6753)
g g g g g g gr d White Rose by James Crisp's Bull
g g g g g g g gr d Rose by Burley (1766)
g g g g g g g g gr d Young Anna by Isaac (1129)
g g g g g g g g g gr d Anna by Pilot (496)
g g g g g g g g g g gr d Ariadne by Albion (14)
g g g g g g g g g g g gr d Bright Eyes by Booth's Lame Bull (359)
g g g g g g g g g g g g gr d by Shipton (187)
g g g g g g g g g g g g g gr d by a son of Suwarrow (636)
g g g g g g g g g g g g g g gr d by a son of Twin Brother to Ben (88)
g g g g g g g g g g g g g g g gr d by Twin Brother to Ben (660)

32

BEEF & MUTTON FROM AMERICA.

As it is now a fact beyond dispute that beef and mutton can be imported from America in perfect condition at 25 per cent. less than European prices, and yet yielding very large profits to the cattle dealers in the West. See the recent reports in English papers;

And in the *Evening Standard* of February 8th and March 7th, 1876, as follows:

AMERICAN MEAT IN THE LONDON MARKET.—There was another consignment of American meat in the market yesterday morning, the quantity being not less than seventy tons, which was sent up from Liverpool on Saturday night. The beef is just as good in every respect as that previously reported on, and it is evident that the salesmen in the market are finding the meat pay, for in some instances they are careful to ticket the meat as "killed in America." The cargo included about seventy sheep, which were all right as to condition, and sold readily, but were scarcely fat enough for the English market. This fault the consignors say they will remedy in the next lot. It is reported from Liverpool that this trade in fresh meat from the United States is assuming considerable proportions. All the steamers of the Guyon Line have now been arranged and adapted for this trade. Each has a room 40 feet long, 28 feet wide, and 9 feet high, at one side of which is an apartment containing forty tons of ice, and a blower worked by an engine, which keeps a constant stream of cold air passing over the meat.—*The Farmer*.

AMERICAN MEAT.—There is a supply of American imported beef in the Metropolitan Market equal in every respect to recent importations. The shippers of this meat have now five vessels prepared for the carriage of the meat, and we may soon expect a cargo almost daily. The noticeable feature in the present supply is that the carcases are larger than in any previous shipment, and that the shippers, challenging attention to what they can do, are exhibiting in the Metropolitan Meat Market a hind quarter of beef, which they have sent as a special joint for the Queen. It is expected that American meat will be in the London market at least three other days this week.—*The Farmer*.

"GLOBE," *September 27th*, 1875.

. . . . In regard to meat the case is very different. Our herds and flocks *do not increase in number* with the increase of our population. On the contrary, they fall off. As compared with 1874, we have in 1875, in the United Kingdom, 1,342,203 *fewer* sheep, and 125,652 *fewer* cattle!

"DAILY TELEGRAPH," *September 10th*, 1875.

PRICE OF MEAT.—The data recently published in the *Daily Telegraph* showed that the present high price of meat will be increased later on in the year. It should be borne in mind that the supply of the London meat market is, under the most favourable circumstances, dependent on importation from abroad to the *extent of more than one-half*.

33

A MORAYSHIRE MAN'S ACCOUNT OF VICTORIA, IN KANSAS.

VICTORIA, ELLIS COUNTY, KANSAS, U.S.
1st February, 1874.

SIR,—Would you be kind enough to give space in your paper to a few things I would like to say in regard to Victoria, the property of George Grant, Esq., which may be useful to any one intending coming here?

The first of this colony (of which I am one) landed here on the 16th May, 1873. There was a large house ready for our reception, in which Mr. Grant kindly allows settlers to stay till their own houses are ready. Mr. Grant was very kind to us, and did everything in his power to make us comfortable. Mr. Grant has a fine property here; the land is of excellent quality, little or no broken ground—all covered with a thick coat of buffalo grass. A person would think it had been painfully sown "out" by human hands. The land is not quite level; it rises in gentle slopes. We can see a great distance, as there are neither hills nor forests. The only wood we have is along the banks of the creeks and rivers. Saline River is within fourteen miles of us to the north, and Smoky Hill River within twelve miles to the south, and between these rivers we have North Fort (on which Victoria is built), and Big Creek, now Victoria River.

Mr. Grant's property extends between Saline and Smoky Hill Rivers. He has every alternate square mile; Government has the other. On Government land any man, after becoming an American citizen, can pre-empt 160 acres, of which he has to pay 80 acres at 2¼ dols. per acre, or he can homestead 80 acres. All he has to pay is 14 dols., and by living on it, and improving all or part, he is entitled to the title-deeds of it at the end of five years, and the land too; if it were in the Laich o' Moray that would be worth at least £3 10s. an acre per year.

But I don't think this will be a good agricultural place, not but the crops will grow, but we are too far from market. The men I think who ought to come here are those who have a few hundred pounds to buy land, and enter into sheep or cattle raising, to which every facility is offered by Mr. Grant, who has imported a few excellent polled bulls, and a lot of very good tups. It is quite easy in summer to store as much hay in a short time as will do a large herd through the winter. Of course the summer is warm, but I never felt it oppressive. There is always a fine cool breeze, which rises with the sun and falls when it sets. The mosquitoes are rather troublesome after the sun goes down. In October and part of November we get very pleasant weather. It is called the Indian summer. This is the 1st day of

February, and I must say we have had a pleasant winter; we occasionally get a *norther*, a few of which have been very cold, but they never last longer than three days, and then we get a while of fine weather.

You must remember, sir, I am not an emigration agent, but really, when I see so many thousands of acres of splendid land without a human being on them, I do wonder at and pity the rack-rented farmers of some parts of Morayshire working hard every day, and even with that can scarcely make daily bread. Still this is not a place as yet to which I would advise any one to come who has not enough means to start with stock. I mean one altogether dependent on his wages for manual labour, as provisions are rather dear in this part, owing to heavy charges on the railways. Tea is from 50 cents to 1 dol. per lb., sugar from 12 cents to 16 cents per lb., butter 35 cents per lb., flour from 2½ dols. to 5 dols. per 100 lb., boots from 5 dols. to 10 dols. per pair. Every kind of woollen stuff very dear.

Please excuse me for taking up so much room in your valuable paper.—I am, yours respectfully, GEORGE PHILIP.

LETTER FROM COL. JOHN T. ROBESON, U.S. CONSUL.

ROYAL HOTEL, EDINBURGH,
13th March, 1874.

GEORGE GRANT, Esq.,

DEAR SIR,—I have read the clippings of the newspapers you sent me yesterday.

I have no hesitation in saying that Kansas is one of the most healthy States of America, and no portion of the said State more so than Ellis County. I have been stationed in Fort Hays, in Ellis County, and know the country well. There is no part of America better adapted for stock raising or agricultural pursuits than what is known as the Smoky Hill country, which lies along Smoky Hill Fork and Grand Saline Rivers, in Kansas.

Should any one doubt my statements, I would respectfully refer them to the Agricultural Reports, published by the United States' Agricultural Department, Washington, D.C.

With high consideration, I have the honour to be, respectfully yours,

JOHN T. ROBESON.

P.S.—I consider the climate of Kansas more pleasant than that of New York. J. T. R.

FROM THE FOREGOING, it will be seen

THAT A MAN MAY HERE BECOME

THE OWNER

OF A

FREEHOLD ESTATE

OF 640 ACRES OF

GOOD LAND

FOR THE SAME AMOUNT HE WOULD HAVE
TO PAY IN ENGLAND FOR A

YEAR'S RENT ONLY,

For a Farm of 500 Acres. The Former a

VIRGIN SOIL,

The Latter, more or less

EXHAUSTED.

Chapter 3
Raising Cattle in Kansas

By Rhonda McCurry and Corinne Patterson

Through the lives and the stories of Kansas's great cattlemen, you can see how an industry was built on the backs of hard-working families who sought freedom in the course of self-employment. Their "bosses" were the land, cattle, horses, and other livestock that sustained their operations. The adoption of good management practices and forward thinking helped make Kansas a beef industry leader 150 years after she officially secured her statehood on January 29, 1861. The evolution of cattle-handling protocols and ranching facilities, coupled with a desire among ranchers to embrace proven management practices, have made raising beef in Kansas a sustainable livelihood.

It is impossible to paint a complete picture of the Kansas beef industry—quite simply, it is too vast to fully describe. Like the crested buttes of the Gypsum Hills, the vast plains of western Kansas, the cobble-stoned tallgrass of the Flint Hills, and the rolling Smoky Hill prairie, there are countless stories of the cattle industry that will forever remain fossilized in the rural holdings on the historical Kansas landscape. Folklore varies from one region of the state to another, but raising beef in Kansas has given its cattlemen a certain pride and progressivism that, together, form a common thread. While not every story can be celebrated or told, this chapter features some of the stories of multigenerational Kansas beef producers. It provides snapshots of the pioneering spirit and the way of life from which Kansas's rich history of raising cattle has unfolded.

Previous: (Source: Corinne Patterson)

Early Claims

Generations of ranchers have a way of defining an operation, with each adding its own mark in attempts at progress. It takes a lot to build on what our forefathers pioneered, and such efforts are not taken for granted by cattlemen such as Bill Paige and his son, Jerry. Since 1912, members of the Paige family have worked to build a solid farming and ranching operation from an original two-acre farm near White City, which Bill's grandfather acquired in exchange for one acre that he owned in Missouri. Today, it is rare to find a family's fifth generation working on the same ground that was staked out by its ancestors, whose "heavy-lifting machinery" was comprised of a horse, a plow, and their own hands. While tractors, a large mixer wagon, a telehandler, and a modern working facility have brought

Processing calves at Paige Farms involves vaccinating, deworming, and implanting. Individual weights are recorded with a scale under the chute, and each calf is given an electronic identification ear tag. (Source: Corinne Patterson)

Charolais cattle are shown grazing pasture under a cloudy Kansas sky. (Source: Lewis Browder)

visible changes to Paige Farms, it is actually an enduring family bond that keeps the operation in fine tune.[1]

The Paige family's story is familiar to many of Kansas's multigenerational ranchers. As the country opened for settlement farther to the west, pioneers struck out to find more opportunity from the land, when they dispersed their current holdings for new stakes out on the open prairie. Leon Mosteller's great-grandfather, George, was a coal miner in Pennsylvania who spent his meager earnings at mine-owned businesses to sustain his family. Realizing that he was essentially bestowing his labor in the coal mine for free, George left in 1879 and made his way to Nemaha County, Kansas, where he bought 160 acres from the man who had laid the homesteaded claim. Today, more than 130 years later, the fourth and fifth generations of the Mosteller family run a small commercial beef herd, and a few purebred shorthorn cows maintain the heritage of a purebred herd that was started by the family during the 1960s.[2]

In Kansas, it is common to refer to land by certain family names—often without truly knowing how to spell or even correctly pronounce the names. Indeed, some native Kansans, including those with ancestral ties to the early settlers of the state, will speak about land in terms of particular names associated with particular pastures or fields. For example, some still speak of *the Klotz, the Guthrie, the Lind, the Griffith,* and *the Grother,* where Howard Blender and his family have grazed cattle since his grandfather first came to Chase County from Illinois in 1924. Parcels of ground and places where cattle still roam today are often tied to an original homesteader's or early resident's name.[3]

Kansas territory opened for settlement after the Kansas-Nebraska Act was passed in 1854. Although this act contributed to the outbreak of the American Civil War,[4] it also encouraged the settlement of Kansas's open prairies. Before there were those who laid claim to a homestead, Kansas served as an open range for grass-finishing cattle that were bound for markets in Chicago and Kansas City. Today's modern-era cowboys—like Norm Wilson of Strong City—are cut from the same cloth as the early-day cowboys who drove the cattle trails. With almost seventy years of experience, Wilson still ships tens of thousands of cattle from the ranges in east central Kansas each year. Although they do not possess any cattle of their own, the cowboys who handled thousands of cattle during the early 1900s and the cowboys of today are a significant part of Kansas's beef history. They prove handling cattle is a genuine work of art.[5]

Western Kansas remained unorganized for many years after it opened for settlement. Cattle were ranged over thousands of acres without the restrictions of fences. The bovines were joined by only a handful of cattlemen, who used the range for nothing more than to grass and water their stock. The cattlemen's supremacy was short-lived, and as quickly as the railroad brought cattle to the country, settlers flooded the area to lay claim to the land.[6] According to distinguished historian Dr. Jim Hoy, "The days of open range that followed the Civil War were numbered once a practical and economic barbed wire was patented in 1878."[7]

Hoy said that under open-range conditions, farmers bore the responsibility for fencing crops to protect them from foraging livestock, but fence laws required livestock owners to confine their livestock. When the Kansas Legislature later passed a fence law, it was on a county-option basis; therefore, "In some Flint Hills counties cows could roam freely, while in others they had to graze behind fences."[8] Cattlemen and farmers were resourceful with their fencing options, and limestone is a good example. The skill that was required to build a limestone fence can be seen today on the scenic Native Stone byway, which winds through Morris and Wabaunsee Counties. The Post Rock byway, which connects communities in Ellsworth and Russell Counties, is named for the unique native limestone rocks that were used for fence posts in that area.[9] The remaining hedge rows that are scattered across Kansas remind us that the land was once sectioned off to mark one man's hopes for prosperity. Leon Mosteller's father told stories about the early settlers who

Cattle are transported across the country via semis and cattle pots. Busy feedlots and backgrounding lots receive cattle at all hours of the day. (Source: Corinne Patterson)

protected their crops or confined their cattle by erecting fences during the 1890s and early 1900s, when thick rows of Osage Orange trees were trimmed every five years or so to maintain a blind that was approximately five feet in height. Many of those trees have since been removed or repurposed into hedge fence posts for stretching barbed wire.[10]

Even with the introduction of fencing, the vast Kansas prairie remained home to roaming cattle, which often varied in age. Norm Wilson recalled seeing cattle with their tails dragging the ground as they neared their fourth year. When Wilson was just nine years old, he would saddle his horse and ride before sunrise to meet his neighbor Clod; together, they would spend the day rounding up cattle to ship by rail to Kansas City. Wilson and his neighbor would gather between 250 and 400 head of cattle and sort out which ones were fattened enough and ready for market. It took 30 to 33 head of cattle to fill a railcar; once a couple of railcar-sized loads were sorted off, the boys would trail those cattle fifteen miles to the Cassoday depot. Wilson recalled,

> One summer we got close to five thousand head of calves that came in by train out of Mississippi. They weighed anywhere from three hundred pounds to probably four and a quarter. We didn't do anything but doctor, doctor, doctor. They came in at the Neal stockyards, and we'd have to drive them out. At the time, the trains weren't good about being on time. They'd tell you when they were supposed to be there, but they might not get there till the next day and at night.[11]

Cattle shipping revolved around the railroad stockyards until the 1960s. As integral as livestock scales are today in the beef industry, they were few and far between during the railroad days. Wilson recalled that scales were kept in good working order by inspectors whose only job was to keep gates swinging easily so the cattle kept moving. Many of the cattle owners were not much for "cowboying," yet they were sure to be on horseback come shipping day. "For a lot of them you'd lead their horse to the pasture and they'd pull up a car and get out and cut what cattle they wanted," Wilson recalled. "Then they'd get back in the car and you had to lead that horse back home."[12]

In twenty-first-century Kansas, most grazing cattle are confined with miles of fences. Galvanized barbed wire that is stretched from metal T-post to metal T-post and eventually tied into a solid pipe corner is preferred. However, in some rural areas of Kansas, there are small portions of what is still considered open-range country. For example, large sections of prairie in the Gypsum Hills near Coldwater are grazed with few divides and a couple of cattle guards. In the Flint Hills open range south of Matfield Green, a handful of different brands share common grazing.

Sustaining a Lifestyle

Small tracts of fertile land once offered a hard day's work for early settlers, but today, forgotten homesteads and run-down barns dot the Greenwood County prairie, where Hal Luthi and his family run stockers and manage their purebred Simmental herd. "There were three or four times more family-type operations when I was a kid growing up," Luthi recalled. "Maybe a generation prior to that there was a family farm on almost every quarter or half section, and now a quarter or half section won't support much of a lifestyle."[13] In 1900, Kansas recorded 173,000 farms with an average size of 241 acres. According to the U.S. Department of Agriculture (USDA) National Agricultural Statistics Service, the definition of a farm has changed over time. Before 1959, a farm was defined as three or more acres with annual sales of $150 or more, or fewer than three acres with sales of at least $250. From 1959 to 1975, a farm was defined as ten or more acres with $50 or more in annual sales, or fewer than ten acres with sales of $250 or more. After 1975, a farm was defined as any place with at least $1,000 in annual sales of agricultural products. In 2009, the USDA reported 65,500 farms in the state with an average size of 705 acres.[14]

These bulls are shown near the working facilities of Sylvester Bull Development in Wamego, Kansas. (Source: Blair Tenhouse)

As each family member has been incorporated into the business, the Luthis' operation has grown in both size and efficiency. Luthi said that the beef industry as a whole has lent a hand in their expansion.

> When my grandfather fed cattle, he kept them until they were two years old and then put them on full feed. We got them tremendously big. Once you bought a set of steers you kind of lived with them, and you knew them individually before you sold them. Now you can turn over a calf by the time it is 15 months old without hardly any problems, just by the improvement in the genetics.[15]

Today, the Luthi stocker operation receives calves that weigh 300 to 800 pounds. Tom Ballard, a longtime employee, handles the new cattle that generally come in weekly during the busy winter receiving months. Larger calves are sorted off and may head straight to a commercial feed yard, but the majority of calves are backgrounded to get ready for summer grazing. Like many background and stocker operations in Kansas, the Luthis purchase cattle from southern states such as Alabama, Georgia, and northern Florida (Tennessee has also shown promise as a buying source). Thirty years ago, most calves were purchased locally; however, in today's market, it is cheaper to ship the cattle to the feed source rather than ship the feed to the calf, so most cattle movement in the country is northward. According to Luthi,

> The reason we buy southern cattle is because of less conditioning. The cattle maybe aren't as fleshy so we can put on a pound of compensatory gain a little faster than the local calves. Presentation is a lot of good marketing, and the cattle that we buy are not presented in their best clothes. They are usually under-conditioned, probably not castrated, not dehorned, and not weaned.[16]

Luthi said cattlemen who want to do the background work and group calves uniformly will be more highly compensated.

> But we've found that we have a niche where we can baby a lot of those higher-stressed cattle along and do pretty well with them. If you would take those kinds of calves and put them directly into a dry lot, there's potential to have a lot of disasters. It takes a little more labor. We group

> them in grass traps where they have a healthier environment versus laying in dust and mud. We kind of pamper them along until they get the background, the immunity, and get themselves built up to where they can stand with the calves that have already had the work done.[17]

Like many early Kansas cattlemen, Philip Blender had a herd that was started by his father. Philip's son, Howard, recalled that they ran hogs behind the cattle and broke ear corn over the edge of the wooden feedbunks to sustain the stock. By the time Howard was old enough to do chores with his father, they were feeding the cows with a tractor hitched to a two-wheel wagon that today rests on the family's land in a haven of rusted equipment from days past. Howard remembered, "I'd ride on the wagon with him to feed the cows that were wintering on the stocks at the Lind place. He'd pull the wagon with the tractor, and I'd fork silage to feed the cows."[18]

The Blenders dispersed their herd during the mid-1950s when landlords demanded that only stockers graze

Controlled burn of native grasses. (Source: Lewis Browder)

their holdings in the heart of the Flint Hills. Double stocking or early intensive grazing of yearlings for ninety days had started to become the norm during the 1960s and 1970s. However, Philip and his son had witnessed years where the grass would come on too late, or various other factors outside of their control would leave the native prairie in disrepair after intensive grazing. It was determined that on the Blender Ranch, cattle would graze full season from April 15 to October 15 so as to "take half and leave half." The remaining half was used to fuel the fires that restore the tallgrass prairie each spring.[19]

Indeed, burning in eastern Kansas is as much a part of the beef industry as cattle feeding is in western Kansas. A million acres or more go up in flames just a few weeks before turnout in the Flint Hills each April. Ranchers have long practiced prairie burning as a way to restore and to maintain a healthy, native landscape; however, Kansas cattlemen may eventually face regulation of this practice, which has sustained one of the last remaining tallgrass prairies in the world. Air quality issues in metropolitan areas have prompted the U.S. Environmental Protection Agency and other organizations to draft a burn management plan.[20]

Herd Health

When the change was made to import yearling cattle into Kansas, Philip Blender began working with an order buyer to secure what Howard Blender recalled were "very plain cattle."[21] When they reached the ranch, many were still bulls with horns, and they were in various conditions as they jumped off the crowded trailers. Managing these cattle that ranged in age from yearlings on up was often referred to as "straightening them out" by the cattlemen and cowboys who handled Flint Hills stockers. Howard got the job of riding pens at age eleven. As soon as he got home from school, Howard would saddle Trigger, a palomino quarter horse his father had purchased for the solid price of $125. Trigger was

(Source: Corinne Patterson)

really more horse than the youngster needed, but doctoring was a big job before preventative vaccinations.

During this time, penicillin was top of the line in the medicine box. When it lost its effectiveness for treating sick cattle, products such as Liquamycin—a forerunner to LA 200—and Tylan were used on the ranch. A combiotic and penicillin sulment drench was a popular choice for doctoring sick cattle. Howard remembered coccidiosis outbreaks as a top threat to herd health. If sick calves were caught in early stages, treatment with a particular bolus pill would stop it in its tracks. When vaccines and early treatments were developed, Howard said they were commonly injected directly into the muscle. Before vaccines became combined subcutaneous injections, cattlemen had to use the hindquarters to vaccinate or treat sick cattle, because there was not enough muscle in the neck region.[22]

Howard recalled a blackleg vaccine that was one of the first preventative vaccines being used on the Blender Ranch. Blackleg was a concern all across the beef industry; it was a virulent and invariably fatal disease that could destroy 25 to 35 percent of a herd of calves overnight. In 1916, Oliver Morris Franklin, a veterinarian who is credited with developing the first reliable preventative for blackleg, founded the Kansas Blackleg Serum Company in Wichita. Many of the company's investors included prominent Texas ranchers, so the company moved to Amarillo, Texas, shortly after its inception. Dr. Franklin made his mark on the beef industry before he left Kansas. He received his doctor of veterinary medicine from Kansas State Agricultural College in Manhattan on June 13, 1912, and he worked at the experiment station as an assistant in the Department of Veterinary Medicine.[23]

STATE OF KANSAS
DEPARTMENT OF
STATE LIVESTOCK SANITARY COMMISSIONER

RECORD OF CALF VACCINATION

(Owner) (Address) (County)

Tube No.	Breed Shorthorn Tattoo, Vaccination Tag, or Registered Name and Number	Sex	Date of Birth			Agglutination Test: +, Complete agglutination; I, Incomplete agglutination; —, No agglutination; R, Reactor; S, Suspect; N, Nonreactor					
			Mo.	Day	Year	Test Dilutions 1-50	1-100	1-200	1-400	Test Result	Reactor Tag No.
1	Tattoo 65	F									

Veterinarians making optional prevaccinal or required post vaccinal tests are directed to fill in and sign the following statement:
I hereby certify that I bled the above described calf/calves and agglutination blood test was completed by .... (Veterinarian or Laboratory)
.... (Address) on the .... day of ...., 19........
(Authorized by State L. S. S. Commissioner)

At time of vaccinating either tested or untested calves veterinarians are directed to fill in and sign the following statement:
I hereby certify that on the 16 day of Aug, 1943, I used 6 cc. of strain 19 Brucella abortus vaccine Serial No. 316 produced by Norden laboratory Lincoln Neb. (Address)
in vaccinating the above described calf/calves
"Send to State Livestock Sanitary Commisioner, Topeka, Kansas"
(Authorized by State L. S. S. Commissioner)
19-3254-a 4-42—15M Sets

State of Kansas record of calf vaccination, circa 1943. (Source: T&M Ranch, Argonia, Kansas)

A common health management protocol is to vaccinate calves before they are weaned. Insulated coolers keep vaccines cool to preserve their effectiveness. (Source: Corinne Patterson)

Vaccines for other diseases have been developed throughout the years, including vaccines for pinkeye, foot rot, and mycoplasma, just to name a few. Today, cattlemen regularly vaccinate their herds to help prevent bovine respiratory disease complex (BRD), which is commonly called shipping fever. BRD vaccines generally include several strains that may cause respiratory disease alone or in combination, including bovine viral diarrhea, infectious bovine rhinotracheitis, parainfluenza virus, bovine respiratory syncytial virus, and a host of other lesser-known respiratory viruses. In 1999, a USDA study that was conducted by the National Animal Health Monitoring System investigated the top-twelve cattle feeding states, including Kansas, for the presence of BRD. Producers reported that 14.4 percent of all cattle placements developed BRD while at feedlots, which was nearly five times the percentage of cattle placements that developed the next most commonly reported disease, acute interstitial pneumonia.[24] Many of today's cattlemen vaccinate calves for BRD before weaning and then give booster shots at weaning or when the calves are received at a backgrounding or finishing yard elsewhere in the state.

Hormone supplementation also has played a role in the commercial beef industry. One of the first implants that the Blender family used was diethylstilbestrol. It was approved in 1954 by the U.S. government for oral administration, and implants were approved for use in cattle in 1957. The use of diethylstilbestrol in growing-finishing cattle was rapidly adopted; however, health concerns associated with this product eventually led the U.S. government to remove oral diethylstilbestrol from the market in 1972, and implants were removed one year later.[25] Howard recalled sitting in Dr. Dan Upson's anatomy and physiology class at Kansas State University when announcements where first made that diethylstilbestrol was banned for use in cattle. Today, approved growth implants are still used in backgrounding and finishing programs. Implants have been designed specifically for heifers and steers to enhance their rate-of-gain performance, and their regulated use includes proper withdrawal periods before harvest.

Veterinarian's Certificate of Vaccination

This is to Certify that I, a duly licensed, graduate veterinarian, have this day vaccinated the following described calves with

Norden Certified BRUCELLA ABORTUS VACCINE

Owner: Andrew McIntire Address: Argonia
No. Vaccinated: 1 Helen Louise Breed: Sh. 2108752
Age: 4 Mo. 5 Mo. 6 Mo. 7 Mo. ✓ 8 Mo. ✓
Tag Nos. Tattoo 61 L E
Serial No. 597A 3976 Expiration Date 1-1-43
Date 9-21-42 Signed [illegible]
Address Harper

Certificate of vaccination for *Brucella abortus*, circa 1942. (Source: T&M Ranch, Argonia, Kansas)

Since the mid-1990s, the Blenders have moved away from receiving cattle that were considered to be high risk, i.e., those without pre-weaning vaccinations and preconditioning. Today, the majority of stockers that the family purchases are age- and source-verified, and most have been preconditioned for forty-five days before they are delivered to the ranch from December through March. Heifers that weigh 500 to 550 pounds are vaccinated and branded 24 to 48 hours after arrival, and they are fed a mixed ration so that they gain about a pound per day until they are turned out to grass around April 20 (on the summer range, heifers typically average gains of 2.5 to 3 pounds per head per day).[26] Vaccines are given in front of the shoulder, which is the injection triangle area as defined by the Beef Quality Assurance (BQA) program.[27] Sick cattle are treated according to appropriate protocols, including the administration of antibiotics or other injections subcutaneously in the BQA-defined injection triangle.[28]

The chore of doctoring sick cattle is not what it used to be when Howard and Trigger were a pair during the 1950s. Today, Howard and his trusty palomino, Mike, rarely pull "sicks" from the backgrounding lots. Together, they cover the pastures each summer to monitor health and thus get a good head count.

Building a Cow Herd

Kansas is a genetics leader and boasts numerous seed stock producers. From Angus to Gelbvieh to Horned Hereford, Kansas producers have used recordkeeping and technology to offer top herd sires for commercial beef production.

Bill McIntire of Argonia, Kansas, was born into the registered shorthorn business. Bill's grandfather, John, and his father, Andrew, encouraged him as a youth to show cattle through 4-H and FFA events across the state. Bill said his family taught him to appreciate the cattle and ranching environment, which would one day become his career: "We're in the cattle business because we enjoy it. We want to make a profit, but we like it and really understand the cattle. A cow may run over you and stomp you into the mud, but she'll at least tell you about it before she does."

Growing up, Bill said most registered cow herds were small, between thirty and fifty head. This size of an operation required face-to-face networking and personal interaction among fellow ranchers in order to sell seed stock. The main reason Bill's family exhibited cattle was to meet fellow breeders and to market their herd at county and state fairs and livestock events. After attending Kansas State University, Bill went to work at the Thompson Ranch, a commercial Angus operation in Barber County. A Duquoin, Kansas, native, Bill and his wife, Linda, moved back to southern Kansas in 1975. Linda's father, Olen Taton, wanted a partner for his small registered Angus herd, and Bill thought it was a good opportunity to move closer to home and still manage his own cows.

(Source: Blair Tenhouse)

Under a new name, T&M Angus Ranch, LLC, Bill began to use such popular technologies as expected progeny differences and artificial insemination to improve and grow the operation. Bill explained, "Linda's dad had 15 registered cows that basically were his kids' 4-H projects. We wanted to add value, and I saw at that point in time an opportunity for herd improvement. I went to [artificial insemination] school to learn how to expand our cow herd."

Bill now has 150 registered Angus and 150 commercial cows on 4,000 acres. Bill and Linda have two sons: Andrew, who works in agribusiness, and Travis, who makes his home at the ranch as the fourth generation of McIntires to raise their family and cows together. "I wanted my kids to have an opportunity to raise cattle but was never sure it would work out," Bill said. "But since the time he was knee-high, Travis has said 'cow, cow, cow' repeatedly; it was his calling."[29]

In 1958, the American Angus Association began offering the Angus Herd Improvement Record system.[30] During the 1990s, the T&M Angus Ranch made the move to a personal computer for maintaining cattle records. Before he bought a computer, Bill McIntire had been using his neighbor's computer knowledge and Internet access to enter herd improvement data. Bill credited his neighbor for coaching him through tasks that ranged from turning on and off the machine to performing keyboard shortcuts. Bill can laugh now about his previous lack of computer knowledge, but he said it was pure intimidation that kept him from entering the computer age earlier. He now realizes neither he nor the ranch could function without the ability to maintain accurate records, input data, and send e-mails:

> I equate it to the way my dad started farming with horses and then moved into using machinery and four-wheel tractors. In this day and age, the ability to go in and click buttons and find unlimited amounts of data is absolutely necessary to maintain farm and ranch records. I still request paper copies of my American Angus Association registration papers, but my son Travis might someday be the one who chooses to keep registration information stored online.

Other technologies have helped the T&M Angus Ranch move forward into the digital age, including electronic animal identification. If Bill and Travis added 300 cows to their operation, the cost for animal ID tags would be $3 per head. Bill said that figure adds the right amount of value for the lowest amount of cost. He also said the ranch is slowly transitioning into a radio frequency

Left to right: Arden Oleen, Aaron Oleen, Josh Patry, Jesse Oleen, and Jan Oleen. (Source: Tom Patterson)

identification device system, which uses scales that work with electronic scanners and laptop computers to instantly record information. For many years, Bill used a red pocket notebook, commonly called a herd book, to write down birth weights and tattoo numbers. "When a calf is born the first thing we do is tie its feet, then tattoo, tag, and weigh it," he said. "This info all gets written down in my herd book."[31]

When Robert Oleen started Oleen Brothers near Dwight, Kansas, he took a chance on a business and a breed of cattle. Having grown up near Falun in Saline County, Bob's first venture into cow-calf production was purchasing twenty-five bred Hereford heifers from his brother. It was the 1930s, and times were tough. Bob's son, Arden, said he has heard many times how the economic Great Depression instilled a sense of survival in his family:

> Dad would tell the story of when he would go hunting with a 22-rifle, that he wasn't just shooting at tin cans. His shot was to bring a rabbit back home to eat. A lot of people in that generation learned to save and waste not. In our family, you don't just throw something away; rather, you fix it.[32]

Today, Oleen Brothers is a forty-year-old cow-calf operation comprised of 350 registered Horned Hereford and black Angus cows and 400 commercial cows. The cattle and

Cowboys on horseback still move and gather cattle throughout the state. Some cattlemen use ATVs to check the pasture and complete other chores. (Source: Tom Patterson)

horses carry a JAK brand, which stands for brothers Jan, Arden, and Kent (who is a silent partner). Their cows calve in both the spring and the fall, and the Oleens base their cow herd decisions on fertility, calving ability, disposition, and overall performance. To measure the quality of their cattle, the operation uses ultrasound technology—a process whereby sound waves from an electronic scanner capture an image with a high degree of accuracy to measure back fat, ribeye area, and marbling on live cattle. Each year, the Oleens hire a technician to ultrasound their entire calf crop as yearlings, and both heifers and bulls are evaluated. Collecting ultrasound data is a much cheaper and a more efficient means of evaluating carcass traits on breeding animals relative to progeny testing and carcass data collection. Additionally, fat thickness, ribeye area, and marbling are moderately to highly heritable, which suggests that the differences found between animals are expected to be passed on to their offspring.[33]

Ranch Equipment and Tools of the Trade

One hundred and fifty years ago, Kansas cowboys relied on their trusted steed—the horse—to work cattle. A horse was the most reliable way to drive cattle away from a rocky hillside or to chase after a stray through the open range. Although some ranch equipment will never be totally replaced, new tools have found their way into everyday cattle management.

Feeding hay used to be done by hand from the back of a wagon; today, modern bale beds allow cattlemen to roll out hay during unfavorable grazing conditions. (Source: Corinne Patterson)

In 1980, the first ATV was used at Oleen Brothers. Even though the four-wheeler is an easier way to take a newborn calf back to the barn (and a heck of a lot of fun to drive), Kansas cowboys still rely on their four-legged best friend. Arden Oleen explained,

> No one will ever invent a four-wheeler or mechanical device that can replace a horse. A four-wheeler can help with the rocks we go through in the Flint Hills, and we often use it and horses in combo. But you can't send four wheelers through the rocks to gather cattle. And sometimes the only way you can get a cow out is at the end of a rope—better on a horse than an ATV.[34]

At the McCurry Brothers Angus Ranch in Sedgwick, the cowboys use both horses and ATVs, but they are used for different tasks. Twenty years ago, McCurry Brothers purchased a three-wheeler to survey cattle, irrigation pivots, and fences. Greg McCurry said the three-wheeler made jobs faster and easier, but the three-wheeler was more dangerous than a horse. The rancher upgraded to a four-wheeler, which was a more stable and safer piece of equipment. Greg said the ranch has always used horses to sort cattle, which is a daily task on the ranch: "In a seed stock registered Angus business, horses are a must . . . Riding horseback is pleasant and easier on the cattle."[35]

One of the most appreciated pieces of equipment at Oleen Brothers is a hydraulic bale bed on a pickup—a necessity that has since been installed on four pickup trucks at the ranch. As kids during the winter, Arden and Jan Oleen used to feed thirty small square bales a day: they loaded the bales onto a pickup each morning, drove through the pasture, and then furiously cut the strings and threw bales to the ground for hungry cattle. Before four-wheel-drive pickup trucks were common on ranches, Arden remembered how his dad would often plow ahead full speed. Today, Oleen Brothers uses six four-wheel-drive pickups, four of which are outfitted with bale beds.

In a corral along Kansas Highway 177, just south of Council Grove, there stands a unique loading chute. Its

(Source: Corinne Patterson)

purpose is to load cattle from the lush pasture in the Flint Hills onto semi-trucks. The corral itself was built with heavy steel pipe when the Oleen Brothers purchased the land; however, they did not particularly like the layout and design for handling cattle and soon made plans to change it. They cut out the large, overhead pipes and replaced the rusty pipes with hedge posts and steel panels. Everything was replaced except for the rock-loading chute, which is a unique and often photographed element in the Flint Hills.

Arden Oleen said a solid working facility is absolutely essential to raising cattle: "Cattle pick up a lot of bad habits when they are handled in poor facilities. When they get by you once, they remember it and will try again. We've built a lot of corrals in our pastures because we believe we are better managers with good facilities."[36]

Cowboys like Norm Wilson also have seen more than their fair share of various working facilities. From the used railroad ties and woven wire of the past, he said there is no comparison to the solid-pipe construction that ranchers build today. Cattle once had to be trailed for miles to reach the stockyards and a scale. Today, individual pastures house their own gathering facilities that are large and allow for easy sorting and shipping of cattle, which can now be sold directly from the ranch because of the readily available scales.[37]

Branding is a popular way to identify livestock. Today's electric iron is an efficient tool that maintains a consistent heat to create legible brands. (Source: Corinne Patterson)

Roping sick cattle and treating them on the range was a way of life before working facilities started springing up along the countryside. When vaccinations and other management practices were implemented, many facilities were outfitted with headgates and squeeze chutes. Howard Blender claimed that the scars on top of his head were the result of the guillotine-style manual headgate. Commercial feeders and larger operators had hydraulic chutes long before, but the Blender Ranch still uses a Trojan chute that was purchased in 1972 for processing several hundred head each year.[38]

Cattle Identification

Cattle that came up the trails from Texas carried brands that were registered to their owners. Typically, calves were roped and then branded with a running iron that seared the identification into their hides. The practice of ranch identification branding has changed little through the years, but the tools used to apply brands have evolved. Since the 1990s, an electric iron, with its ability to maintain a consistent heat, has replaced irons that were heated with open fire. When today's cattle are run through the hydraulic squeeze chute, the simplicity of plugging the iron into an electrical outlet allows the cattle to be branded effectively.[39]

Freeze branding is a method of livestock identification that applies a chilled branding iron to the hide of an animal.[40] When the iron touches the skin, pigment-producing cells are destroyed or altered. When the hair grows back, it comes in white and creates a legible, permanent form of identification. Since 2006, the McIntires have used this technology as their main method of identification. Every registered cow at the T&M Angus Ranch has a white ear tag with a four-digit number that is also tattooed on the right ear. When cows are one year old, Bill and Travis freeze brand them with the same four-digit number on their right hip. The ranch also places its own TM state-registered brand on the left hip.

A freeze brand is more legible than a hot-iron brand throughout most of the year, and some believe freeze branding causes less damage to the hide. On branding day, the McIntires fill an ice chest with dry ice, 99 percent alcohol, and three-inch copper irons. Bill ensures that the animal is in good condition, because he believes a layer of fat helps to freeze the hair follicles quickly, and the hair on the animal's hip is clipped short to allow better and more instant contact with the skin. On a good day in January or February, ten head can be branded per hour. On rare occasions, Bill said freeze branding not only gives a rancher a visual identification method but also advertises the herd: "We had an instance where a neighbor saw the brand on our bulls and called us. We sold several bulls to them because they liked what they saw and were able to find out where the bulls came from."[41]

Progressive commercial producers try to capture premiums through identification of their calf crop. Bill Paige's son, Jerry, has really taken the cow herd to the next level. He is open-minded about technological advancements, including a third-party verification program for identifying in their calf crop. Jerry shared his experience talking with cattle buyers:

> It was kind of frustrating. I would take all of our age and source verification information to the sale barn the last couple of years, and we put all that information together. If we didn't have some third-party verification where they came out and looked at your cattle and verified that you were doing what you say you were doing, that source and age information wasn't worth the paper it was written on. They just told me that if you are going to reap the rewards for doing all this extra work, you are going to have to get involved with a company like ABS.[42]

For the first time in the fall of 2008, the Paiges took their recordkeeping through third-party verification and enrolled their calf crop with the Power Genetics program (http://www.powergenetics.com). This marketing alliance, Jerry said, allowed them to market cattle to a national audience via the Big Blue Sale Barn website (http://bigbluesalebarn.com), which many cattle buyers monitor weekly. The calves are sold directly through the website for a listed price or through a sale on a specific date.

Jerry has made many friends and partners in the beef industry who have helped him improve his herd. Pat Koons, owner of Kearny County Feeders, is one such friend who took Jerry and his wife, Dianne, under his wing. The Paiges were able to get carcass data back on groups of calves that were fed at Koons's feed yard. Although they were not focused directly on carcass traits, the Paiges were interested in how their calves performed. They even went to the packing plant to watch calves get harvested. "We were really surprised at how good the carcass information was and how they feed," Jerry said of the data they received from the feed yard. "It's a 30,000 head lot, and they were in the top 10 percent of all the cattle that they fed the years that they had them."[43]

Another new tool that the Paiges are using is a scale under their squeeze chute, which helps them track individual weights. They already track birthdates on each calf; they can now correlate that information with an individual weight to identify top producers and thus be more selective with poor performers. However, no matter how advanced the tools get, come December "preg-check time," the cow better be carrying next year's paycheck, or she is out the door. According to Jerry, "They don't get a second chance around here. If they aren't bred, they are gone. It costs too much to keep them around. Some of them we probably ought to roll over and make fall cows out of them, but they don't get second chances."[44]

Certified scales allow cattlemen to both measure performance and sell calves directly from the ranch. (Source: Corinne Patterson)

(Source: Corinne Patterson)

Technology on the Ranch

Oleen Brothers hosted its first production sale in 1987. Today, the sale serves as an annual marketing event that features their best cattle for sale to the public: Horned Hereford and black Angus bulls, foundation-bred quarter horses, F1 black baldy heifers with calves at side, and fall-bred Angus, Hereford, and black baldy heifers. To prepare for that first sale, Arden Oleen sat at his brother's kitchen table and addressed two hundred catalogs by hand that were mailed to potential customers; in 2010, the ranch mailed fourteen hundred catalogs across the country with labels generated by their home computer. This is just one example of how computer technology has changed the way the Oleens manage their ranch.

Today's ranchers also use computers for online video auctions. At one time, Arden used to take photos of bulls before a sale, send the film via mail for development, and then spend numerous hours packaging the sale catalog material by hand before he sent the final product to a printer. He now uses a digital camera to photograph the bulls, uploads those photos to his desktop computer, and then e-mails the selected pictures to the printer—a much faster

Sale day at the Eby Ranch near Emporia, Kansas, 2010. (Source: Blair Tenhouse)

process. With video auction technology, customers who live an hour away from the ranch or those who live several states away are able to purchase top-selling bulls. Arden explained, "If a customer comes to the ranch and looks at our cattle, we have a pretty good chance of selling them something. If a customer knows our cattle and that our word is good, they also feel comfortable bidding online. It's not as good as having people in the sale bleachers, but it is better than a called-in bid."[45] Arden said the family still believes that a live animal walking through the sale ring is better than a four-foot video screen. An old-timer once told him, "If I wanted to watch TV, I would have stayed at home." However, the live-video auction has significantly expanded their customer base and has been well worth the added cost, especially when bad weather hits the area on sale day.

Although many things have changed at Oleen Brothers during the past forty years, the tradition of raising quality cattle is still the same. Arden readily admitted, however, that he never thought there would come a day when he would talk on a cell phone while riding a horse.[46]

Cattle Feeding

The early forerunners to today's modern cattle-feeding industry in Kansas were many small farmer-feeder operations sprinkled across the state. Warren Weibert, owner of Decatur County Feed Yard, grew up near Durham in Marion County during the mid-twentieth century, and his family was one of those small feeder operations. His father fed cattle that were sorted and then shipped to the stockyards in Wichita and Kansas City. On the way to Kansas City, they would pass through Strong City, where Weibert remembered being in awe of the vast numbers of cattle that E. C. Crofoot had on feed. The Crofoot yard was one of the state's earliest commercial yards and had a capacity for 15,000 head during the late 1940s and 1950s.

Feed trucks similar to this Harsh 502 are used across Kansas to deliver feedstuffs to the cattle in feed yards. (Source: Blair Tenhouse)

During his lifetime, Weibert has witnessed the evolution of feed delivery to cattle. From the back of a four-wheeled wagon being pulled behind a tractor, and with a scoop shovel to the evolution of the automatic systems of a feed truck, feed delivery dictated the number of cattle an individual could handle. Weibert recalled,

> Back when I was a kid, we just fed cattle and they were what they were. With commercial cattle feeding, that changed in a big way. Through the sixties and into the mid-seventies, there was a tremendous evolution of the way feed was delivered to the bunk using large automatic mixing trucks. Obviously, before then it was scoop-shovel driven. As farmer feeders got bigger and commercial feed yards started, they advanced the mixer truck in a very big way, and there were those businesses that sprung up through Kansas. Two or three different manufacturing companies in Kansas soon developed successful versions of a feed truck. Over time, scales were added, cards were stamped, or numbers from the scales were written down to make it easier to track and bill commercial customers for the type of feed that was fed to a particular pen of cattle.[47]

While a handful of small farmer-feeders still fatten cattle, the commercial feeding industry in Kansas today ranks second in the nation.[48] More than five million head of fed cattle are marketed in Kansas annually.[49] Weibert explained how things have changed:

> A farmer-feeder is more diversified. Farmers had residue crops available on some of the rougher land that they could utilize to cheapen the cost of gain, and utilize their labor and resources through the winter during their off season. That really helped farmers in earlier generations like my parents. When feedlots came along, they were more management intensive and less land intensive and ran a pencil quite sharply in looking

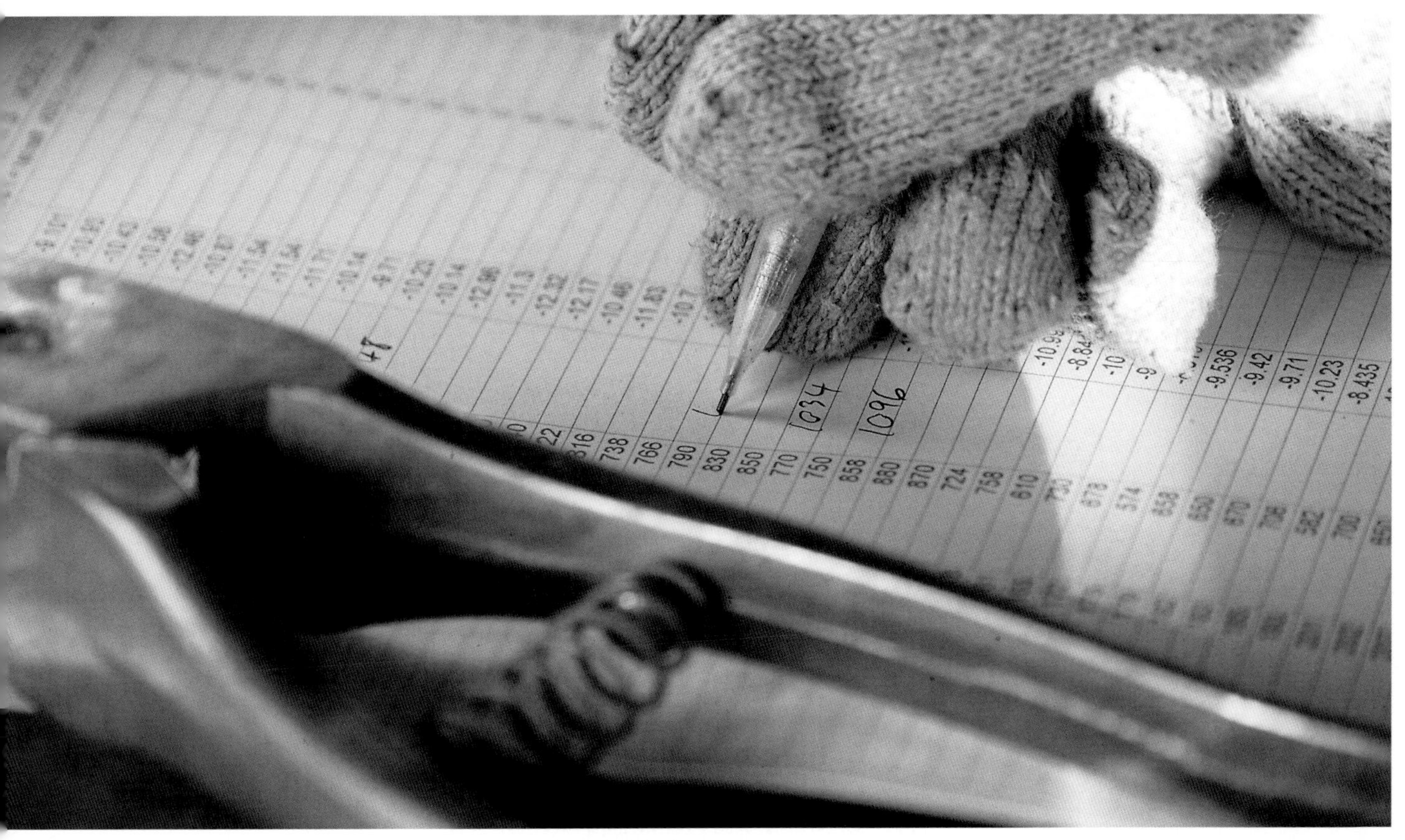

Recordkeeping is common throughout the beef cattle industry. (Source: Blair Tenhouse)

> at input costs and output, trying to figure out what was making money, and what wasn't.[50]

Weibert graduated from Kansas State University in 1969 with a bachelor of science degree in animal science. He took a job with a telephone company out of college and married his wife, Carol, in 1970. He spent seven years working in Wichita before Carol's father, Milton Nitsch, asked them to move to Oberlin to take over the management of Decatur County Feed Yard. They moved to Oberlin in 1977, and they now own the feed yard that boasts a one-time capacity of nearly 40,000 head. Weibert spoke about some of the changes he has witnessed:

> Recordkeeping and automation were a major change through the seventies and into the eighties and allowed the feeding industry to really grow in Kansas. The packers came along when the feeders expanded through the western part of the state. The packers were building new plants throughout the eighties, and that really helped spark the commercial cattle feeding industry in Kansas.[51]

From early on, Weibert was involved in the Kansas Livestock Association, the National Cattlemen's Association (present-day National Cattlemen's Beef Association), and CattleFax. Much like his first fascination with the scale of the Crofoot yard, Weibert made cattle feeding an obsession. He said the story of Decatur County Feed Yard really took root in about 1980, when the yard starting weaning calves for beef producers in Wyoming. These producers questioned how they could make more money out of their calves: specifically, they asked how Weibert was going to

Brian Hagedorn, manager and co-owner of Sylvester Bull Development, believes that consistent and accurate feeding provides a strong foundation for bulls to thrive under his care. (Source: Blair Tenhouse)

sell their cattle for a higher value—a so-called premium. Weibert recalled,

> As they came out and visited the yard, they would ask me, "How are you going to sell those cattle over there, and how are you going to sell mine?" And, of course, many of those people thought they deserved and should get some sort of a premium for those cattle. The fact unfortunately was they all kind of brought the same prices. We lumped 8 or 9 or 10 pens together and they brought the same price. Any thinking and discerning person would see that there's something wrong with that kind of

> marketing. All cattle were not created equal or surely not all equal in value. The problem was getting true value out of cattle.[52]

Marketing fat cattle thirty years ago was quite different from marketing cattle today. On the day that Weibert walked in as the yard's manager in 1977, he was greeted by thirteen different cattle buyers. In the following years, smaller packers began to disappear: "You had to be a little bit mindful every week or month or year, who were still the buyers and whether they had the money to pay for the cattle when they actually took them. There became attrition of buyers, and it had to do with efficiencies of operation and the ability to sell meat to their buyers as well out in the country."[53]

Many improvements in animal health came about as the cattle-feeding industry grew. Kansas has profited from this growth, and today Kansas and Missouri boast the "animal health corridor" of the twenty-first century. The animal health corridor, with its center in metropolitan Kansas City, is home to many of the world's animal health companies, and companies in the Kansas City area account for one-third of sales in the $19 billion global animal health sector.[54] Weibert has observed the growth of animal health–related businesses in Kansas:

> With the advent of commercial cattle feeding when it grew on such a large scale, it enhanced possibilities and also expanded problem areas with confinement. The use of consulting veterinarians and consulting nutritionists became a big business in the late seventies and early eighties and has progressed right up to the present day. Feedlot operators had access to a very talented group of people that had a very broad view of the industry because they traveled geographically between states and got to see a lot of successful operations. Some things worked better than others throughout time.[55]

Around 1986, Decatur County started to use ultrasound technology as a way to sort cattle, a project that was developed with Patsy Houghton when she served as the Northwest Kansas Beef Cattle Extension Specialist for K-State Research and Extension. In 1990, a video-imaging study with 700 cattle was conducted. Weibert recalled that, initially, the program did not work extremely well, but the goal was to explore the implementation of technologies that were coming down the line.

The first version of the electronic ear tag, which was marketed by Allflex, debuted in 1993; in 1994, Decatur County worked with Micro Beef Technologies to track cattle individually through the feeding operation. They secured a packer who worked with them to develop a value-based marketing grid; the tracking data were then used to target various points on the marketing grid.[56]

John Atwater's family has fed their cattle with Decatur County Feed Yard since the 1970s. Dating as far back as the 1880s, the Atwaters have raised beef in Jackson County for several generations. Their relationship with the feed yard has allowed them to use data collection when they make decisions about herd improvement. Value-based marketing has changed the way Atwater is paid for his cattle. Data collection (e.g., through ultrasound technology) produces individualized data on each of his calves: "It pays you for improving your herd," Atwater said.[57] He also noted that preconditioning calves is one of the biggest management improvements that has come from owning both fall and spring calves in the commercial feed yard. Calves are vaccinated before weaning begins, and follow-up booster shots are given during the weaning process. With facilities and available feedstuffs, Atwater said they background calves on the ranch for at least thirty days before the calves are sent to the Decatur County Feed Yard.[58]

Conclusion

Although there are no living beef producers from 150 years ago, many of the changes and improvements from that time still remain. A cowboy from 1861 would recognize the beef animal that is raised today, but technology has certainly changed the tools that are used to care for and manage the herd. What continues over time is the belief that if you take care of each animal to the best of your ability, you will in turn sustain a livelihood in the cattle industry. Warren Weibert noted the nature of Kansas and its people—and the shared progressivism that brings unity in the beef industry:

> Kansas is blessed with talented people, and I would say open-minded people, who are open as much to new ideas as anyone. We are a diverse ag state. There's a lot of difference in rainfall from one end of the state to the other. With the different types of crops we have, the weather that we have, and the people that we have, our producers are probably as open minded as any around. In my own experience through the Kansas Livestock Association, there's quite a bit of cooperation between individuals in this state. Whether it's your neighbor or people across the state, there's a progressive attitude among the people.[59]

(Source: Corinne Patterson)

Chapter 4
Buying and Selling Cattle: The Kansas City Livestock Exchange and the Development of the National Cattle Trade

By K. C. Olson

After the Civil War, drovers brought cattle overland to the railhead towns of Kansas for shipment to the market centers in Kansas City and Chicago. The cattle shown here were shipped from Abilene, Kansas. (Source: Kansas State Historical Society, Topeka, Kansas; reprinted with permission)

The nineteenth century was characterized by rapid change and reorganization in American agriculture. The affluent, urban populace of the eastern United States needed food and textiles that were plentiful, varied, and of high quality. As rail transportation and telegraph communication developed, national markets for agricultural commodities emerged. Before the 1840s, the traditional mercantile firm marketed and distributed the nation's goods. Early large-scale trade across state lines was chaotic and inefficient; there was little organized commerce. Out of this chaos, a new class of merchants appeared. The commodity broker—a middleman between the producer and the buyer—arranged the purchase, transport, and delivery of agricultural products across the nation, and he brought an organizational revolution to agriculture. Within a generation, the modern commodity dealer replaced the mercantile firm. This new form of administrative coordination reduced the number of transactions in the flow of goods, increased the speed and regularity of flow, and, consequently, lowered costs and improved the productivity of the nation.[1]

Within a short period of time, the commodity broker came to totally dominate the movement of grain. These commodity brokers organized the Chicago Board of Trade in 1848,[2] and in so doing, they provided a new opportunity for the nation's agriculturalists. Farmers and ranchers were no longer relegated to subsistence living, bartering, and extracting a living from limited local markets. Even the most isolated of farmers and ranchers were able to distribute their products nationally via the commodity broker, the railroad, and the telegraph.

THE

KANSAS STOCK YARDS

AT KANSAS CITY, MO.

IS THE

LARGEST ESTABLISHMENT

OF THE KIND WEST OF THE MISSISSIPPI.

RAIL COMMUNICATIONS EAST, WEST, NORTH and SOUTH.

EVERY FACILITY IS AMPLY FURNISHED FOR

Receiving, Yarding, Watering,
Feeding, Resting, Selling,

WEIGHING AND SHIPPING

LIVE STOCK

OF ALL DESCRIPTIONS.

YARDS FOR CATTLE FLOORED AND COVERED PENS FOR HOGS AND SHEEP AND GOOD BARNS FOR HORSES AND MULES.

All necessary conveniences at hand for transacting Stock Business. No expense, labor or pains will be spared to make the

KANSAS STOCK YARDS

THE LARGEST AND BEST POINT FOR CONCENTRATING LIVE STOCK IN THE MISSOURI VALLEY.

Here has sprung up a complete Stock Mart of great magnitude. Here are responsible Commission Firms, also Banking Houses and Railroad offices. Patronage solicited. Fair dealing and satisfaction guaranteed.

JEROME D. SMITH,
Gen'l Supt.

This handbill is an early advertisement for services at the Kansas City Stockyards. (Source: Kansas State Historical Society, Topeka, Kansas; reprinted with permission)

This organizational revolution was not unique to the grain trade. It also occurred with the distribution of livestock throughout the United States, which was enabled by the advent of rail transport. The Kansas-Pacific Railway pressed into Kansas and Colorado during the 1860s, and the Missouri, Kansas, and Texas Railway opened the Texas overland trade during the early 1870s.[3] The potential for railroads to bring animals and animal products from the production centers of the west to the population centers of the east was quickly recognized.

Unlike the grain brokers who never left the mercantile exchange offices, and who only saw samples of the commodity that was traded, cattle traders of that time traveled between the farms and ranches of the Southwest and the trading centers of the Midwest, where they personally monitored the movement of specific animals on the railroad. Eventually, these traders helped establish markets that were similar in scope to the Midwestern grain exchanges. The Chicago and Kansas City Live Stock Exchanges were organized in 1884 and 1886, respectively, almost forty years after the Chicago Board of Trade was organized.[4]

Grains graded and standardized easily and were sold through futures markets that the Chicago Board of Trade introduced in the United States during the mid-1900s. Conversely, livestock could not be standardized easily for the purposes of trade; thus, they were not sold through the futures markets. Each animal was unique in both weight and product yield, which made it necessary to trade livestock on a spot cash market.[5] Live animals required human oversight at every step of the marketing process, and there still had to be personal contact between buyers, sellers, and merchandise. Not until 1974 did modern cattle-feeding practices, selective breeding, and veterinary science bring enough standardization to the cattle trade to enable the establishment of futures markets.[6]

The Livestock Trade in the Nineteenth Century

After the American Civil War, drovers moved market-ready cattle out of the Southwest to the railheads of storied Kansas cowtowns such as Abilene, Ellsworth, and Dodge City. During the late 1860s and the 1870s, one Texas drover could efficiently move about 3,000 animals at a time. The job required a substantial investment—a large crew of cowboys, a trail boss, food, horses, equipment, and an entire summer's wage.[7] In most cases, Southwestern drovers purchased cattle in their home states immediately before the overland drive to Kansas. This proved to be a profitable venture despite the time, trouble, and financial investment associated with a trail drive. In 1867 Texas, a two-year-old steer could be purchased for $8.70. A few months later, that same steer could be marketed in Kansas for $23.32—a price difference of 268 percent.[8] The rancher was paid roughly a third of the final sale price, while the drover received about two-thirds. This resulted in some resentment toward the drovers.

Southwest cattlemen were anxious to retain a greater share of the value of their animals and were therefore quick to seek more equitable methods of marketing. As railroads extended further into the Southwest, ranchers were presented with the possibility of being able to transport cattle overland more rapidly and without the involvement of a middleman. The railroads provided ranchers with direct market access and the means to capture a greater share of the value of their product. By the early 1870s, railroads were collecting thousands of cattle off of the Southwest grasslands and were bringing them into Kansas City for national distribution.[9]

A trip from Las Animas, Colorado, to Kansas City via the Santa Fe Railroad took two days in 1873; the trip to Chicago took five days.[10] Moving animals great distances by train introduced livestock owners to the concept of *shrink:* confinement for long hours without feed or water caused livestock to experience significant weight loss. Injury and death also were significant problems. Producers closer to the stockyards were at an advantage, because their livestock experienced relatively little loss during the trip.

Cattlemen from all over the Southwest quickly learned this lesson, and many moved their herds to the rangelands of eastern Kansas before the herds were sold. When the price of cattle in Kansas City rose to an adequate level, the cattlemen would quickly load their herds aboard stock trains and ship them to take advantage of the price change. Grazing transient cattle on the grasslands south and west of Kansas City became common in the ranching business. Each summer, the bluestem pastures of the Kansas Flint Hills around Emporia supported thousands of transient cattle that had been shipped from west Texas for a few

The Phillip Armour packing plant is shown in 1870. It established Kansas City as a major meatpacking center. (Source: Kansas State Historical Society, Topeka, Kansas; reprinted with permission)

months' growth and fattening. In fact, the 101 Ranch in the Texas panhandle purchased 75,000 acres in the Flint Hills just for this purpose.[11]

Kansas City was a logical center of operations for the new and burgeoning national market in livestock. It was a town of only 32,000 residents when Phillip Armour built a packing plant in 1870; however, it was closer to the source of Southwestern beef than the primary competing market in Chicago was. Just sixteen years later, Kansas City had seven packing plants that employed 2,234 workers.[12] Because of a lack of refrigeration, meatpacking was seasonal until 1877, when the Armour packing plant installed a chill room that enabled year-round work. That same year, Kansas City became the primary supplier of packed beef to the United States.[13] After that time, beef packing in Chicago dwindled into comparative insignificance because of the large number of cattle that were processed in Kansas City.[14]

It would be incorrect to suggest that the Chicago market did not play a significant role in the advance of the national livestock trade, because it certainly did. However, the Kansas City and Chicago markets dominated fairly distinct, non-overlapping regions of the country because of the pattern of railroad development. In 1890, a railroad inspector who was stationed in Kansas City along the Santa Fe Railroad recorded 8,988 railcars of cattle shipped. Bills of lading for these cars showed that 85 percent stopped in Kansas City, while only 15 percent went to Chicago. These figures can be interpreted to suggest that the Kansas City market dominated the trade out of southwest Kansas, Colorado, Indian Territory, the panhandle of Texas, and New Mexico. By contrast, the Missouri, Kansas, and Topeka Railroad shipped the majority of eastern Texas cattle because of its more easterly routing. A railroad inspector in southeastern Kansas reported that of the 8,500 cars shipped from northeastern Texas in 1890, 69 percent went to Chicago, 17 percent went to St. Louis, and only 14 percent went to Kansas City.[15] The pattern of regional dominance that was established during the early days of the national cattle trade persisted into the twentieth century.

Initially, Kansas City lacked an organized livestock market, and packer representatives had to travel into the countryside to buy livestock. That changed in 1871 when a number of Kansas City businessmen chartered the Kansas City Stockyards Company[16] to provide a location for buyers and sellers to congregate and transact business. It was complete with rail lines, animal handling facilities, and livestock scales. By 1893, the stockyards covered about one hundred acres along the Kansas-Missouri border; the entire area was floored with three-inch cypress plank; the daily capacity was 20,000 cattle, 35,000 hogs, and 15,000 sheep; and 300 men were needed to handle, feed and care for all the stock.[17]

It took little more than twenty years for Kansas City to be reborn as the major livestock marketing and meatpacking center in the United States. When this rebirth was combined with deeper penetration of the railroads into the grasslands of the West, cattlemen were able to transport their animals to a competitive national market, and for the first time, they were able to negotiate face-to-face with a number of potential buyers. The new system also enabled cattlemen to capture a greater share of the value of their animals. As long as the Kansas City Stockyards remained relatively small, a rancher could find a buyer without any assistance, but the volume of animals soon increased to the point that the stockyards became large, impersonal, and confusing.[18]

The Kansas City market offered a wider selection of more reasonably priced animals than competing markets to the east did. This circumstance prompted large numbers of eastern commodity merchants to travel to Kansas City relatively soon after the organization of the Kansas City Stockyards Company. The large concentration of buyers and sellers at the stockyards caused livestock prices to fluctuate wildly—up to 30 percent in a single day. A producer who attempted to market his own livestock frequently sold for less than market price. No one but an expert could determine when to make a sale for optimal return within the Kansas City market. An additional complication was the fact that cattle alone had fourteen different classifications for the purposes of marketing: fancy cattle, choice cattle, good shipping steers, medium shipping steers, common to fair steers, common to choice bulls, good to choice cows, poor to medium cows, stockers and feeders, northern range steers, Texas steers, Texas cows, veal calves, and milk cows. The plethora of classifications meant that the seller had to find multiple buyers for each load of stock that was brought to Kansas City. Finding all of these buyers was difficult; only someone intimately familiar with the market knew all of the buyers and what types of cattle they sought. These factors meant that a cattleman could no longer market his own livestock without encountering a significant price risk.[19]

The Livestock Commission Firm

As the size and the scale of the Kansas City Stockyards increased, it became increasingly difficult for individual ranchers and farmers to effectively market their own animals. Farmers and ranchers lacked market savvy and familiarity with stockyard personnel, which reduced the likelihood of a suitable financial return. It became clear that an intermediary was needed to work between the buyers and the sellers at the Kansas City Stockyards.

The livestock commission merchant—an entirely new class of trade broker—appeared to fill this void. Commission merchants arranged the sale of livestock for a consignment fee. They carried out this task not only within the Kansas City Stockyards but also throughout the Southwest, where they often traveled great distances to locate, solicit, and sell livestock. In 1887, forty livestock commission firms operated in Kansas City; that number

R. C. WHITE, Kansas City. L. A. ALLEN, Bent Co., Col. M. M. EVANS, Kansas City.

WHITE, ALLEN & CO.,

Commission Merchants for the Sale of

Stock

Kansas Stock Yds. KANSAS CITY, MO.

REFER TO

WESTERN LIVE STOCKMEN

OR TO

BANKS DOING BUSINESS WITH WESTERN STOCKMEN.

An early handbill advertised the commission services of White, Allen, and Company. (Source: Kansas State Historical Society, Topeka, Kansas; reprinted with permission)

eventually increased to three hundred, a number that remained constant well into the twentieth century.[20] Because the commission merchants operated continuously in the Kansas City Stockyards, they were expert observers of price trends and were acquainted with the numerous order buyers and packer buyers who frequented the stockyards. Commission merchants reduced much of the risk associated with marketing livestock in Kansas City because the livestock owners chose opportune marketing windows and arranged for the simultaneous sale of multiple classes of livestock.

It is significant that the commission merchants did all of this at a relatively small cost. Unlike the drover, the commission merchant did not accept ownership of animals because one commission merchant could not take title to the thousands of animals that were sent to the market daily. Instead, they minimized their investment risks by operating on consignment. The standard consignment charge was

An aerial view of the cattle pens at the Kansas City Stockyards during the early twentieth century. (Source: Kansas State Historical Society, Topeka, Kansas; reprinted with permission)

initially $0.50 per head for cattle; later, an equivalent charge of $12 per railcar was used.[21] These commission rates remained constant between 1886 and 1921, a span of thirty-six years. During that same period, the price of cattle varied from a low of $3.65 per hundredweight in March 1889 to a high of $9.60 per hundredweight in August 1912.[22] The attractive and eminently fair rate scale of the commission merchants soon made selling livestock on consignment the industry's preferred method of marketing.

The early livestock commission merchants of Kansas City sought ways to distinguish themselves from their primary competitor—the drover. First, the commission merchant provided more services than the drover. The drover traded market-ready beef cattle only. Railroads enabled the transport of not only market-ready cattle but also young cattle that needed to be finished, beef cows, dairy cows, breeding bulls, sheep, and hogs over long distances. Commission merchants happily accepted the task of marketing all classes and species of livestock; however, cattle dominated the trade receipts in Kansas City. An average of 2.2 million cattle were shipped annually to Kansas City between 1906 and 1915, which represented 75 percent of the total weight of livestock marketed during that period.[23] Second, commission merchants operated on a grander and speedier scale than drovers did. In October 1885, a single commission firm owned by A. J. Snider took 19,000 cattle off the grasslands in the Southwest and marketed them in Kansas City. Only a few days passed from the time that the cattle were loaded onto a railcar to the time that the ranchers received payment. Third, commission merchants were quick to exploit the value of the telegraph as a marketing tool on behalf of their customers. The telegraph was the first reliable source of price discovery—for the first time in the history of the cattle trade, the telegraph instantly supplied information on price changes in the stockyards. Commission merchants kept their customers, who were

A present-day view of where the stockyard pens were located. (Source: Holli Kepner)

far away in the countryside, informed of changing prices with regular market quotes that helped customers pick an advantageous marketing window for their stock. The prepaid telegram was the means by which this information was communicated. On February 9, 1889, the *Texas Live Stock Journal* reported the importance of one such prepaid telegram. Several days before, the cattle market had advanced $0.25 per hundredweight. Thousands of telegrams that quoted the price increase were issued simultaneously from the Kansas City Stockyards. Producers who were holding cattle on the grasslands west of Kansas City loaded nearly 17,000 animals onto railcars and rushed them to market in time to profit from the information. The marketing capabilities of a typical drover paled in comparison to those of the commission merchants, and the drovers were quickly driven from the marketplace. In the end, there was little or no protest from ranchers at the passing of the drover.

The American Royal began as a national Hereford show in a tent at the Kansas City Stockyards. This show has grown into the annual American Royal Livestock Show, which draws an audience of nearly 70,000 from a thirty-state region. Attendees include owners/breeders, future business and agriculture leaders (high school and collegiate), and animal enthusiasts. (Source: Holli Kepner)

Daily operations at the Kansas City Stockyards were highly ordered. When cattle arrived, stockyard employees unloaded them from the railcars. The livestock were then delivered to the assigned alleys and pens of the commission firm that was designated by the cattle owner to handle the sale. Employees of the commission firm then fed and watered the stock until they were sold. In 1886, the February 6 and March 13 editions of the *Texas Live Stock Journal* featured fee scales for the stockyards: feed charges at the Kansas City Stockyards were $1 per bushel of hay and $0.75 per bushel of corn, and the yardage charge was $0.20 per head per day for cattle. These prices were substantially cheaper than in Chicago, where the fees were $1.50 per bushel of hay, $1 per bushel of corn, and $0.25 per head per day yardage.

The morning after the livestock arrived, the commission merchant transacted business in the alleys and pens that housed his client's cattle. Packer buyers, order buyers, and cattle feeders rode horses through the alleys or walked along the catwalks constructed over the pens to make their selections. Contracts struck between the commission merchant and the buyers were private treaty affairs and were transacted orally. Many commission men kept the figures in their heads until they returned to the commission office in the exchange building. Once terms of the sale were agreed upon, the livestock were herded to scales and weighed, and the first printed record of the transaction appeared when a scale ticket was attached to the bill of lading from the railroad. This was subsequently delivered to the stockyards company office, and the buyer was free to collect his purchase. The commission merchant received payment for the livestock at the Kansas City Stockyards Company office, where the commission firm was charged by the stockyard for the rail freight, feed, and yardage fees. The commission firm then paid the seller after commission, freight, feed, and yardage fees were deducted from the total sale price. This process took place twenty-four hours per day, seven days per week.[24]

A large amount of money changed hands during the course of an average day at the Kansas City Stockyards Company, which created financial opportunities for both the financier and the cattleman. Kansas banks leaped at the opportunity to do business with the patrons, and many opened offices adjacent to the stockyards—or within the stockyards building itself—to handle the deposits of cattlemen. By the same token, ranchers had the convenience of being able to borrow operating capital from these banks. Kansas bankers routinely took risks on cattle that originated from areas all over the Southwest. One such institution was the Emporia National Bank; in 1900, it loaned 16 percent of all its operating capital to ranchers in Roberts County, Texas. That same year, Kansas banks loaned 50 percent of the cattle money that was borrowed in the Texas panhandle.[25] When making operating loans to ranchers who lived far from Kansas City, the stockyard banks relied heavily on the commission merchant's judgment of the rancher's character and business acumen. Stockyard banks also depended on the commission merchant to report on the location and the disposition of any cattle that were under mortgage. In this way, the commission merchant became the critical link between the Kansas banks and the ranchers in capital-poor areas of the West.

Banks were not the only institutions to finance the new cattle trade. The Kansas City livestock commission merchants themselves provided much needed cash flow to ranchers. They often provided financial advances on consigned cattle, and they quickly learned that they could direct more business to their firms by giving these advances. It soon became customary for cattle owners to draw upon commission firms for at least part of the purchase price of the animals that they shipped.[26]

Crisis and Reform

It was inevitable that the freewheeling economic conditions associated with the development of the national market in livestock would lead to controversy. Before 1900, and particularly in the frontier areas, the federal and local governments were not a regulatory force; government agencies rarely imposed rules upon the industry. It was in this laissez-faire trade environment that unique biological and ethical problems first challenged the national livestock trade.

The biological problem that confronted the trade was an unintended consequence of a nationwide transportation system. As livestock were moved by rail

The American Royal is not only one of the oldest and best-loved traditions in Kansas City but also one of the largest combined livestock and horse shows and rodeos in the nation. (Source: Holli Kepner)

from the four corners of the country to the national markets of the Midwest, they brought with them endemic diseases. Outbreaks of Texas fever, pleuropneumonia, and tuberculosis were only regional in scope before the era of rail transport; by 1883, these diseases were national epidemics, and the livestock markets in Kansas City and Chicago were the focal points of disease transmission.[27]

The U.S. government was ill-equipped to deal with the crisis. There was no stated policy on the control of animal diseases before 1884, and no federal agency had executive power to deal with such a problem. Livestock producers demanded that the U.S. Department of Agriculture act, and their demands reached new levels during outbreaks that were brought about by rail transport in 1883. The government responded by creating the Bureau of Animal Industry in 1884. This agency was given broad powers

Bull Wall was designed by sculptor Robert Morris and is a tribute to Kansas City's agricultural heritage and the American Royal Livestock Horse Show and Rodeo. This sculpture connects two 120-foot-long steel walls that feature cutouts of running bulls. (Source: Holli Kepner)

to identify and destroy diseased animals, and it had the power to quarantine any stockyard that was determined to be infected.[28] The scope of this power threatened the livelihood of the livestock commission merchants.

The bureau's veterinarians wanted to shut down all stockyards in 1884 when pleuro-pneumonia was diagnosed in Chicago; however, their views inspired little confidence among ranchers and commission merchants. Government veterinarians at the time were not held in high regard.[29] Furthermore, disputes arose between and among veterinarians, commission merchants, and others about such issues as disease transmission and the appropriateness of closing stockyards.[30]

In response to the disputes with the Bureau of Animal Industry about disease-related issues, the Chicago livestock commission merchants organized the Chicago

Livestock Exchange in March 1884. As an association, they successfully lobbied the U.S. Congress to reduce financial appropriations to the bureau and to limit the bureau's executive powers over the stockyards. In effect, the association won the right to regulate stockyards commerce privately and free of government interference.[31] The commission merchants in Kansas City faced similar pressures, but they resisted organizing. The Chicago Livestock Exchange had effectively quashed government attempts to regulate the livestock trade's animal disease issues, so the need for the Kansas City merchants to form an organization of their own was no longer urgent from that perspective. There was, however, an impetus behind the development of a livestock exchange that came from a wholly different source: the cattlemen and the customers of the Kansas City Stockyards Company.

The Golden Ox Restaurant opened its doors in May 1949. The restaurant's purpose was simply to provide the ranchers and farmers who brought livestock to the stockyards with a good place to dine. It featured the then and now world-famous Kansas City steak. (Source: Holli Kepner)

Although epidemic livestock diseases threatened the national livestock markets externally, unethical behavior by the commission merchants and the packers threatened the markets internally. The volume and the anonymity that was typical of the newly reorganized livestock trade made the sellers totally dependent on the integrity of their trading partners; however, it was unavoidable that corrupt people would take advantage of what was essentially an honor system. Certain business practices of the commission merchants and the packers began in good faith but were later used to routinely defraud sellers. Naturally, the livestock producers were angry and sought justice.

Within the stockyards of Kansas City and Chicago, all contracts between buyers and sellers were oral. The integrity of the commission merchant was the only thing that protected livestock owners from fraud. It was not until the commission merchant rendered the bill of lading and the scale tickets that there was an identifiable contract for a given sale. It was relatively easy for a dishonest commission merchant to falsify the account of the sale so as to collect a larger consignment fee than was originally agreed upon. One historical example illustrates this problem. A Chicago commission merchant, J. S. McFarland, sold sixty-six heavy cattle on behalf of D. P. Taylor of Avoca, Iowa. The cattle weighed 93,030 pounds and sold for $4.70 per hundred pounds. McFarland only returned $4.60 per hundred to Taylor and pocketed the difference. An audit of McFarland's books by the Chicago Livestock Exchange uncovered the fraud, and he was immediately expelled from the exchange.[32] Kansas City, however, had no such mechanism to police its own merchants. Crimes like this, if discovered at all, were instead subject to the purview of local courts, which were notoriously slow to prosecute offenders.

The telegraph also was subject to misuse. Some commission merchants, in an effort to encourage livestock sales during periods of low prices, purposefully misquoted

This mural was painted in about 1958 by Necati Alkis, a recent immigrant from Turkey. The mural is more than 20 feet long and about 5.5 feet high, and it covers most of the south wall of the Golden Ox Restaurant's bar. (Source: Holli Kepner)

the market in telegraphed price circulars that were widely distributed. The fact that the Kansas City Stockyards initially operated twenty-four hours per day, seven days per week, encouraged this type of behavior. Some merchants conducted their business during the evening hours simply because fewer people were around to report misdeeds. The fact that livestock were shipped in on night trains also provided the potential for deceit because these cattle were often sold before buyers had accumulated the next morning. This situation afforded some buyers an unfair advantage, and it suppressed competition for the seller's animals.

Unscrupulous livestock buyers also were a problem. Some were known to take advantage of the time lag between the oral contract and the delivery of payment. Deceitful buyers sometimes denied agreements that were made with the commission merchant in the stockyards after the price of livestock declined during the day. A second problem was known in the trade as the *dockage swindle*. Animals were sold by the pound, and any animals with a physical imperfection were subject to packer-mandated price discounts. The *price dock* was applied by discounting the weight of an animal that was recorded at the time of the sale (by some predetermined amount

Kansas City has a rich restaurant tradition. Two KC-area brothers, Herb and Marvin Quick, founded Quick's 7th Street Bar-B-Q in 1956. Today, cousins Scott (Marvin's son) and Steve (Herb's son) run this Kansas City–based family business. They are moving this restaurant to the Shawnee Mission Parkway area, where it will continue to be family owned and operated. (Source: Holli Kepner)

and according to the nature of the physical imperfection). Conveniently, the dock was imposed after the time of sale. Packer buyers quickly learned that the dockage system was a convenient way to reduce the price they paid for animals; unfortunately, there was no way a rancher could appeal the dockage decision of the packer buyer.[33] In principle, ranchers, commission merchants, and packers agreed that there was a legitimate reason to discount the price that was paid for imperfect animals; however, it was clear that the packer buyers were using the so-called dockage swindle to control their costs. Volatility was inherent in the cattle markets—prices could change as much as 30 percent in one day. Packer buyers were able to easily lower the price of cattle bought early in the day by applying the dock late in the day. Stockyard customers noticed that the applied dockage seemed to be more fair if hog prices went up during the day. It did not take long to figure out why; the buyers were manipulating prices.[34]

Several business practices of the commission merchants, while inside the letter of the law, were ethically questionable. From the beginning of the commission trade, firms employed solicitors who lived at locations that were remote from stockyards and were usually close to a large concentration of cattle. Solicitors often held other influential jobs within their local livestock industries; railroad agents and cattlemen's association representatives were the favored solicitors of the commission firms. Their job was to convince ranchers to consign their stock to specific commission firms; in return, solicitors received one-half of the consignment fee. Inside information provided by its solicitors gave the commission firm a competitive edge. However, the fact that blatant conflicts of interest often occurred did not escape the notice of cattlemen. Solicitors became such a controversial part of the business that commission firms rarely identified them publicly. Furthermore, solicitors had little loyalty to any single commission firm, and they frequently switched employers for more pay.[35]

Another questionable practice of the commission merchants involved independent speculation by commission firm employees, who were free to purchase livestock on their private accounts with the hope that prices would rise before resale. Although legal in practice, speculating was ethically questionable. Commission firm employees received discounts on yardage and feed at the Kansas City Stockyards. While waiting for a market upturn, they could afford to keep livestock at the stockyard for longer periods than a livestock producer could. Inevitably, some poor-quality stock came into the possession of commission firm employees who traded on their own accounts. Unscrupulous employees were easily able to substitute their own inferior animals for a customer's high-quality animals. The employee was required only to report the number of livestock sold; a few inferior animals placed in a large load went undetected.[36]

Quick's 7th Street Bar-B-Q slow-cooks their briskets for twelve hours. (Source: Holli Kepner)

A third ethical problem involved the rebates of consignment fees. Before 1886, commission firms began to rebate one-half of the consignment fee for especially large shipments of livestock. This practice became a useful way for commission firms to win customers. Unfortunately, it escalated into a form of abuse. Vigorous competition made the commission business difficult to enter because the larger, wealthier firms operated on narrow financial margins. They could afford to offer continuous commission rebates while the smaller, newer firms could not. In 1885, four commission firms dominated the cattle trade: A. J. Snider and Company marketed 44 percent of the cattle that was sold through the Kansas City Stockyards, and three other firms accounted for 30 percent of the cattle that were sold through the stockyards.[37] In an effort to eliminate competition, unscrupulous commission firms advertised commission rates that were below operating costs. In some instances, these merchants made up their losses by falsifying the account of sale and defrauding their customers.[38]

By 1885, poorly handled animal disease outbreaks and unethical business practices made the Kansas City Stockyards the primary subject of industry derision. Kansas City commission merchants once again felt pressure to organize a regulatory exchange similar to what the Chicago market had organized in 1884, but they procrastinated until 1886 when powerful cattlemen's associations in the Southwest forced the commission merchants to act. As

Left to right: Scott Quick, Beth Quick (Scott's wife), Janelle Quick (Scott and Beth's daughter), Candy Quick (Marvin's daughter), and Steve Quick. (Source: Holli Kepner)

early as 1867, these private associations (e.g., the Bent County Colorado Cattleman's Association) were organized to police their respective local ranching industries. Increasingly intolerant of the ineptness and graft in the stockyards, a number of western cattlemen's associations collectively created the International Range Association in January 1886. The New Mexico Territorial Cattle Growers Association and the State Livestock Association of Texas were the primary organizers, and they invited cattlemen from other states, Mexico, and British Columbia to join.[39]

Believing that the U.S. Bureau of Animal Industry would act too late to save western cattle herds from such diseases as pleuro-pneumonia, the International Range Association established a quarantine against livestock shipped to the West from markets east of the Missouri River. Moreover, they developed a series of executive committees. One committee timed, adjusted, or regulated the shipments of cattle from different regions to prevent a surplus in the markets and to bypass the commission merchants, and another committee formulated quarantine regulations. A third committee was charged with controlling stock thieves by prosecuting violators throughout the west. That same committee also created a blacklist of cattle industry personnel who were viewed as dishonest, and association members were forbidden from employing or cooperating with the blacklisted personnel.[40]

The activities of the International Range Association deeply disquieted the Kansas City commission merchants. They were especially concerned with the threat of bypassing the commission firms, and the association controlled enough cattle to make that threat a reality. The Kansas City commission merchants were forced to acknowledge that the livestock commission business had major problems, and without some form of self-regulation, it was certain that regulation would be imposed from the outside.

Organization and Operation of the Kansas City Livestock Exchange

The threat that was posed by the International Range Association brought about the organization of the Kansas City Livestock Exchange in February 1886.[41] Its relatively rapid organization after a twenty-three-month period of indecision was based on two key points: mitigation of animal disease outbreaks and restoration of integrity to the livestock trade.[42] Despite this apparent unity of vision among commission merchants, it is inaccurate to portray the organization of the Kansas City Livestock Exchange as the result of a peaceful consensus. From the beginning, there was disagreement about setting standard consignment rates. In order for all commission firms to

The Kansas City Livestock Exchange building, 1887. (Source: Kansas State Historical Society, Topeka, Kansas; reprinted with permission)

The exchange was structured around a nine-member board of directors (three-year terms), a president (a one-year term), and four appointed committees that were tasked with accomplishing specific goals of the exchange. The Kansas City Stockyards Company and a consortium of local packers were allowed to appoint one representative each to the board of directors, who met a minimum of twelve times per year but could hold special sessions when needed. Ten members of the Kansas City Livestock Exchange could force the directors to call a special meeting by signing a petition; exchange members also had the power of referendum over decisions made by the directors.[43]

compete effectively, the policy of offering commission rebates had to be abolished; however, two competing factions emerged. The free-trade faction consisted of well-established, large commission houses that wanted to reserve the right to offer commission rebates as they saw fit. The regulator faction, which was made up of newer, smaller commission firms, opposed the free-trade faction; rather, they wanted to set uniform consignment rates for all commission houses.

In spite of these philosophical differences, the organization of the Kansas City Livestock Exchange went forward. Members of the regulator faction were at the forefront of organizing the exchange; however, they invited all commission firms to join and, eventually, all of the firms accepted. They also invited livestock traders, packers, and the Kansas City Stockyards Company to provide input into the operation of the fledgling organization.

Initially, the board of directors appointed an Executive Committee, an Arbitration Committee, and an Appeals

The Kansas City Livestock Exchange building, 2010. (Source: Holli Kepner

The general manager's office in the Kansas City Livestock Exchange building is shown during the late nineteenth century. (Source: Kansas State Historical Society, Topeka, Kansas; reprinted with permission)

Committee; in 1899, they added an Investigating and Judiciary Committee to prosecute rule breakers within their own ranks. There were five members of the exchange on each committee, including one representative from the Kansas City Stockyards Company and one from the packer consortium.

A series of self-imposed taxes and fines were used to fund Kansas City Livestock Exchange operations. There was a fee for each railcar of livestock received, fees for membership, fees for arbitration, and fines for rule infractions. The exchange was relatively inexpensive to operate. In 1908, exchange income amounted to $14,559.54, while cash outlays totaled $12,028.54. Sixty-three percent of this income was from railcar taxes, 4 percent was from arbitration fees, 7 percent was from membership dues, and 21 percent was from fines levied against members.[44]

The constitution of the Kansas City Livestock Exchange granted the board of directors the power to prosecute and to discipline rule breakers. The board was empowered to fine, censure, suspend, or expel offending commission merchants; however, it could not discipline a member without itself convening a commercial trial. Through their Investigating and Judiciary Committee, the board identified and prosecuted rule violators. Even a rumor of impropriety was enough to precipitate an investigation. The committee also was responsible for presenting charges and evidence against accused exchange members during commercial trials. Accused commission agents or commission firms were allowed to defend themselves, but they could not employ professional legal counsel. Refusal to appear before the committee or the commercial court brought an automatic suspension from the exchange.

The rules of the exchange expressly forbade any member from taking commercial disputes into a court of law. Commission merchants were required to submit all commercial disputes to the Arbitration Committee; independent livestock traders and ranchers could force a dispute into arbitration simply by notifying the board of directors. If a customer was unsatisfied with the decision of the Arbitration Committee, the Appeals Committee reviewed the dispute. Decisions by the Appeals Committee were final and were binding. Commission merchants who refused to pay an award as ordered by either the Arbitration or Appeals Committees were guaranteed expulsion from the exchange by the board of directors. If a member sought an injunction in any court of law against the exchange, that membership was considered to be forfeited.

Once the administrative and revenue-gathering structures of the Kansas City Livestock Exchange were established, members addressed the specific biological and ethical dilemmas that were facing the commission trade. To exercise control over the spread of livestock diseases, the Executive Committee hired public inspectors to work at each of the twelve scales located within the stockyards. The inspector's job was to monitor all animals as they crossed the scales and to mark or quarantine any that were diseased, injured, or otherwise imperfect. The decision of an inspector could be appealed to a chief inspector who was in charge of adjudicating all disagreements. The chief inspector's decision could not be appealed.

The inspectors hired by the exchange were not necessarily qualified to spot a diseased animal. Because of this, the U.S. Congress passed legislation that required all of the major markets to have qualified veterinarians. These officials did not replace the exchange inspectors; they were merely stationed themselves near the scales to watch for diseased animals. The first government inspectors appeared at the Kansas City Stockyards in 1894. The exchange initially opposed the presence of the government veterinarians on the grounds that they were redundant; however, that attitude changed over time, and the government veterinarians eventually convinced the exchange members that their services were essential. In fact, when the appropriations for government stockyard inspectors were drastically reduced in 1920, the exchange hired their own veterinarians to serve that function.

The public inspectors also provided a solution for the dockage swindle problem. The exchange's board of directors immediately removed the packer-paid dockers from the stockyards and replaced them with their own inspection personnel. Each inspector was paid by the exchange and favored neither the producer nor the packer. The fact that the decision of an inspector could be appealed was well received by both livestock producers and traders alike. The board of directors acted quickly to eliminate fraud from the Kansas City commission trade. All member firms of the exchange were required to provide written authorization that empowered the board of directors to audit their books and telegraph messages on an as-needed basis. The threat of a surprise audit made it very difficult for the commission merchants to falsify sales accounts. As a result, livestock producers and traders had greater assurance that they would receive all of the money that they were due. Any merchant who was found guilty of fraud during a commercial trial was expelled from the exchange. Members of the exchange who telegraphed false market reports to stimulate livestock shipments during periods of low prices were easily caught. The board also limited market quotes by commission firms to those firms whose sales were transacted by the firm itself. The exchange reduced other opportunities for fraud by redefining the role of the commission firm solicitor. The exchange forbade commission firms from paying solicitors a percentage of consignment fees; moreover, they required solicitors to register with the secretary of the exchange by name and address. The exchange also required all solicitors to be employed on a full-time basis; a solicitor could no longer serve as a railroad livestock agent or a cattlemen's

These steps inside the Kansas City Livestock Exchange building show many years of use by both businessmen and cattlemen alike. (Source: Holli Kepner)

association representative. By imposing restrictions on the nature of the solicitor's position, opportunities for conflict of interest in livestock sales largely disappeared. The exchange also limited the time that the stockyards and the commission firms could remain open for transacting sales to daylight hours only. This policy ensured that all sellers were able to market their livestock during periods of acceptable buyer concentration. It also made the commission business more public because it became nearly impossible to carry out a misdeed in the stockyards without someone else witnessing it.

The Kansas City Livestock Exchange recognized early that livestock producers and traders had no security against crooked transactions or the insolvency of commission firms. To remedy this situation, the exchange used membership fees as a form of surety bond. Starting in 1893, the exchange membership of any commission firm that defrauded a customer or defaulted on a transaction was revoked, and that merchant's membership fee was used to reimburse the customer. Despite these measures, fraud and insolvency continued to be a problem. In 1912, the exchange created a collection agency that was tasked with verifying all buyers and sellers had the necessary cash to complete their proposed transactions. Later, the exchange provided a blanket bond for all commission merchants; the blanket bond was first used to reimburse customers during the economic depression of 1919.

The final internal reform enacted by the Kansas City Livestock Exchange was to set uniform consignment rates, and $0.50 per head or $12 per railcar was established for cattle. The free-trade faction within the exchange membership was furious, as were some of their larger customers; however, the members who belonged to the regulator faction outnumbered them and thus controlled the vote. Simply publishing standard consignment fees did not guarantee their enforcement. Firms that belonged to the free-trade faction initially ignored the rules governing consignment fees and openly rebated commissions on large shipments of livestock that were consigned to them. To obtain evidence of these violations, the exchange hired detectives to board stock trains that were moving in and out of the Southwest. When evidence was found to confirm that rebates had been offered to certain customers, the board of directors levied heavy fines against the offenders. Failure to pay resulted in immediate expulsion from the exchange and the Kansas City Stockyards. Over a period of months, the Kansas City Livestock Exchange successfully asserted its right to regulate commission rates and to discipline violators of the rules. It also prevented the financially powerful commission firms from undermining the functions of the exchange.

As the Kansas City Livestock Exchange grew in prestige, it began to exert an influence on interstate commerce. It sought, for example, to get new rail mileage into Kansas City. In 1886, the year the exchange was organized, ten railroads funneled traffic into Kansas City; by 1893, there were sixteen. The number of railcars that were received in 1893 was 27,483 more than in 1886, and more than one-half of this increase came from the new railroad mileage. The exchange also used its influence to affect national agricultural policy. In 1892, the federal government ordered cattlemen out of the Cherokee Strip grazing allotment in Indian Territory no later than October. The exchange directors condemned this action in a letter that was sent to Agriculture Commissioner J. M. Rusk. The letter explained that the Cherokee Strip contained between 125,000 to 170,000 cattle, and expulsion of all these cattle would glut the Kansas City market. Cattlemen in that area had no other market option because Kansas, Colorado, and Texas forbade the movement of cattle into their grazing lands before December 1 to prevent a Texas fever outbreak. Moreover, the exchange directors argued that if Cherokee Strip cattlemen were forced to dump their animals on the Kansas City market, it would cause a price depression for all cattlemen in the Southwest. Another mitigating factor was that the summer had

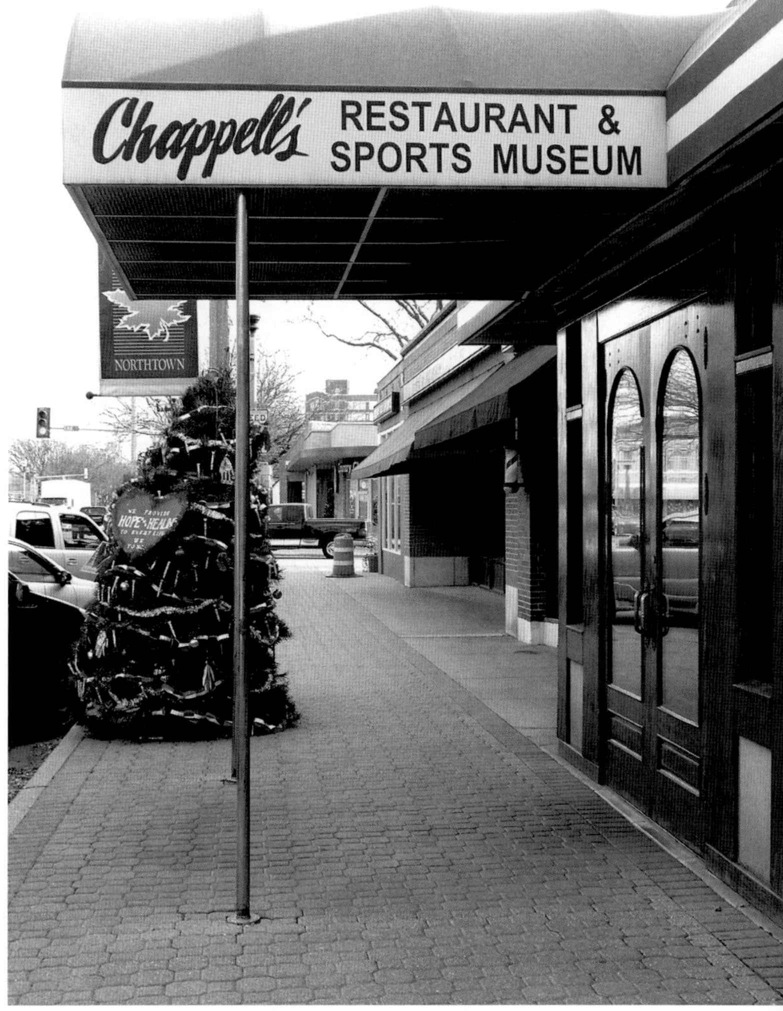

Today's Kansas City restaurants draw beef-loving customers much like the Kansas City Livestock Exchange did for decades. In 1986, Jim Chappell created a cozy little restaurant that was both comfortable and casual. From the beginning, this political/sports bar and grill has served really good food, and the restaurant has had to expand as a result. (Source: Holli Kepner)

been unusually dry in Indian Territory, which resulted in a rather sparse forage supply. Therefore, the cattle in that area were thin and not readily marketable. The exchange's petition was successful—the government delayed execution of the order for two months.

During the course of the next thirty-five years, the Kansas City Livestock Exchange became a model of self-

The original "cozy little bar" now seats 235 people, and it serves everything from the popular Zagat-recommended half-pound hamburger to steaks. People get excited about the memorabilia, but the good food is what brings them back. (Source: Holli Kepner)

regulation in American business. It effectively policed animal disease outbreaks and restored integrity to the livestock trade. Moreover, the exchange gave all of the commission firms a voice in the regulatory process and prevented monopolistic business practices by the largest, wealthiest firms. The International Range Association was not as durable and only lasted two years because it and the other cattle associations lacked effective power. State legislatures were often unsympathetic, and without them, the associations could do very little. This fate forced the cattlemen to accept the commission merchant's ideas about internal regulation.[45]

The Case for Private Regulation of the Livestock Trade

One of the most significant sidebars to the organization of livestock exchanges was the establishment of the right of private business associations to regulate trade. The livestock market in Kansas City grew from a local concern during the 1860s to a national trading center during the 1880s. The volume and the anonymity that this brought to the industry forced participants to invent new trading methods, which, in turn, created a need for some system to administrate disputes in a fair and expeditious manner. When the Kansas City Livestock Exchange was first organized, the commission merchants were fearful of government influence in the livestock trade; therefore, they voted for self-regulation rather than regulation by the state. Because of their subsequent ambivalence toward the new system, the cattlemen tacitly lent support to the proponents of self-regulation.[46]

Although a few historical precedents existed, the move toward self-regulation was bold. When they organized a commercial exchange, the Kansas City commission merchants assumed certain rights from the government and its constituency, including the right to organize their trade, the right to put a stop to anarchy, and the right to promote uniformity in business conduct. They also assumed they had the right to set commission rates for all livestock that were consigned to the Kansas City market. Furthermore, the exchange assumed they had the right to settle business disputes outside of the normal judicial system.[47] In effect, the Kansas City Livestock Exchange performed all of the functions of a government regulatory agency from 1886 to 1921. In 1892, political commentator Edwin Snyder wanted to know why the livestock industry allowed a voluntary association of fewer than two hundred men to dictate the terms upon which thousands of livestock producers sold their animals.[48] Historian J. W. Hurst provided the answer in 1956, when he noted that under a federal constitution committed to limited government, Americans routinely loaned the organized force of the community to private planners. Furthermore, the courts sustained the rights of these planners to act.[49]

State governments of that period developed narrowly defined views of their own power,[50] and this limited-power view of government paved the way for private associations such as the Chicago Board of Trade. The state of Illinois gave the Chicago Board of Trade the right to regulate the grain trade through a corporate charter. The Kansas City merchants followed the Chicago example when they organized an exchange in 1886, but did not apply for a corporate charter. Although they had no stipulated authority from the state government, they simply assumed that they had the right to regulate the livestock trade of the Southwest as an unincorporated private association.[51]

Modern legal experts have expressed amazement at the audacity and the breadth of power that was exercised by the Kansas City Livestock Exchange. For example, the exchange determined who could and could not be a commission merchant, it conducted commercial courts outside of the county and district courts, and it fined members enormous sums of money for rule violations. A mere rumor was enough to have a member summoned before the board of directors and investigated for improprieties. The board even used anonymous witnesses to convict members, and the accused was not permitted to have an attorney in the commercial courtroom.[52] Geography was perhaps one reason why the exchange was able to conduct its business relatively free of government interference. The state legislature in Kansas was fairly hostile toward the Kansas City commission merchants; at times, it attempted to pass laws that imposed on the autonomy of the exchange. Unfortunately for the Topeka politicians, the exchange building was situated exactly on the Kansas-Missouri state line that was within the stockyards. When necessary, the exchange membership simply moved their meetings to the Missouri side of the exchange building, which made it easy to ignore Kansas law. The Missouri legislature

Cattle that were sold at the Kansas City Stockyards by Major Andrew Drumm on behalf of E. R. Lehmann of Eureka, Kansas. Drumm *(inset)* was a prominent member of the free-trade faction and a promoter of the American Livestock Commission Company. (Source: Kansas State Historical Society, Topeka, Kansas; reprinted with permission)

rarely interfered with the operation of the exchange and, reportedly, even encouraged the rebellion against Topeka. In spite of the seemingly high-handed management style of the exchange, its activities were found to be consistent with the trade rules of its day. In both 1889 and 1906, the right of the Kansas City Livestock Exchange to regulate its trade was sustained when cases were brought before the State Supreme Court.[53]

The most significant challenge to the authority of the Kansas City Livestock Exchange began in 1889 when a number of individuals who belonged to the free-trade faction of the exchange membership formed a new business venture, the American Livestock Commission Company (ALCC). This new venture was intended to destroy the commercial and the political influences of the exchange. Members of the ALCC wanted to return to the nonregulated environment that existed before the exchange was created in 1886. The ALCC charter made it clear that the ALCC would ignore the no-rebate rules of the exchange and would use the Wyandotte County, Kansas, district court and the Kansas legislature to try and force the exchange out of business. In anticipation of opposition from the Kansas City Livestock Exchange, the ALCC directors incorporated their organization in Illinois. Most historians have interpreted the ALCC as a part of the late-nineteenth-century cooperative crusade, because the ALCC portrayed itself as a farmer's alliance. In reality, the officers and the managers of the ALCC were not poor farmers; they were old-style Kansas City free traders who posed as poor farmers. W. F. Peters was the only member of the ALCC's management team who was not a large dealer in cattle; rather, he was a commission merchant. Peters, who was already a member of the Kansas City Livestock Exchange, was needed by the ALCC so that ALCC cattle could be sold at the Kansas City Stockyards.

Once their presence in the Kansas City market was announced, the ALCC threatened to boycott the Kansas City exchange if the commission firms and packinghouses discriminated against it. The new organization promised to ship their cattle to Chicago instead. The ALCC relied on its contracts with every stockholder that required stockholders to transact all of their business through the ALCC. By the terms of its charter, the ALCC rebated 65 percent of net earnings back to consignors in proportion to the number of cattle that each consignor marketed through the ALCC. Remaining earnings were distributed among the stockholders in proportion to the number of shares each held. Members of the ALCC controlled enough cattle that the Kansas City Livestock Exchange perceived them as a threat. If successful, the ALCC would have vastly undercut the services of exchange members and

The Hereford House was established in 1957. The restaurant only uses the top 12 percent of all graded beef from Sterling Silver Premium Meats. (Source: Holli Kepner)

would have driven most of the members out of business. Furthermore, the authority of the exchange would have been undermined to the point of making it a laughingstock. The board of the Kansas City Livestock Exchange resolved to take action.

On June 11, 1890, the exchange notified the ALCC that charges had been filed against it for violating commission rebate rules. Upon receipt of the charges, the ALCC sought the intervention of the Kansas legislature. Eli Titus, general manager of the ALCC, had enough influence that the 1891 legislature, which was largely controlled by the Populist Party, passed the Roe Bill. This bill declared the regulation of commissions on the sale of livestock unlawful in the state of Kansas and thus outlawed the Kansas City Livestock Exchange. The exchange board of directors responded by revoking the memberships of W. F. Peters and the ALCC, because seeking legal injunctions against the exchange was against exchange rules. The board also expelled all members of the exchange who were associated with the ALCC. The board stood firmly on the Missouri side of the exchange building when they announced that they would have nothing to do with the Roe Bill or with any laws enacted by the Kansas legislature. They further adopted a new rule that gave them a new disciplinary power over members: the authority to blackball, which was immediately invoked against the ALCC. Previously, the board could do nothing more than refuse to inspect any animals at the Kansas City Stockyards that belonged to the offending firms. Now, it could prevent commission merchants, traders, packers, and railroad officials at the Kansas City Stockyards from doing business with an offending firm by threatening a boycott. The implementation of the blackball rule against the ALCC effectively shut it out of the Kansas City market—no packer would buy ALCC livestock, and no trader would buy ALCC animals for speculation.

In retaliation, the ALCC sought an injunction from the Illinois Supreme Court against the Kansas City Livestock Exchange to prevent the expulsion. The Illinois court ruled in favor of the livestock exchanges on October 31, 1892. The opinion of the court was that the livestock exchanges had the right to regulate their own memberships, which included the right to expel members. However, the court did concede that there was a strong basis for declaring the stockyards a public market, which would make private regulation impossible under federal law. Nevertheless, the court declared that until the U.S.

Feedlots, and privately finishing-out cattle, have become a major part of the beef industry in Kansas. Source: (Blair Tenhouse)

Congress specifically declared that stockyards were public markets, the livestock exchanges had the right to regulate the livestock trade. The ALCC never recovered from its challenge of the Kansas City Livestock Exchange. In the end, the exchange regulated the commission trade in Kansas City for another thirty-two years.[54]

The Livestock Trade in the Twentieth Century

The Kansas City Livestock Exchange's influence on the national livestock economy lasted until 1918 when the packers began to construct private stockyards and once again went into the countryside to purchase livestock directly from producers. This practice effectively bypassed the Kansas City Stockyards and the regulation of the Kansas City Livestock Exchange. Merchants of the exchange were not allowed access to packer-owned stockyards to either buy or sell livestock. Additionally, competing markets in Denver, Oklahoma City, and Fort Worth began to circumvent the flow of cattle away from Kansas City.[55]

The development of irrigation on the Great Plains brought large-scale feed grain production to the region. Feedlots soon followed, and cattle finishing began in earnest. In turn, the packers moved west to be near this new supply of finished cattle, and much of the livestock industry shifted out of the Corn Belt and onto the Great Plains. By the mid-1980s, packers within a 250-mile radius of Garden City, Kansas, slaughtered four out of every ten cattle in the United States.[56]

The invention of the motor truck and its ability to transport livestock brought about a second organizational revolution to the livestock trade. The use of trucks encouraged the trend of selling directly to packers by increasing market flexibility for livestock producers, who were no longer bound to the railroad. The motor truck also spawned new marketing methods. As the livestock trade continued to decentralize during the 1940s and 1950s, livestock auctions emerged on the Great Plains, which made the livestock exchanges and their associated stockyards much less convenient by comparison.[57]

On August 15, 1921, the U.S. Congress passed the Packers and Stockyards Act, which declared the major stockyards of the nation to be public markets. The American Farm Bureau and the National Farmers Union—groups that had lost confidence in the free market and thus wanted regulation from the federal level—lobbied for this new legislation. At that point, a federal bureaucracy assumed the function of a regulatory force in the stockyards. Ironically, it employed a system of operation that was already institutionalized by the livestock exchanges: it supervised the setting of commission rates, it regulated the membership, it disciplined commission merchants and traders, and it conducted audits—all at taxpayer expense. The new bureaucracy innovated very little beyond what the exchange had implemented during the previous thirty-five years, with two exceptions: it permitted the cooperative commission firms back into the stockyards, and it permitted livestock traders and livestock producers to provide direct input into the operations of the public market.[58] According to O. J. Hazlett, economists generally agree that the original iteration of the Packers and Stockyards Act was a failure[59] because the packers escaped regulation under the statute through lengthy litigation in the courts until 1932. Furthermore, the motor truck eventually made federal regulation of the stockyards meaningless. As the livestock trade decentralized, the marketing of livestock bypassed the major stockyards. Nevertheless, the Packers and Stockyards Act brought an era of private regulation in the livestock trade to a close. After nearly 120 years of operation, the Kansas City Stockyards Company closed in 1988.

The Kansas City Stockyards, which are shown during the height of its commercial prestige, closed in 1988 after 120 years of operation. (Source: Kansas State Historical Society, Topeka, Kansas; reprinted with permission)

Conclusion

Explanations as to why and how the U.S. livestock business evolved the way it did are all too often absent in historical accounts of our Western heritage. A perceived lack of romance may have discouraged some from researching the livestock exchanges, stockyards, and packinghouses for historical perspective because these enterprises were a symbol of industrialization in the

American West that may not have fit into the idyllic image sought by some historians.[60] However, there is much in the accounts of these institutions to enlighten us about the modern beef industry.

During the nineteenth century, livestock commission merchants were a new economic institution born of innovations in transportation and communication. Commission merchants marketed livestock in a faster and more efficient manner than the drovers did. They also became a source of operating capital for ranchers in the capital-poor areas of the West. In some areas, the commission merchants were involved in 90 percent of all funds that were executed, and many ranchers in the nineteenth-century West could not have operated without the financial aid of the commission merchants.[61]

The rise of livestock exchanges fundamentally altered the beef industry in the American Southwest. The regulatory efforts of the Kansas City Livestock Exchange went far beyond the expectations of livestock producers. The exchange increased productivity by coordinating the activities of the railroads and the stockyards. It acted as a lobbying force and pressured the federal government to modify its decrees. The exchange also reformed aspects of the trade that were little understood by outsiders. It recognized that unless market information and business hours were controlled, the activities of unscrupulous livestock commission merchants would discredit the Kansas City market. It even assumed the power to audit the books of commission merchants against their will in an effort to ensure that livestock producers and traders received a correct return of funds. Finally, the exchange pioneered the concept of a surety bond to protect customers from fraud and insolvency.[62]

The Kansas City market reached its peak of influence in 1918, when cattle receipts totaled approximately $3 million—the equivalent of $42.4 million in 2009. From 1886 to 1921, the Kansas City Livestock Exchange assumed and responsibly executed regulatory powers over the livestock trade of the Southwest on behalf of the government and its constituency. The Packers and Stockyards Act of 1921 ended that allowance of power,[63] but the Kansas City Livestock Exchange continued to operate until its closing in 1988.

Note: Some segments of this chapter appeared in the October 2001 issue of *Rangelands* magazine (vol. 23, no. 5, pp. 3–21; Allen Press Publishing Services). The authors and editors thank Allen Press for reproduction permission.

Paul Hibbard, Greenwood County

Paul Hibbard epitomizes the honesty, hard work, integrity, and dedication of the Kansas farmer/stockman. Hibbard's family moved to south-central Kansas in 1869. "They scouted for a year before they found the poorest ground around," Paul used to kid. Hibbard backgrounded mixed steers in the Flint Hills for nearly a half century. (Source: John Schlageck)

The Flint Hills of Kansas

Next page, top: The Flint Hills are the world's largest remaining tract of continuous tallgrass prairie. The tract stretches two hundred miles from north to south and sixty miles from east to west, and it is primarily in Kansas. Each year thousands of young weaning calves spend several months out on this native rangeland where they feed on the calcium-rich Flint Hills native tallgrass. (Source: John Schlageck)

Sea of Grass

Next page, bottom: The Flint Hills region has almost four million acres of tallgrass prairie pasture and rangeland. Chase County in the Flint Hills is still almost 80 percent tallgrass prairie, the highest percentage of any Kansas country. Rainfall averages between 30 and 42 inches per year, and the best quality forage is produced from late April through early July when cattle from Texas and the southeastern section of the United States graze on the tall, lush sea of grass. (Source: John Schlageck)

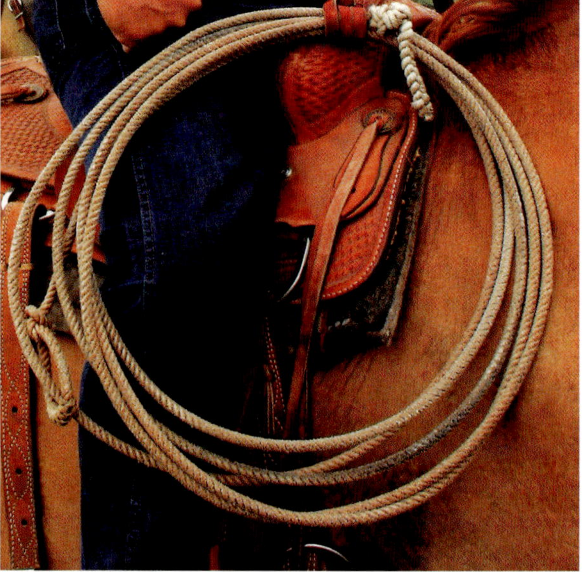

Tools of the Trade

Top left: Every Kansas rancher, stockman, and farmer wears a pair of pliers on his belt or drops one in his back pocket. Coupled with a little bit of sweat and elbow grease, this handy tool is without a doubt their best friend. (Source: John Schlageck)

One on Every Horse

Above: The lariat (or lasso) is the long rope with a noose that stockmen use to catch and secure cattle and horses. The lariat can then be wrapped round the horn of the saddle. The word lariat comes from the Spanish word *lareata* (rope). No ranch hand worth his salt would ever ride out in the pasture without one. (Source: John Schlageck)

The Real McCoy

Bottom left: Historically, cattlemen have always needed protective footwear, and they prefer boots with a higher heel. The origin of the cowboy boot that we know today comes from a variety of boot styles, including the Coffeyville-style cowboy boots that originated in Coffeyville, Kansas, around 1870. This style of boot is normally black leather with a low Cuban heel. Part of the enduring appeal of cowboy boots is that they can be worn by anyone—even real cattlemen. (Source: John Schlageck)

Dennis Ricke, Barber County
Ranchers such as Dennis Ricke have tended cattle in Kansas since the first cattle drives of the late 1800s. This deep family tradition takes time, common sense, and hands-on know-how to ensure mama cows and their calves stay healthy and bring in top dollar. (Source: John Schlageck)

Cow Skull Art

Left: This sun-bleached cow skull in Wabaunsee County recalls the nostalgia of an Old West cattle drive and the pioneer spirit of the American West. The cow skull has become a symbolic icon of the struggles, death, and dedication of our early pioneers. Cattle and livestock were a huge part of taming the Old West, from the oxen that pulled many a Prairie Schooner across the endless prairies to the rugged terrain of the Rocky Mountains and the harsh desert of the southwest. (Source: John Schlageck)

Food Source

Below: Round bales can weigh a ton or more, and they are well-suited for modern large-scale livestock operations. In 2009, Kansas placed fifth in all hay produced with 7,225,000 tons and sixth in alfalfa hay produced with 3,665,000 tons. This beautiful hay field is located in Rooks County. (Source: John Schlageck)

Beef on the Hoof
Beautiful black Angus cattle on the Paul and Judy Wasko farm/ranch in Hodgeman County. Plenty of grass and a rather mild summer brought out the best in these fleshy, good-looking stock. (Source: John Schlageck)

Summertime and the Livin' Is Easy
Just like kids, cattle know water is a good place to be when the mercury tops 100 degrees. All across Kansas, cattle take dips in farm ponds during the blistering heat of July, August, and September. These cattle found relief in a Barton County pond. (Source: John Schlageck)

Hess Family Tradition
The Hess family settled in Wabaunsee County, Kansas, more than 150 years ago. Today, Alan and his brother, Jim, still live on the same property that their great-great-grandfather settled in 1858. As Alan said jokingly, "You'd think we could have figured out a better way of making a living by now." Truth be known, neither Alan nor Jim ever wanted to do anything else but raise cattle in the Flint Hills region. The brothers are shown pushing a cow-calf herd toward the working corral during a rainy spring roundup. (Source: John Schlageck)

Breakfast Is Served
Lip smackin' good, the sweet smell of silage draws cows and calves to feed bunks during a tough Kansas winter. Kansas cattlemen are dedicated to the care and feeding of their livestock year-round. This dedication produces the healthiest livestock and also provides the highest-quality and safest food on the planet. This photo was taken on the Bert Schlageck farm/ranch in 1973. (Source: John Schlageck)

Angus Pioneers
Longtime Ford County rancher John Warner is shown moseying through his herd of black Angus cows and calves. The Warners were one of the first families to run Angus in southwestern Kansas. John Warner ranched in southwestern Kansas for nearly seventy of his eighty-nine years. (Source: John Schlageck)

Post Rock Country
A sprinkling of wild flowers adds a touch of yellow to this pasture, which is located in Post Rock Country northeast of Russell. When the pioneers staked their claims on open range during the 1870s, they found little timber available and an abundance of limestone; thereafter, limestone became their building material. Fence posts like the ones in this photo are still in use today. Limestone posts have supported barbed wire fences around thousands of prairie acres in north-central Kansas for more than 120 years. They reflect the sturdy character of the early Kansas pioneers. (Source: John Schlageck)

Dana Pieper, Rooks County

Above: Fifth-generation livestock producer Dana Pieper is shown with one of her favorite Hereford cows, Annie. This eight-year-old cow was raised as a bucket calf, and as Dana said, "She throws a big, healthy good-looking calf." The Piepers are part of the Sutor Hereford Farms, which belong to her maternal grandparents. Herefords have been raised on this farm since the 1880s. (Source: John Schlageck)

Picture-Perfect Day in the Flint Hills

Left: Blue sky, fresh grass, and a sunny day are the perfect complement to this windmill in the Flint Hills west of Manhattan in Riley County. Source: (John Schlageck)

Way Out West

Above: Red Angus cross cattle are shown grazing on short grasses in Cheyenne County. Part of the Shawna and Shawn Blanka herd have farmed and ranched in this far northwestern county for four generations. They run 400 head of momma cows and finish out the majority of their calves each year. (Source: John Schlageck)

Family Plan

Right: Four-and-a-half-year-old Repp Braun is shown with his father, Mike. Just like his father, Mike Braun raises cattle for a living because he knows their product will nourish families around the globe. The care they provide to their animals is part of a carefully constructed business plan. (Source: John Schlageck)

Animal Care
Momma cows are shown looking on as one of the baby calves is gently and carefully led through a roaring spring creek by Jim Hess, a rancher in Wabaunsee County. For generations, Kansas farmers and ranchers have cared for and nurtured their livestock. These dedicated men, women, and children work from sunup to sunset—every day of the year—to ensure that the animals on their land receive the best care possible. (Source: John Schlageck)

(Source: Bragg/Coffman Families)

Chapter 5
Families of the Kansas Beef Industry

By Barbara Bragg, H. Hurst Coffman, Susan Russell, Lois Schlickau, Justin Kastner, and Blair Tenhouse, with Podcast-Related Research Files from Gabe Webb, Tiffany Lee, Tiffany Ebert, and Kelsey Rezac

Nearly fifty years ago, a distinguished scholar cautioned fellow historians about the danger of missing the human element when chronicling the past: "The various aspects of life which in history we inevitably divide are to the living all one."[1] His warning is worth heeding, because the history of the Kansas beef industry indeed has multiple and distinct economic and social components, and it must be remembered that people experienced these components in combination. From the nineteenth-century cattle drives to modern-day beef production enterprises, this chapter features a number of family stories that show some of the integrated "lives lived" in the history of Kansas beef. Some of these stories are also available in audio form; please visit www.beefcattleinstitute.org for the *150 Years of Kansas Beef* podcast series.

From Tanning Hides to Sewing Shoes to Producing Beef: The Story of the Bragg Ranch

The citizens of Whitehaven in Northumberland, England, specialized in trades such as textile production and tanning of leather. For generations, the Bragg family had been experts in tanning—a process where tree bark was used to produce the chemicals needed for curing hides into leather. In 1832,

James Bragg. (Source: Bragg/Coffman Families)

the Bragg family immigrated to the United States and settled first in Ludlow, Kentucky, and then in Sinking Springs, Ohio. The Braggs operated tanneries in both locations.

James Bragg, who was born in 1830, came to America as a small child and grew up in Ohio. During a Methodist youth gathering, James met Salome Hoffman of West Milton, who was the daughter of the minister and the granddaughter of the bishop. After they married in 1852, they had three children—Thomas, Henry, and Katherine. During the early 1880s, Henry Bragg moved to St. Joseph, Missouri, where he established the Bragg Leather Goods Company. A few years later, James, Salome, and Katherine moved to St. Joseph, where James joined his son in the leather business and established the Specialty Shoe Machinery Company, while Thomas stayed in Ohio to operate the tanneries.

As leather goods production flourished, there was greater and greater demand for a steady and reliable supply of cowhide. At that time, the natural next step was to push westward. By the early 1890s, James Bragg and his sons had accumulated enough wealth to consider purchasing their own cattle ranch, so they contacted a group of land brokers and lawyers in Ford County, Kansas, to discuss land acquisition. Much land had become available for purchase after the economic depression of 1893, and between 1898 and 1905, James and Thomas Bragg acquired an almost contiguous spread of 14,000 acres that were located twelve miles south of Dodge City along the banks of Mulberry Creek. Father and sons agreed to sell the tanneries in Ohio, and Thomas relocated to Dodge City.

The Braggs had been tanners and businessmen who enjoyed every comfort, so taking on the operation of a ranch was a new venture that, quite simply, proved challenging. It was necessary for a member of the family to oversee the ranch, so it was decided that Thomas Bragg would go live on the ranch. Thomas and his wife, Margaret, who passed away in 1892, had four children—Harry, James,

Minerva, and Eva. Thomas and his small children moved west: the two daughters stayed in St. Joseph with their grandparents; the two boys stayed with Thomas on the ranch in Dodge City.

The pastures on the ranch were not all connected, so it was necessary to drive cattle from one place to another. One especially important feature of the Bragg Ranch was a dammed reservoir that was built next to a spring and served as a watering place for herds

Above: This spring fed the reservoir at the Bragg Ranch. A ranch hand can be seen taking a rest. (Source: Bragg/Coffman Families)

A water reservoir that was used for watering livestock at the Bragg Ranch. The reservoir was a key feature of the ranch because a common spring provided clean, fresh water for the cattle, ranch, and passersby. (Source: Bragg/Coffman Families)

of cattle that were being driven to Dodge City for shipping. The spring and the reservoir were important factors when the Braggs selected the property, and each contributed to the success of their operation. Also, the ranch's close proximity to the train lines in Dodge City provided a way for shipping cattle to St. Joseph, Kansas City, and other exchanges.

In 1895, Thomas returned to Sinking Springs to marry Blanche Belleson, with whom he had reestablished a relationship. Blanche then accompanied Thomas back to the ranch in Dodge City, where they had four

Cattle are shown drinking at the Bragg Ranch reservoir. (Source: Bragg/Coffman Families)

were known for their hospitality. In 1905, Charles Faulds came to Ford County from Sinking Springs, Ohio, to work at the Bragg Ranch. He and his wife, Susan Pearl Allen, lived on the ranch for fifty-one years; their son, Jim Faulds, and his descendants continue to farm much of the land that was part of the Bragg Ranch.

The Braggs achieved great success building factories, raising cattle, and farming. This was a family of astute genius, strong faith and character, and an unrelenting work ethic. They were educated and inventive. The Braggs also had a sense of style, culture, and refinement that they brought with them from England. After church on Sundays, the table was draped

children—Cecil (who was known as Frank), Thomas Jr., Low (who died as a young child), and Margaret.

In 1898, Thomas and Blanche built a good-sized sod house on the ranch that was close to the spring and the holding pens. The house was built into a rise in the ground for cooling in the summer and protection from the north winds in winter. The walls were three feet thick, the windows were wood-framed with glass panes, and the roof was made of wood shake. The house had several bedrooms and finished floors, and there was fresh spring water piped inside. Thomas and Blanche lived in the sod house with their family and

Thomas Bragg and his family are shown with hired hands in front of the sod house in 1898. The house was located in Ford County, Kansas, just south of Dodge City. (Source: Bragg/Coffman Families)

The Braggs had a sense of style, culture, and refinement that they brought with them from England. Thomas's sister, Katherine, was an artist who specialized in painting china, quilting, and creating cloth pictures with shaded silk embroidery threads, which is now a lost art. This is an example of Aunt Katie's work in shaded silk embroidery. (Source: Bragg/Coffman Families)

with crisp white linens, china, and silver, and servants cooked and served the meals. Each linen napkin was two-feet square, monogrammed, and had hand-rolled edges. Today, more than a hundred years later, some of those napkins are well preserved and are still used for special family occasions.

Handmade quilts were popular during this era. Thomas's sister, Katherine (who was known as Aunt Katie), was an artist who specialized in painting china, quilting, and creating cloth pictures with shaded silk embroidery threads, which is now a lost art. The family still owns and prizes many samples of her artistry. The development of culture was further reflected in Blanche Belleson Bragg, who

Linens and silver were featured at the Bragg family's weekly Sunday dinners. Shown here are two-foot-square one-hundred-year-old monogrammed linen napkins that were rolled by hand. Also shown are monogrammed sterling-silver spoons that were brought over from England by Thomas Bragg's grandparents. (Source: Bragg/Coffman Families)

studied piano at the Cincinnati Conservatory in Ohio. When she moved into the sod house, she brought her piano and began to give lessons. These talents have been carried down for generations, and today there are many Braggs who are painters and musicians.

From early childhood while growing up on the ranch, Frank Bragg was always drawing and painting, mostly with watercolors. When he was a youth, his parents encouraged him to draw pen and ink cartoons. A friend of the family, who was an editor for the *Kansas Farmer,* visited the Bragg home and saw the young man's work. He then asked Frank to submit his work to the paper. Then, around 1923, Frank left Dodge City to attend the prestigious Chicago Academy of Fine Arts.

A cartoon by Frank Bragg that was titled The Evolution of Kansas. It appeared in *The Topeka Daily* on January 29, 1926, to celebrate sixty-five years of Kansas statehood. The images depict the difference in the state from its first birthday to its sixty-fifth birthday. (Source: Bragg/Coffman Families)

During his studies in Chicago, Frank's cartoons appeared in the *Capper's Weekly Journal.* He also applied to *King Editors Feature*, a newspaper in New York, and successfully achieved that goal. Later, Frank studied art and journalism at the University of Missouri, where he received a bachelor's degree in journalism. He then trained to be a newspaper cartoonist, and he began to create cartoons about the trials and tribulations of ranching. Having been inspired by his own early ranching days, Frank gained much recognition for his acute cleverness and his subtle portrayal of ranching life, which included addressing issues associated with the U.S. government and its approach to agriculture and the growth of the beef industry.

After he graduated from the University of Missouri in 1931, Frank returned to the ranch to face the Great Depression and the Dust Bowl Days. Frank did everything in his power to hold the ranch together. The shoe and tanning factories were still in production but were waning in prosperity with the demise of James and then Thomas Bragg—the brains behind the leather goods operations. Frank had never worked in the shoe factory, but he knew all about western Kansas and cattle. The future of beef production appeared to be promising, so Frank went back to what he knew best—cowboy life—to save the family spread. As he struggled to keep hold of the ranch, his first-hand experiences inspired his cartoons, which documented the signs of the times. The greatest challenge for Frank was the collapse of the economy and the disastrous heat of 1934 and 1936. Frank stuck it out at the ranch, where he faced dust, grasshoppers, rabbits, and blizzards that destroyed both crops and livestock alike.

A group of Hereford cattle grazing on the range, circa 1930. (Source: Bragg/Coffman Families)

Frank was married to Vara Bragg, who lived with him in the sod house as they continued to operate the ranch. They told stories about the dust blowing so much you could not see more than three feet in front of you. They also recalled that the grasshoppers were so thick, they swarmed a fence post and ate it. Many cattle were lost to the drought of the 1930s, and much of the land was lost because there were no crops and cattle.

Beginning in 1937, the overall situation began to improve, and the ranch produced a welcomed good harvest of wheat. After that, both beef and wheat production continued to improve, and Frank was able to pay off the debts he had incurred while borrowing money to pay taxes and ranch hands during the bad years of 1931 to 1936. During the 1940s and later, the tanning and shoe factories were closed, and a new direction became evident: produce beef cattle for sale to the beef packers in St. Joseph, Kansas City, Oklahoma City, and Chicago.

Thomas Bragg's older sons, Harry and James, continued the family tradition on the ranch; Harry's son, Joseph, was a rancher in Kansas and Colorado. Thomas's daughter, Minerva, married Clyde Coffman of Overbrook, who was highly regarded as a political figure primarily interested in organizing farmer and rancher cooperatives across the state. Daughter Margaret Bragg Baird spent her life in Dodge City, where she owned and operated the Fairytale Shop, a children's clothing store located on Front Street. Son Thomas Jr. also spent his life in Dodge City.

One of Thomas Bragg's grandsons, Roger Coffman, D.V.M., operated a ranch veterinary clinic and feedlot out of Limon, Colorado, and a great-grandson, James Coffman, D.V.M., is a professor emeritus at the College of Veterinary Medicine and the provost emeritus at Kansas State University. Great-grandson Thomas Alison Bragg, who was his namesake, purchased the famous Bay State Ranch in Scotts Bluff County, Nebraska, and formed the T. A. Bragg Ranch Company in Centennial, Wyoming. Between the two ranches, he operated and ran some 2,000 head of cattle at any given time. Another great-grandson, Larry Coffman, and his wife, Karen, raise registered organic Angus in Meeker,

Close-up of one of the herd bulls, circa 1930. (Source: Bragg/Coffman Families)

Colorado. Other descendants still raise beef for a living in Colorado, Wyoming, and, of course, Kansas. The only piece of the original Bragg Ranch still owned by family members is a half section owned by four of the grandchildren of Minerva Bragg Coffman.

Ark Valley Cattle Company. (Source: Blair Tenhouse)

A Family Committed to Growth and Innovation: Fink Beef Genetics*

In 1977, Galen and Lori Fink began their breeding cattle operation with just one cow. They eventually built up a herd that was entirely Angus cattle; in 1999, they expanded the herd when they added the Charolais breed. Today, the Finks' operation—Fink Beef Genetics—sells approximately 600 bulls to customers in Florida, Montana, and California, among other places. In addition to the bulls, the Finks typically wean around 900 calves each year. Until just a few years ago, the entire operation was headquartered on forty acres; today, the operation has just over six hundred acres.

A Charolais cow and calf relax in the sun. (Source: Lewis Browder)

Because of the relatively small acreage, the Finks do not invest in equipment or produce their own crops; instead, they purchase what they need from local farmers. Fink Beef Genetics has no hired help, but the Finks' daughter does help when she is home. Although the operation is not necessarily a large one in terms of land, Fink Beef Genetics has been ranked among the top twenty-five largest seed stock producers in the National Cattlemen's Beef Association. In 2008, the Finks were named Certified Angus Seedstock Producers of the Year. Fink Beef Genetics has incorporated artificial insemination into its breeding for the past thirty-three years. In 1988, the Finks started to put embryos in customers' cows, which was rare at the time because embryo transfer was a fairly new technology. Today, they put in about 1,000 to 1,200 embryos a year.

In addition to running their operation, Galen and Lori are involved with numerous state and national cattle

*Go to www.beefcattleinstitute.org and listen to a full audio podcast entitled, "Growth and Innovation in the Cattle Industry," and learn how, over the course of thirty years, Galen and Lori Fink built a large bull operation from just one cow.

Hays House, Council Grove, Kansas. (Source: Justin Kastner)

organizations, extension work, fair boards, and 4-H. Both serve on the federation side of the National Cattlemen's Beef Board, and both serve on the Kansas Beef Council Board of Directors. The Finks' other business ventures include partial ownership of Manhattan's Little Apple Brewing Company, which they started in 1992, and the Hays House in Council Grove. The Finks maintain that restaurant ownership makes them better beef producers, because it helps them to realize the importance of producing a consistent product. Although Fink Beef Genetics does not directly produce Certified Angus Beef for its restaurant customers, many producers use the Finks' commercial bulls to produce the products that then go into the Certified Angus Beef program.

Bill and Kathy Hogue and the Mission Valley Ranch†

Located eight miles west of Topeka, Mission Valley Ranch is one of the many examples of Kansas ranches that were built with passion and diligence. Its owners,

Bill Hogue is shown on top of the hay bales that he donated to fellow ranchers. (Source: Mission Valley Ranch)

†Go to www.beefcattleinstitute.org and listen to a full audio podcast, entitled "The Mission Valley Ranch: Built by Hard Work and a Passion for Agriculture."

Black beauties along the fence line in one of Mission Valley Ranch's Flint Hills pastures. (Source: Mission Valley Ranch)

Bill and Kathy Hogue, moved to their current location in 1957; today, there are two other divisions: the bull division, which is located one mile north of headquarters and houses one of the two ranch hands, and the Flint Hills division, which is situated another twenty-three miles to the west.

The Hogues began with approximately twenty-five head of cattle. Twice they have depopulated their herd to introduce new genetics, and on one occasion they kept only one bull from which they built up their herd. Today, the Hogues have 556 head of cattle, and their ranch totals 3,500 acres. Despite the obvious change in size, the Hogues' operation has in many ways remained the same. When they started, they wanted to build the best herd they could and thus provide breeding stock for the neighbors. The quality of their cattle has improved significantly since they began; yet their philosophies on raising and selling cattle have remained the same. Major improvements in the quality of the herd are due largely to artificial insemination, a technology the Hogues adopted around 1978. The Mission Valley Ranch does a lot of embryo transfer work, too.

In addition to running the ranch, Bill builds homes in Topeka. He has been doing this for fifty years, and despite being in his late seventies, he still runs both businesses. Typically, Bill works in Topeka until around noon and then comes home to the ranch. Although he works two professions, agriculture is his passion. Kathy has said her husband is a builder by trade but a cattleman at heart. Along with this passion comes involvement in outside organizations. Both Bill and Kathy have been members of

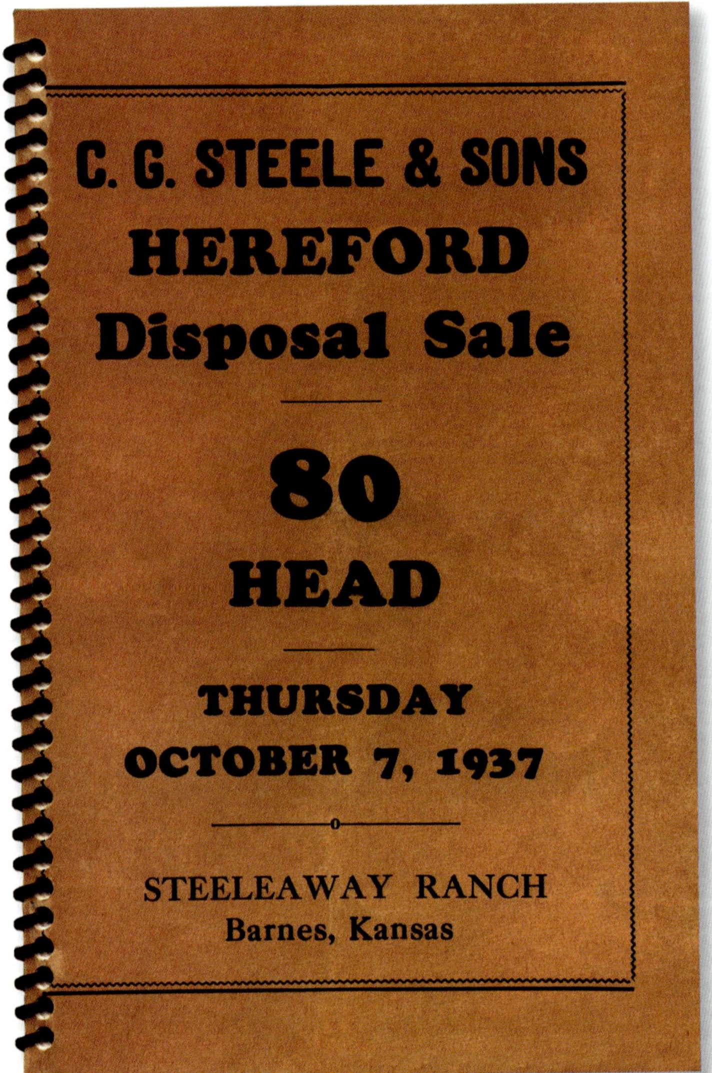

The sale catalogue from the Steeleaway Ranch disposal sale in 1937. (Source: Killingsworth/Steele Family)

the American Angus Association since 1960, and they are both members of the Kansas Angus Association. When their children were involved in the 4-H beef project, the Hogues were as well. With more than 500 head of cattle, and with more than fifty years in the business, the Hogue family continues to produce high-quality cattle with dedication and a love for what they do.

A Family Farm 150 Years in the Making: The Steele Family Ranch‡

A century and a half ago, Samuel Benton Steele was traveling in Kansas when he hit the end of the railroad line in Waterville. Almost immediately, Samuel bought 1,200 acres of land and settled down with his family to farm. He purchased a prize-winning Hereford bull from Great Britain to incorporate quality into his herd, and his farm soon became one of the largest in Marshall and Washington Counties. Within three decades, it was formally known as the Steele Ranch.

Hard work and hardship are no strangers to Kansas agriculturalists, and this certainly can be said of the Steele Ranch. After renting out the farm for several years, Samuel's youngest son, Charles G. Steele, eventually took over the remaining 1,100 acres, a house, and some barns. Starting off with little capital and little experience proved to be a challenge for Charles. However, the timely purchase of good registered Herefords helped his bottom line, until he was forced to sell them during the drought of the 1930s. Charles pioneered other business ventures (e.g., renting an ice box, selling seed, and developing a dairy business), but further hardship hit the Steele Ranch when different diseases ruined the herds.

Having weathered many economic storms during the course of their life, both Charles and his wife were highly respected members of the community, and Charles was often called on as a promotional speaker at local and county events. After Charles G. Steele II took over the family business, the economic lot of the Steele Ranch improved as pasture rotation and the higher prices generated by World War II enhanced the profitability of the ranch. Today, many of the original barns still stand at

‡Go to www.beefcattleinstitute.org and learn more about the Steele family farm, in an audio podcast entitled "A Family Farm 150 Years in the Making."

the Steele Ranch. Animals are no longer raised there, but a full century-and-a-half of family pride still runs deep at the family homestead.

The Gardiner Angus Ranch

The Gardiner Angus Ranch is a family-owned seed stock operation that specializes in Angus cattle. It dates back to the late 1800s; today, Henry Gardiner runs the ranch with his wife, sons—Greg, Mark, and Garth—and his sons' families. Henry's grandparents lived in a dugout for nine years on the initial 160 acres of homesteaded land southwest of Ashland. Henry's father, Ralph, is credited with assembling the present-day Gardiner Ranch during the 1920s.

Henry attended Kansas State University and took much of what he learned back to the ranch, including artificial insemination technology. Mark explained, "Everything here has been totally [artificial insemination] without the use of clean-up bulls since 1964. Dad, when he was in school at

Henry Gardiner, Gardiner Angus Ranch. (Source: Julie Tucker, Graphic Arts of Topeka, Inc.)

Yearling bulls graze in a pasture southeast of the Gardiner Angus Ranch headquarters. (Source: Blair Tenhouse)

A western Kansas sunset over the artificial insemination barn at the Gardiner Angus Ranch. (Source: Blair Tenhouse)

K-State—and that would have been from 1948 to 1951—he learned how to [artificially inseminate]."[2]

The Gardiner Angus Ranch today encompasses 21,000 acres in southwest Kansas. The ranch breeds 1,500 head of registered and commercial females annually, and it collects and transfers more than 2,000 embryos each year. Mark spoke about the progression from Gardiner Angus Ranch's previously smaller commercial herd to its substantially greater size today.

> We have always been commercial cattlemen. Dad wanted to improve cattle. We tried as hard as we possibly could to make changes to the cattle. In 1964 to 1980, he was using what was believed to be the best Angus bull of the time. He said we are only going to use high-accuracy, progeny-proven bulls for the traits of economic importance. And that's what we have done.[3]

One thing that Henry has always done—even when the operation was just beginning—is guarantee that every animal is functionally fertile to the best of his knowledge. Cattle are guaranteed for one full year. Mark discussed one of the biggest challenges they faced in recent years:

Mark Gardiner and Doug Tenhouse are shown discussing the current bull crop. (Source: Blair Tenhouse)

recessive genetic defects. When the defects first came about, worried customers needed answers to questions. It was during this time that the ranch relied on one of Henry's rules—always guarantee the animal. As a result, the Gardiners believe that the relationships they now have with the majority of their customers are better than ever. According to Mark,

> When you have a product to sell and then, all of a sudden, it becomes a discounted product, it displays as a challenge to your business. The recessives have proven to be an opportunity and a challenge. My brothers and I were sitting here in the office of the [artificial insemination] barn, and I said that this will be an opportunity to prove who we are; so we stood by our decision to make things right with our customers. That hasn't been easy, but the flip side is it was the right thing to do . . . It's a people business. It's a good business.[4]

Respect for the cattle and the land that they forage is not lost at the ranch. The Gardiners believe that cattle are the byproducts of good land stewardship. In this respect, the ranch has witnessed much prosperity and growth over the

(Source: Blair Tenhouse)

decades. The ranch's growth has enabled the Gardiners to involve their own family members, including future generations. As Mark explained, "People from my great-granddad to my granddad, through their own trials and perseverance, gave us the opportunity to have a chance at this business. I view my one and only mission is to make sure this is a sound business, so that my kids or my brothers' kids have the opportunity for a future in the ranch."[5]

From 1873 to the Present: The Schlickau Family's Cattle History

In 1868, the first George Schlickau came to the United States from Germany at the age of eighteen. He arrived by ship with his older brother, Henry, and Henry's family.

George Schlickau first purchased a cow-calf pair in 1873. His family is shown circa 1901. *Front row, from left:* George, Helena, Esther, and wife Clara. *Back row:* Albert, Martha, and Walter. (Source: Schlickau Family)

Albert Schlickau's family all participated in the diversified agricultural farm and ranch. Albert and his wife, Florence, were married in 1905 and raised a family of four children. The family is shown in 1929. *Front row, from left:* Albert, George, and Florence. *Back row:* Harry, Opal, Mildred. (Source: Schlickau Family)

George Schlickau is shown in 1933 with his first Kansas State Fair club calves. (Source: Schlickau Family)

After settling in Wisconsin for a few years, George moved for a short time to Illinois before he eventually homesteaded in Reno County, Kansas, in 1872.

Kansas was open territory, and his homestead was situated in present-day Haven Township. In addition to homesteading, George was employed. With part of his earned money, he purchased two oxen, which he used to break additional prairie sod. After working for a lumberman near Wichita for about four months, George acquired enough money to purchase a horse; several months later, he bought a cow and a calf. These were the first cattle on George's claim, and from this commercial pair, a herd was started in about 1873. As this herd increased, George found a market among the settlers who came to lay their claims and build their homes.

The Schlickau family has raised cattle and crops continually since these homestead years. In 1900, George's oldest son, Albert, bought his first purebred Hereford bull to use with his commercial cows. Albert and his wife, Florence, were married in 1905 and raised a family of four children—Mildred, Harry, Opal, and George. With two sons joining the operation, Albert purchased additional pasture south of his ranch in what longtime residents of the area call *red jaw country.* It is called that because of the red soil, which was unsuitable for cultivation but produced good native pasture grasses. In 1913, Albert and his brother, Walter, bought sixteen head of registered Hereford cows. In 1919, Albert's ten-year-old son Harry exhibited his home-raised heifer at the Kansas State Fair, where he won overall 4-H champion female. Nearly every year since then, the Schlickau family has shown Hereford cattle in either open class, 4-H, or FFA divisions at the Kansas State Fair.

In addition to being a progressive rancher, Albert was a leading farmer. In 1926, the Kansas Wheat Improvement Council and the Kansas State Agricultural College presented Albert with a trophy for being the wheat-growing champion in Kansas. As Wheat King of Kansas, Albert shared agricultural tips during whistle-stops on the back of the special Kansas Wheat Festival Train. The Schlickau family sold its cattle far and wide, even during the Great Depression. In 1935, Albert had nineteen head at the Great Western Livestock Show in Los Angeles, California. In 1939, the Los Angeles show catalog listed Albert as exhibiting twenty-five head. In

George Schlickau is shown with one of his FFA projects in August 1945. Schlickau earned an American Farmer Degree and expanded his registered cattle operation. (Source: Schlickau Family)

A group of sale bulls near the Schlickau's barns, circa 1950. Annual production sales began in 1943. (Source: Schlickau Family)

1940, Albert's son, George, had a champion pen of heifers. At first, young George thought the heifers had sold for more than $500 for the pen. He was shocked to find out that the winning bid was per heifer. During the 1940s, the Schlickaus consigned cattle to the Haven Hereford sale, the Reno County Hereford Association sale, and the Kansas Hereford Association (KHA) sales. George was quoted as saying that he and his siblings were "hooked on Herefords almost from infancy." In 1932, for example, Albert offered his ten-year-old son the pick of any heifer from the herd. A later magazine article noted that George's selection did well in three states, including winning as grand champion at the Kansas State Fair in 1933.

In 1942, Albert bought the Moffett Ranch, which had both pasture and cultivation land and was located near Argonia. Son Harry and his wife, Lucille, managed this ranch while they raised two children, John and Jane Ann. The land expansion helped fulfill the diversified agricultural needs of Albert's family. In 1943, Harry held a cattle, sheep, and hog sale in Harper County. Albert passed away in April 1967, but George, Harry, and their families have continued the successful business.

A. R. Schlickau & Sons partners were (*from left*) Harry, Albert, and George. They are shown displaying a pen of bulls near their new Quonset hut buildings in 1955. (Source: Schlickau Family)

In 1951, when the national 4-H club fat stock show at Wichita was closed because of a lack of event space, the Schlickau family helped the 4-H'ers set up a judging school on their Haven ranch, where they created several judging classes of their Herefords. Additionally, neighbors brought classes of sheep and hogs for the 4-H'ers to evaluate, and a K-State judging team officiated the event. For the next forty-four years, the Schlickau family continued to host that same judging school for youth at their Haven ranch.

George's progressive outlook was found in his herd management practices.

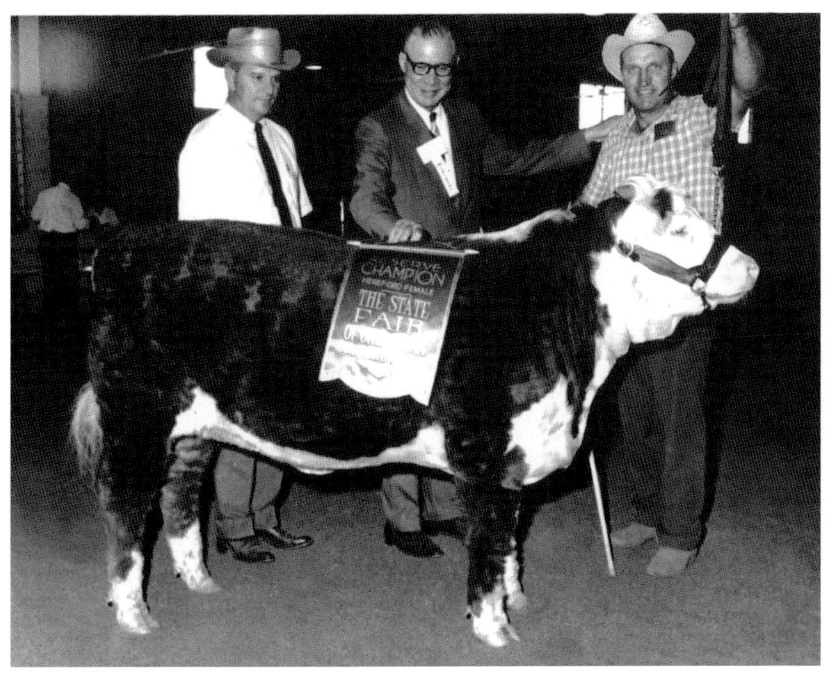

The Schlickaus showed cattle throughout the country at many state fairs and major stock shows. Kansas State's Dr. Miles McKee and the judge are shown behind a top Schlickau heifer, with George Schlickau at the halter, circa 1960. (Source: Schlickau Family)

In 1954, George sought out Hobart Fredrick, Reno County's K-State cooperative extension agent, to help him set up a Total Performance Records program. Together, the pair measured, weighed, and recorded data for many years, and George continued to gather and use the performance information thereafter. George commented, "If an animal can perform on paper, he deserves a second look to show what else he has to offer." During one of George's terms as president of the KHA, he helped form the nation's first junior breed association. "During the [Kansas Junior Livestock] show at Wichita in 1959, a group of kids cornered me and asked what they could do to help," George stated in a magazine article. "So we got our heads together in a hotel room and the organization [Kansas Junior Hereford Association] was born." Four years later, the national Junior Hereford Association was born, and it was the first such group of any beef breed. George served as the American Junior Hereford Association advisor for six years.

George served on the first committee to organize and implement the central Kansas bull test station at Beloit. He also was on councils for the Kansas Livestock Association, Kansas Junior Livestock Association, and American National Cattlemen's Association, among others. George and wife, Lois, raised four children—Bruce, Susan, James, and Nancy. In 1969, the Kansas Hereford

Schlickau Herefords, a partnership between Harry and George, exhibited pen bulls in the yards and on the hill at the National Western Stock show, circa 1965. (Source: Schlickau Family)

George Schlickau removed the halter from Tex Real Onward 178 when he released the Super Register of Merit Sire into the pasture in 1969. (Source: Schlickau Family)

During the 1970s and 1980s, the Schlickaus were honored to be recipients of several awards, including the induction of George into K-State's Block and Bridle Hall of Fame and being named Kansas Seedstock Producer of the Year. From 1969 to 1975, George served on the American Hereford Association board; in 1975, George served as the association's president. During his six years of service to the national board, George had the opportunity to represent the breed and the U.S. cattle industry on many levels. While serving as a cattle ambassador to Brazil in 1975, George observed how the breeders had tipped horns instead of using horn weights. George began the practice on his cattle and encouraged his friend, Alfred Meeks of Nebraska, to try it on his

Auxiliary was formed, with both Lois and Lucille as charter members. Lois served on the original board and was involved in subsequent service to the industry. The auxiliary later was renamed Kansas Hereford Women. In 1981, Lois was elected president of the American Hereford Auxiliary. That year marked the 100th anniversary of Herefords in America, and George and Lois were the first husband and wife team to serve as national association presidents. Lois was honored to receive the first annual American Outstanding Hereford Women Award in 1985. Lois was eventually elected to the Kansas State Board of Agriculture. She was the first woman president and served two terms (1988–1989 and 1989–1990), a unique accomplishment.

This Tex Real Onward 178th son was renamed Denver Domino 1971. That same year, he was named Grand Champion Hereford bull at the Oklahoma State Fair (above), the National Western Stock Show, and numerous other major shows in 1971. (Source: Schlickau Family)

This 1975 photo shows George Schlickau *(third from left)* while he was in Brazil with the U.S. cattlemen delegation. They served as beef ambassadors and Hereford representatives to several South American countries. (Source: Schlickau Family)

commercial herd. Others soon followed suit, and the new trend of tipping made horn weights obsolete nationwide.

While George traveled internationally as a cattle ambassador, foreigners also sought him out at the Haven ranch. His family hosted individuals and delegations from twenty-seven countries, including the Soviet Union's Evgeniy Primakov and his entourage in October 1989, the Governor's Farm Tour in 1991, and delegates from ten former Soviet blocs in April 1996.

In 1983, George and his family became the sole operators of Schlickau Herefords. Harry and his son, John, incorporated a separate firm and marketed their cattle under the new entity. That same year, the Kansas State Fair's Hereford show was dedicated to Harry in recognition of his numerous years of service as KHA treasurer. Other family members have served in many offices, including John as KHA president and executive secretary; Bruce as KHA president; Lois, John's wife Jane, and Bruce's wife Tracy as Auxiliary/Hereford Women

The Schlickau family is shown in 1985. *Front row, from left:* wife Lois, Nancy, and Susan. *Back row:* George, James, and Bruce. (Source: Schlickau Family)

These sale bulls enjoyed supplemental feed before they were sold at the Haven ranch in 1989. (Source: Schlickau Family)

George Schlickau (left) is shown with Kansas State's Dale Blasi and their Mexican host, during one of George and Lois Schlickau's trips to Chihuahua, Mexico, in 1991.

presidents; and Lucille as KHA treasurer. George was named KHA Breeder of the Year in 1976, and Harry was selected as an honorary member in 1978. John, Bruce, Susan, James, and Nancy served as Kansas Junior Hereford Association directors and officers. Susan and Nancy also represented the breed as queens. Along with the next generation, they have continued to serve in cattle, 4-H, and FFA organizations.

In an effort to improve grass and water quality, James implemented a rotational grazing system in the fall of 2005. Nearly seventy conservation specialists and visitors from twenty-two states and five countries have toured the rangeland to observe its benefits.

Harry died in January 2000, and George died in August 2005. Bruce and Tracy established a separate operation under BTS Herefords; Lois, son James, and wife Charlene have continued to operate Schlickau Herefords at the Haven ranch headquarters, where they also continue their commitment to quality genetics by Schlickau Herefords.

U.S. and international conservation specialists came to see Schlickau Herefords' grassland improvements in 2005 and 2006. James Schlickau is shown explaining the rotational grazing system to a tour group. (Source: Schlickau Family)

Andy Schuler Jr. and Fairview Angus

Andy Schuler Sr. was born in December 1873—three years after his father Martin came to the United States from Austria. After settling briefly in Rochester, New York, Martin moved to Kansas and purchased a homestead in 1875. With his initial herd of "black cattle," Andy took some of his animals to Kansas City in 1895, and a few years later, he started his registered Angus herd. For many decades during the twentieth century, Andy Schuler Jr. showed cattle and won numerous awards. In 1939, he showed the reserve champion Angus at the Kansas City Royal, and the granddaughter of one of his heifers was the international champion female in 1973.

Andy Schuler Jr. attended Kansas State University for one semester before he returned to home to the same homestead his grandfather had settled more than a century ago. Today, Andy Schuler Jr. runs about eighty head of cattle, all of which are registered Angus, and he still sells breeding stock.§

§Go to www.beefcattleinstitute.org and listen to a podcast, entitled "Fairview Angus: 115 Years in the Making," and learn about the multigenerational Schuler family's efforts to raise Angus cattle.

Andy Schuler Jr. and his pen of yearling bulls during the fall of 2008. (Source: Blair Tenhouse)

Eby Ranch. (Source: Blair Tenhouse)

A Legendary Kansas Showman: George Crenshaw

George Crenshaw began showing cattle in 1937 at the Kansas Junior Livestock show, which still takes place in Wichita. By age nineteen, George had shown three consecutive reserve champion steers in Wichita, one of which he took to the National Western Stock Show in Denver, where his steer was overall grand champion and sold for $1 per pound. Following his success as a youth, George went to Kansas State University and was hired as the college herdsman—the last person hired for this position who did not hold a college degree. While working from 1946 to 1950, he continued to show cattle, and many of his accomplishments are chronicled in the Beef Cattle Institute audio podcast series.**

**For more on the legendary accomplishments of George Crenshaw, go to www.beefcattleinstitute.org and listen to the podcast entitled "A Kansas Showman: George Crenshaw."

Dalebanks Angus. (Source: Blair Tenhouse)

Dalebanks Angus and the Perrier Family

Less than a decade after Kansas was granted her statehood, an Englishman by the name of Robert Loy and his wife, Alice, homesteaded near Eureka, Kansas. They named their homesteaded farm Dalebanks, which was a reference to their own ancestral farm and region in England. During the early twentieth century, the Loys' daughter, Amelia, married E. L. "Bert" Barrier, who was a capable farmer-stockman from a neighboring ranch. Angus cattle were introduced around 1904, and more than a century later, Dalebanks Angus continues the Barrier tradition of focusing on economically relevant traits in beef cattle production.

Family descendants of the Loy and Barrier families—Tom and Carolyn Perrier—continue to manage, operate, and own Dalebanks Angus along with their children and their children's families. Tom and Carolyn's son, Matt, lives on the ranch with his wife, Amy, and their children—Ava, Lyle, Hannah, and Henry. Matt superintends the breeding, marketing, and customer service operations of Dalebanks Angus. Before the Barrier family brought Angus cattle onto the ranch, the cattle

The Perrier family. *Front row, from left:* Lyle, Hannah, and Ava. *Back row:* Tom, Carolyn, Amy, and Matt. (Source: Blair Tenhouse)

Dalebanks cow-calf pairs roam the pastures of Greenwood County in the southeast region of Kansas. (Source: Blair Tenhouse)

were probably so-called "cracker cattle"—a mix of longhorns, shorthorns, and possibly Herefords.[6] After the addition of the Angus breed, and after many decades, the Dalebanks operation has changed tremendously. However, Tom stressed that much remains the same:

> I don't think the basic philosophy has changed at all. My grandfather was in the business to raise bulls for the commercial cattleman, and he wasn't interested in the fads and fantasies of the time. He didn't show cattle much; we have done very little of that over the years. We are more interested in what the commercial man needs.[7]

The Perrier family's enduring commitment to economic value continues to this day. When mission statements were being developed by private companies, Dalebanks Angus simply adopted a saying that was attributed to Tom's grandfather, Bert Barrier: "We've always tried to produce an animal that would profit its owner through its production."[8] Indeed, the Perriers are committed to this mission, and Matt explained how it fits into his and his family's vocation, which he finds both meaningful and fulfilling:

> We produce food that feeds the world. I don't know a more noble profession than production agriculture. And yet we get so downtrodden with the challenges that we have during the day, or the markets, or the weather, or whatever else. But sometime we have got to step out and realize that . . . it takes a long time to get a calf clear to market weight and, yeah, it takes a lot of cost and a lot of heartache. But instead of just making something, like . . . electronics that somebody could do with or without . . . this is something where we keep people alive. [It helps us] realize that, yeah, I actually did something today.[9]

Four-year-old Lyle Perrier is shown checking on the cows while he worked with his dad. (Source: Blair Tenhouse)

Chapter 6
Kansas State University's Department of Animal Sciences and Industry

By Miles McKee

Kansas State Agricultural College was founded just two years after Kansas became a state. The first college circular, dated September 2, 1863, featured the following declaration:

> Every possible effort will be made to make the facilities for acquiring a full and thorough education in this institution equal to those of any other in the country. Its government will be firm, but mild and parental. Its aim will be to promote the highest welfare of the student, physical, mental and moral. Females as well as males will be admitted to all the advantages

> of the institution. Special instruction will be given to those preparing to teach. All proper attention will be given to subjects relating to the Department of Agriculture. A course of lectures on practical farming and kindred subjects from competent men may be expected during the term.[1]

On July 10, 1901, the Farm Department of the college was divided into the departments of Agriculture and Dairy Husbandry, because the founders and the early administrators had a strong commitment to providing a practical education. This commitment has endured, and the Kansas beef industry has played a role.

(Source: Special Collections, Hale Library, Kansas State University)

The Founding of Kansas State Agricultural College

The settlers of present-day Manhattan, Kansas, arrived in 1854 and 1855, nearly ten years before Kansas gained its statehood. During the 1850s, eastern Kansas was the scene of conflict between the advocates of slavery and those who favored its abolition. The repeal of the Missouri Compromise and the introduction of the principle of *squatter sovereignty* gave control of the region to its inhabitants. As a result, emigrant aid companies were formed in New England to promote the immigration of anti-slavery settlers.

Early in this movement, Isaac T. Goodnow, a professor of natural science at the Providence Seminary in Rhode Island, was compelled to work on the side of freedom. Goodnow and his brother-in-law, the Rev. Joseph Denison, who was a minister in Boston, were inspired by Eli Thayer and decided to join the cause of the New England Emigrant Aid Company. Goodnow organized a group of approximately two hundred, and this group left Boston for Kansas in March 1854.

The New England Emigrant Aid Society was neither alone in its efforts nor the first to arrive in the area. In June 1854, Colonel George S. Park of Parkville, Missouri, staked a claim just east of the mouth of Wildcat Creek. On the west side of the Blue River below Bluemont, Samuel D. Houston of Illinois, along with several others, located a town that they called Canton during the fall of 1854. These early settlers decided to consolidate their efforts and form one strong company, and in April 1855, Colonel Park was invited to discuss an agricultural school with the company

Previous: Bluemont Central College. (Source: Special Collections, Hale Library, Kansas State University)

A display of the pure breeds of cattle are shown at Kansas State Agricultural College. (Source: Special Collections, Hale Library, Kansas State University)

trustees. This is likely the beginning of what has become Kansas State University.

In June 1855, the Cincinnati and Kansas Land Company, which had been destined for a site that became Junction City, joined those who were already in the area to establish another town. They named it Manhattan. Many of these settlers, unlike those of many western towns, were people of education and culture, and they were interested in establishing an institution of higher learning. Apparently, they discussed a school and its support during their voyage, which ended where the Big Blue and Kaw Rivers came together. As early as 1857, an association was formed to build a college in or near Manhattan that would be under the control of the Methodist Episcopal Church of Kansas and would be called Bluemont Central College.

On February 9, 1858, the charter for the Bluemont Central College Association (originally Blue Mont) was received from the legislative assembly of the Territory of Kansas. The association was given permission to locate a college near Manhattan and was empowered "to establish, in addition to the literary Department of arts and sciences, an

Cattle congregated in front of Stone Barn No. 2 in 1889. Designed by Professor E. M. Shelton, it was completed in October 1877 for $4,000. The barn was a one-story structure with a basement that could hold 40 head of cattle and eight horses. In 1885, a 50′ x 75′ north wing that cost $4,500 was added to the stone barn for experimental feeding. Both buildings were equipped with steam and improved machinery for the shelling, grinding, threshing, and steaming of feed grains. (Source: Special Collections, Hale Library, Kansas State University)

agricultural Department with separate professors to test soils, experiment in the raising of crops, the cultivation of trees, and upon a farm set apart for the purpose, so as to bring out of the utmost practical result, the agricultural advantages of Kansas, especially the capabilities of its high prairie lands."[2]

The incorporators of the Bluemont Central College Association were prominent men, both locally and elsewhere. The men who founded Bluemont Central College have been described as persistent, and their success has been attributed to "the character and the sincerity of those who contributed financially to the support of the undertaking." They selected a 160-acre site about one mile west of Manhattan, and the title was secured by a special act of Congress. The Cincinnati Town Company promised help with town lots and town stock, but before it would do so, the company required the Bluemont Central College Association to have property in the amount of $100,000. To help reach this goal, the New England Town Company gave fifty shares of stock in the north half of Manhattan, which represented one hundred city lots. Goodnow and Denison sold the shares of stock, and they solicited locally and in the East to secure funds for a building.

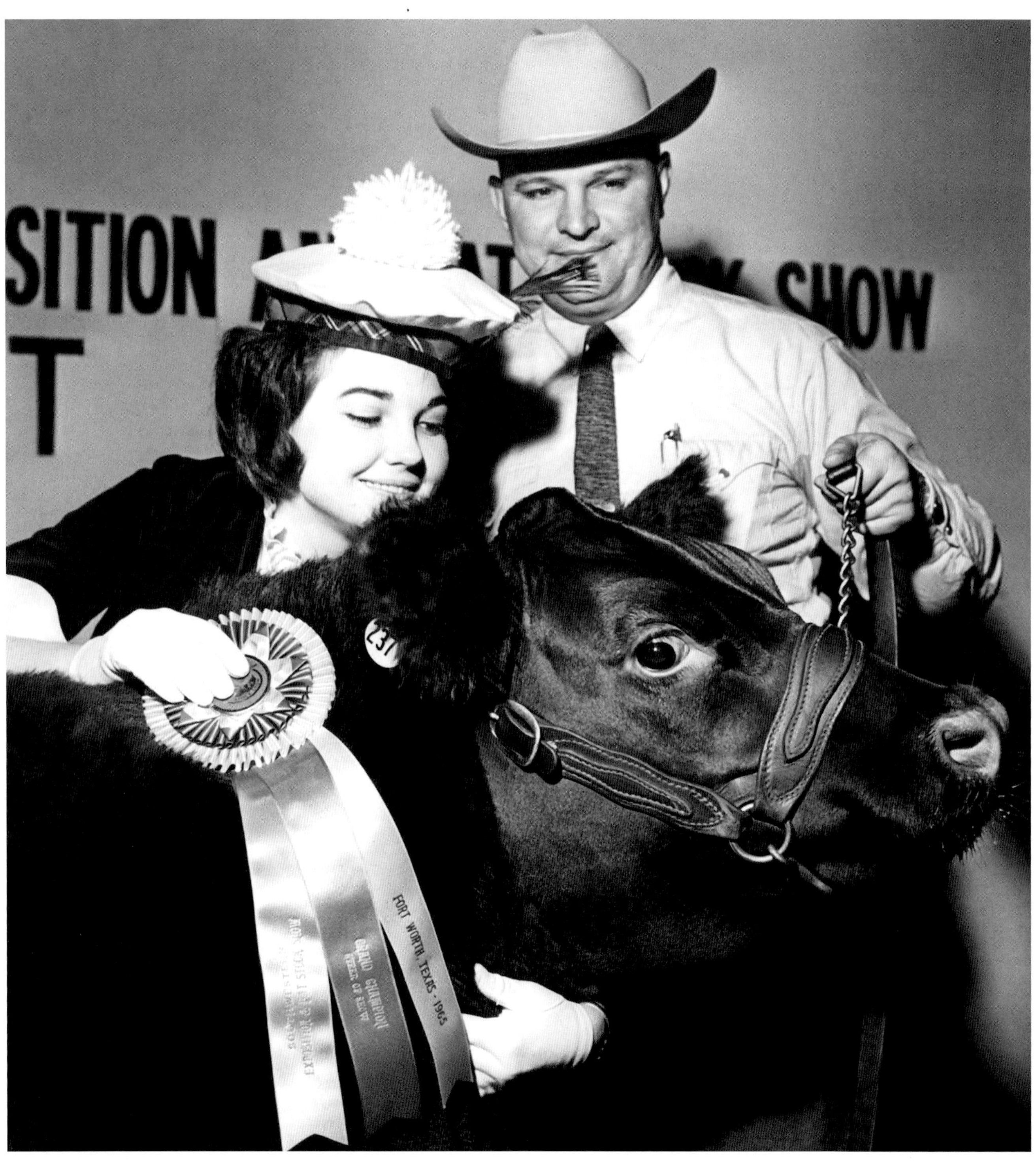

KSU Pete earned champion Shorthorn and grand champion market steer during the 1965 Southwestern Exposition at Fat Stock Show in Fort Worth. Pete also was named reserve grand champion in the carcass show, with a 698-pound average choice carcass. With an 11.2 ribeye area and 0.8 back fat, Pete had a cutability score of 4.8. Shown with Pete are Charlotte Miles, Texas Shorthorn Lassie, and Miles McKee, K-State beef herdsman. The Department of Animal Husbandry purchased the steer from Pete Hawkinson, a Kansas Shorthorn breeder. (Source: Special Collections, Hale Library, Kansas State University)

(Source: Special Collections, Hale Library, Kansas State University)

On May 10, 1859, a cornerstone for a classroom was laid near the present-day intersection of Claflin Road and College Avenue. The three-story limestone building was 44 feet by 60 feet, and it was completed in late 1859 at a cost of approximately $20,000. Bluemont Central College operated until 1863, although college grade instruction was never provided. Apparently, finances became a problem for the school because the only source of revenue was student tuition. Consequently, on February 28, 1861, the school trustees decided to donate the college building, library, and 120 acres of land to the State of Kansas on the condition that the Kansas Legislature permanently locate the state university on this same site in Manhattan.

In February 1863, the provisions of the Morrill Act were accepted, and Kansas State Agricultural College was established on the site of Bluemont Central College at the northwest corner of Claflin Road and College Avenue. It was the first land grant school to be chartered in the United States (it truly was established on a land-based grant from the U.S. government). The Morrill Act gave each state title to 30,000 acres of land for each senator and representative; the income from this land was to endow, support, and maintain at least one college for the benefit of agriculture and the mechanic arts. With two senators and one representative, Kansas received 90,000 acres.

After the Morrill Act was accepted, the trustees of Bluemont Central College once again offered their property, this time for the establishment of the state agricultural college. The Kansas Legislature accepted the proposal, and Governor Thomas Carney signed the act on February 16, 1863. The legislation, which was effective February 19, 1863, gave Kansas bragging rights as the first state to officially accept the provisions of the Morrill Act.

The KSAC Hereford herd are shown foraging in the snow. (Source: Special Collections, Hale Library, Kansas State University)

On March 3, 1863, the Kansas Legislature designated the new school as Kansas State Agricultural College (KSAC). The first session of KSAC had fifty-two students (twenty-six men and twenty-six women) and ended on December 1, 1863. A second session began in early 1864, and the records show a total of just over one hundred students for the school year.

KSAC classes in agricultural subjects got off to a slow start because, at the time, few people were qualified to give college instruction in agriculture and the sciences related to agriculture. The first catalog listed J. G. Schnebly as the professor of natural history and a lecturer on agricultural chemistry that "relates immediately to the practical part of farming." However, it is not clear how much practical instruction in agriculture was given, because there was a vocal faction in the state who insisted that the agricultural needs of the students were not being met. In late 1865 and early 1866, Benjamin F. Mudge became the professor of natural history and natural sciences, and he served the college until 1874. He was much beloved by students, colleagues, and the public. Early histories of KSAC say he was an inspiration to his students and had far-reaching influence.

Unfortunately, no money was available to the new KSAC administration for the purchase of livestock, equipment, or other material necessary for instruction in practical agriculture. During the 1867 session, the Kansas Legislature appropriated $2,120 for fencing and otherwise improving the farm, $1,000 for the salary of a resident agriculturalist, and $1,500 for a residence for the agriculturalist. This was the first appropriation for agriculture work. Additional appropriations of $1,000 and $394 were made in 1868 and 1869, respectively, to complete the dwelling and to provide a cistern and outbuildings (eighty acres of the farm were enclosed by a stone wall in 1867). At this time, "No provision had been made for the grunt of pig, bleat of sheep, low of cattle, or neigh of horse, which might disturb the literary and classical repose of even a more conservative institution of learning," reported one newspaper editor. Additionally, no cropping had been attempted.

True agriculture instruction probably began in earnest in 1868, when J. S. Hougham of Franklin College was elected the first professor of agricultural science. He began his duties in early April 1868, although he was listed as a member of the faculty for the 1866–1867 school year. Hougham taught everything in the college—bookkeeping, commercial science, general and analytical chemistry, and physics—and his title was changed several times to fit his duties. Because he was responsible for so many important courses, it is likely that progress in any of them was slow. In 1870, Hougham planted the first crop, which consisted of oats, barley, and corn. The season was unfavorable—the oats and barley grew only six to eight inches tall, and the corn was all but destroyed by chinch bugs. In August 1870, wheat was sown, and in 1871, the crop yielded 43½ bushels per acre.

During Hougham's four years at KSAC, there were no funds available for the development of his department. He offered to lend some money to the college for necessary expenses, which was accepted, but in the end it brought him much trouble and led to his resignation. Even though Hougham was credited as better qualified in the "sciences related to agriculture" than in the practical aspects of farming, he realized the need for some practical teaching tools. In 1870, he wrote in a report, "We hope to have a wagon and a team of horses to procure some good specimens of different kinds of stock . . . " In 1872, Hougham accepted the chair of agricultural chemistry at Purdue University, where he remained until 1876.

The first reference to the degree Bachelor of Agriculture was made in KSAC's catalog for 1870–1871; the degree was to be conferred on those who "satisfactorily complete the course in Agriculture." The courses with agricultural content were now listed for all four years of study. The following courses were related to animal studies: animal physiology, veterinary science, and breeding and management of stock. This catalog also noted there would be compulsory manual labor for all male students starting in the spring of 1870. Students were to be paid three cents to fifteen cents an hour, depending on the value of the work to the college. Apparently, students objected and there was some relaxation of the policy. Later, during the Anderson administration, students were required to do manual labor for two terms. One term of work was to be on the farm (crops and livestock) and one term for the Horticulture Department.

Between 1870 and 1872, the importance of animal agriculture to the State of Kansas became apparent, and progress was made in the areas of veterinary science and animal husbandry. According to records, "As soon as the resources of the College will permit, choice animals will be obtained. It is hoped the State will provide an ample barn before long."[3] However, the only animals listed as being owned by the college in the 1871–1872 catalog were "three powerful mule teams."

(Source: Special Collections, Hale Library, Kansas State University)

In August 1871, Major Fred E. Miller joined the faculty. Miller had studied at Michigan Agricultural College but had not graduated. He was employed at KSAC as the professor of practical agriculture and superintendent of the farm. During his tenure, the new farm was developed, means were provided for the purchase of stock teams and implements, and foundations were laid for herds of livestock. Miller is credited with laying the foundation for the establishment of KSAC's pure breeds of livestock. In his 1872 and 1873 reports, Miller wrote that "practical progress" had been made.

After the departure of Hougham in 1872, Dr. H. J. Detmers was hired as the professor of veterinary science and animal husbandry. Detmers was a German native whose fluency in the English language was poor. This had a negative impact on his efficiency as a teacher, and he left the college in February 1874.

In 1872, the Kansas Legislature appropriated $15,000 "to fence, improve, and stock the state farm, and to develop the agricultural department of said college"; in June of that year, the KSAC Board of Regents set aside $4,000 of the state appropriation for the construction of one wing of a barn on the new farm. On September 1, 1873, the Rev. John A. Anderson, pastor of the Presbyterian Church of Junction City, became KSAC president. He contributed to the solid foundation for practical education that had been delivered by the Department of Animal Sciences and Industry through these many years. When Anderson's administration ended in September 1879, the KSAC Board of Regents had a hard time finding a replacement. George T. Fairchild, who had left Bluemont Central College and was then vice president of English literature at Michigan State University, agreed to return to KSAC. He served successfully as the school's president from December 1, 1879, until June 30, 1897. One of Fairchild's beliefs that bears citing is the following saying: "Not so much to make men farmers, but to make farmers men."[4]

KSAC cattle were paraded on campus before they were shown at the 1913 American Royal in Kansas City. (Source: Special Collections, Hale Library, Kansas State University)

Animal Husbandry at the Kansas State Agricultural College

Just as agriculture itself is a dynamic and ever-changing industry, the Department of Animal Sciences and Industry has evolved to reflect the unfolding needs of the people and the enterprises it serves. The structure of livestock studies has been reorganized several times in response to those changes and to more effectively realize the school's educational mission.

During the twilight years of the nineteenth century, the offering of agricultural coursework was certainly sparse by today's standards. In 1872, Detmers was the professor of veterinary science and animal husbandry, but his work was limited to veterinary medicine and teaching German. The first actual instructor and researcher in animal husbandry was Professor Edward M. Shelton, who also served as farm supervisor. He

Rufus Cox is shown demonstrating the attributes of a good bull. Cox and the model bull, Royal Husker 3 of the K-State herd, were featured in a herd bull selection article that was published in the *Hereford Journal* during the early 1960s. (Source: Special Collections, Hale Library, Kansas State University)

was appointed in 1874 and served until 1890. The classes related to animal studies at that time were still animal physiology, veterinary science, and breeding and management of stock.

Compared with contemporary times, getting an education in the animal sciences was dramatically unique, and the differences were not limited to the classroom. A severe limitation on instruction in livestock production was the fact that, in its early days, the college farm's stock consisted primarily of mule teams whose most prominent references in the archives include "tearing down the stables on numerous occasions" and dragging their drivers across fields.

Animal science endeavors took another big step forward in 1873, when the Kansas Legislature approved funds for livestock. The executive committee of the KSAC Board of Regents authorized the purchase of four head of three-year-old steers "at five cents" for experiments in "feeding cattle and the relative value of our native and cultivated grasses." The regents also agreed to a two-year lease agreement on some Berkshire hogs, and the regents themselves selected poultry, swine, and cattle for the school, including the nucleus of a purebred Shorthorn herd. Professors Miller and Shelton are credited with laying the foundation for herds of livestock at the school.

This increased involvement in livestock signaled a deeper commitment to animal science, and it reflected the needs of Kansas as well as the nation at large. As the twentieth century approached, there was much concern about the productivity of American agriculture. The nation was growing rapidly, and the rural-to-urban migration was on in earnest as the country's industrial component thrived. Consumers in America's rapidly growing cities both needed and wanted eggs, dairy products, wool, and meat—and it was up to the nation's farmers and ranchers to provide these products. Increased attention to producing better livestock, and to doing it more efficiently, was required. Furthermore, KSAC enrollment was steadily growing. During the 1878–1879 term, there were 207 students, but by the 1891–1892 term, attendance had increased to 584.

In 1899, the Kansas Legislature appropriated $25,000 for a new agricultural building and $5,000 a year for two years for the purchase of blooded cattle. As the focus on animal husbandry sharpened, the structure of the KSAC departments was in a state of flux. On September 26, 1902, the regents voted to divide all of the work being conducted in agricultural education into three departments: agriculture, animal husbandry, and dairy husbandry. In 1903, the two animal-related departments were combined into the Department of Dairy and Animal Husbandry, with Professor Oscar Erf as head of the department and R. J. Kinzer as the assistant in animal husbandry. In 1905, that department was split into Dairy Husbandry and Animal Husbandry, with Erf and Kinzer leading their respective departments. The school herd was divided into dairy and beef breeds, with the dual-purpose Red Polls going to the dairy group.

Emphasis on dairy reflected a rapidly expanding industry in Kansas, but other aspects of livestock production education also were gaining momentum as Kansas agriculture itself was being transformed. In the December 18, 1905, edition of *The Industrialist*, Assistant Professor George C. Wheeler observed that the state's progress in the livestock industry during the previous twenty-five years had exceeded that of any other state or territory in the Union: "From almost insignificant beginnings but a short quarter of a century ago, it is now one of the leading live-stock states of our nation."[5] In particular, Wheeler observed that Kansas was home to a $5,000 Hereford bull, a $22,000 imported cow, several famous draft stallions, and a four-year-old ram that had produced a record-setting fifty-two pounds of fleece.

There were more changes at KSAC during the twentieth century. Highlights include a remarkable communications effort in 1924, when all three livestock departments—dairy, poultry, and animal husbandry—broadcasted college courses over KFKB (and later, KSAC) radio. Known as the *College of the Air,* the program included a poultry course, a dairy course, and a livestock course that were presented weekly by faculty members. The program ran for ten weeks and drew several hundred enrollments. It also drew thousands of other listeners from nearly every state as well as Canada.

In 1968, the Department of Animal Husbandry was renamed as the Department of Animal Sciences and Industry. It was located in the Animal Industries Building, where it had been housed since that facility was constructed in 1957. The structure was later officially named Weber Hall in honor of A. D. "Dad" Weber, a former department head and dean of agriculture. The ensuing years brought dramatic growth to Kansas State University (KSU) in general and to livestock interests in particular. In 1977, dairy, animal science, and poultry were united under the Department of Animal Sciences and Industry. Don L. Good, a veteran educator and head of the department, was selected to lead the new division.

Beef Production Is Cornerstone

The beef teaching, research, and extension programs have been instrumental to the growth, productivity, and success of the beef cattle industry in Kansas. Beef production has grown to become a $6 billion business, with 6.7 million head of cattle at the start of the twenty-first century, and beef production remains a cornerstone within the Department of Animal Sciences and Industry at KSU. Four units support teaching, research, and extension work, which helps keep Kansas beef production in its position of leadership.

KSU's Beef Research Unit is home to about 1,500 research subjects, including horses. Today, novel alternative feed-manufacturing technologies are being developed in collaboration with companies in the United States, the United Kingdom, and Switzerland. Farm-level, pre-harvest food safety research, and work to find natural, plant-based alternatives to eliminate or reduce reliance on antibiotic or steroidal growth promotants, also are emphasized. Projects have included the development of probiotic cultures to help prevent metabolic disorders in horses and cattle, and fat metabolism has been studied to improve cattle health and performance. KSU scientists also have succeeded in increasing the omega-3 fatty acid content of meat.

KSU's Cow-Calf Unit herd consists of approximately 300 Angus x Hereford commercial cows, 70 heifers, and 12 breeding bulls. They graze on 2,000 deeded acres and 3,000 rented acres of Flint Hills pasture. Research is being conducted in the areas of nutritional management, production cycle management (including calving, breeding, and weaning), the effects of calfhood nutritional and health management on carcass quality, and the values and factors that influence grazing distribution.

The KSU Purebred Beef Teaching Unit annually runs nearly 200 breeding-age purebred Angus, Hereford, and Simmental cows on 4,000 acres of native bluestem grass. The mission of the livestock farm is to provide students with practical experience in breeding, feeding, and managing purebred seed stock. It also gives students enrolled in livestock selection and general animal science courses the opportunity to evaluate quality cattle. The herd also provides an opportunity to demonstrate tools for genetic improvement of seed stock, such as performance testing, artificial insemination, embryo transfer, and ultrasound carcass evaluation. Cattle from the herd are used for applied research trials in cloning technology, synchronized and timed artificial insemination, and feed-additive research. Cattle also are employed for teaching purposes in classes such as livestock sales management, pregnancy diagnosis, and bovine calving, and they are used for competition in the annual Little American Royal. Some of the animals from the herd are merchandized during the annual Special K Bull and Female Sale, which is held the first Friday in March. This sale is unique, because it is engineered entirely by students to give them hands-on, practical experience in purebred cattle marketing.

The Beef Stocker Unit consists of 1,120 acres that were acquired by KSU in 1948. The land, which was originally known as the Animal Science Range Unit, was refocused and renamed the Beef Stocker Unit in 2004. The goal is to provide an avenue for backgrounding and stocker cattle research. In 2005, industry needs led to the development of the Animal Identification Knowledge Laboratory. This included the addition of twenty-four receiving pens that are designed to hold 300 head of 500-pound cattle. Current research on receiving cattle and animal identification technology focuses on finding which methods help producers achieve cost-effective results while providing the public with a safe product.

Milestones in Meat Science Studies

No documentation has been found that specifically states when the first class in meats was taught at KSU, although many of the steers that were used for teaching students were taken to Kansas City for slaughter, and their carcasses were subsequently studied. Some records indicate that Professor Kinzer taught the first course in meats to female students who were studying home economics during the winter session in 1905. T. M. Paterson organized and taught the first meats class for animal husbandry students in 1910; the course included a lecture and a demonstration, but in Paterson's words, "It was a mess." In 1915, money was appropriated to build a three-room shack in which to teach meats, but there was only enough money to build a one-room shack. Money was appropriated to finish the shack, but two coolers were built in the Dairy Hall instead. Both instructors and students slaughtered animals in the shack and hauled the carcasses in a wheelbarrow to Dairy Hall, where they were cooled, cut up, and sold.

Lyman Bratzler completed work on a machine that physically tested the tenderness of meat. It became known as the Warner Bratzler Shear. (Source: Special Collections, Hale Library, Kansas State University)

In 1923, KSAC completed an addition to the north end of the east wing of Waters Hall, which provided laboratory space for instruction in meats. Cattle to be slaughtered in the new meats laboratory were penned in the judging pavilion that was located just across the street to the west. A rope halter, with a lead rope long enough to reach from the pavilion to an iron ring in the floor of the new lab, was the means for getting cattle from the holding pen to the slaughter floor.

The meats laboratory in east Waters Hall served the department until the move to Weber Hall in 1957, when a new laboratory was available. During the renovation of Weber Hall in 1988, the meats laboratory was updated to conform to U.S. Department of Agriculture specifications.

Throughout its history, members of the KSU meat science group have provided leadership to the beef industry by serving as presidents, board members, and chairs of national conferences for the American Meat Science Association, the American Society of Animal Science, the Institute of Food Technologists (Muscle Foods Division), the International Congress of Meat Science and Technology, and the Federation of American Societies of Food Animal Sciences. Additionally, they have provided leadership to committees and advisory councils of the USDA, the Beef Improvement Federation, breed associations, the North American Meat Processors Association, and the American Association of Meat Processors.

Note: This chapter features several excerpts from Dr. Miles McKee's book, *Building the Legacy: A History of the Kansas State University Department of Animal Sciences & Industry.* The editors thank Dr. McKee for reproduction permission.

Chapter 7
Looking Forward

By Curtis Kastner and Justin Kastner

During the past 150 years, the state of Kansas has figured prominently in the social, economic, and political history of the United States. Home to a beef industry that, since the mid-nineteenth century, has consistently impressed consumers and investors alike, the state of Kansas has referenced the *150 Years of Kansas Beef* project in its sesquicentennial celebrations.[1] It is good and appropriate for all Kansans to look back and to celebrate the tremendous accomplishments of its agriculturalists during the last century and a half. It is equally fitting to look forward, too. Today, Kansas beef occupies a unique position in the global food system, and future-minded vision will continue to play a role in the state's and the industry's ongoing triumphs.

Many examples are available in recent Kansas history, but one instance of Kansans looking forward involved a State of Kansas Beef Trade Mission to the country of Mexico, a leading importer of Kansas beef. In November 2007, Governor Kathleen Sebelius and the Trade Development Division of

Previous: (Source: Blair Tenhouse)

the Kansas Department of Commerce led a group of beef industry leaders to Mexico City. During the mission, officials engaged in discussions and activities to maintain and enhance the economic linkages between beef producers in Kansas and beef importers in Mexico.[2] Diplomatic, commercial, and educational activities included Kansas political leaders, private ranchers and other leaders in the beef industry, and faculty from Kansas State University (KSU). In addition to meetings with a prominent Mexican beef importers' association, the mission featured a Kansas-Mexico Forum during which cross-border cooperation in food safety was discussed.[3]

Food safety is an area of utmost importance to the future of the Kansas beef industry, and Kansas has a rich record of successes in this area. Since the late 1970s, KSU animal science and food science programs have had a major focus on food safety, with particular attention being paid to the safety of beef and beef products. Because of that focus, KSU has been asked to contribute to a number of beef safety initiatives, including the U.S. Department of Agriculture Food Safety Consortium that began in 1988. Consistent with Kansas's position as a leader in beef safety research, institutions such as KSU have built an infrastructure-rich foundation from which to address traditional as well as emerging food safety challenges. Because of the importance of the beef industry in Kansas, the state's ability to address these challenges is imperative. KSU has a rich tradition of working with industry groups, governments, consumer groups, and other partners. To compliment KSU's leadership role in food safety, the Pat Roberts Biosecurity Research Institute (BRI) has been established in Manhattan, Kansas. This unique facility adds significantly to Kansas's capabilities in beef-related research. Both external and internal partners will collaborate with the BRI at KSU. Significantly, the BRI complements the Animal Health Corridor and the Kansas Bioscience Authority's efforts to help attract the new $650 million federal National Bio and Agro-Defense Facility, which will be devoted to serving the beef industry through the protection of animal health and food safety.

In addition to food safety and animal health, other opportunities face the Kansas beef industry in the twenty-first century, and some of these opportunities may prove to be echoes of the past. As chapters 1, 2, and 3 revealed, Kansas has a history of attracting both interest and investment, thereby fueling the development of a robust trade in beef; as the state looks to the future, the attraction of new investment (from both internal and external sources) will remain vital to the economic health of the state. Chapters 3 and 5 showcased individuals and family ranching operations, and the continual development of a devoted workforce will ensure the future of Kansas beef. Kansans have much to celebrate in 2011, because they can look back on 150 years of triumph in the Kansas beef industry. If history is any indication, the future is bright.*

*Go to the video section at http://frontier.k-state.edu, and watch "K-State: Historically Committed to Food Safety and Security" to learn about KSU's effort to address biosecurity issues for the benefit of Kansas agriculture, including the Kansas beef industry.

(Source: Blair Tenhouse)

Endnotes

Chapter One

1. Kansas Land and Emigration Company, *Emigration to Kansas: The Glory of the West* (London: R.H. Drew, 1869).

2. Kansas Pacific Railway, *Emigrants' Guide to the Kansas Pacific Railway Lands* (Lawrence, Kansas: Kansas Pacific Railway Co., 1871), pp. 3–4, 11.

3. J. B. Edwards ascribed this motive to two fellow early Abilene settlers; *Early Days in Abilene* (Abilene, 1940), p. 1. According to Ray Allen Billington, "the feeling that 'I ought to do better'" was an "uncontrollable passion of the true frontiersmen"; *America's Frontier Heritage* (New York: Holt, Rinehart and Winston, 1966), p. 34.

4. Kansas Census 1860: 103,116; 1870: 361,605; 1880: 995,494; 1890: 1,427,154; *Kansas Statistical Abstract 2004* (Lawrence, Kansas: Policy Research Institute, University of Kansas, 2005), p. 396.

5. Wayne Gard, *The Chisholm Trail* (Norman, Oklahoma: University of Oklahoma Press, 1954), pp. 3–19.

6. Ibid., p. 5.

7. Ibid., p. 12; David Dary, *Cowboy Culture: A Saga of Five Centuries* (Lawrence, Kansas: University Press of Kansas, 1981), pp. 138–9.

8. Joseph G. McCoy, "Historic and Biographic Sketch," *Kansas Magazine* (December 1909), pp. 45–55.

9. Gard, *Chisholm Trail,* pp. 20–56.

10. The symptoms of this infectious blood disease included "high fever, swollen spleen, engorged liver, thick bile, and bloody urine." The disease, caused by a protozoan microorganism, was spread by the tick *Boophilus Bovis;* Cecil Kirk Hutson, "Texas Fever in Kansas, 1866–1930," *Agricultural History* 68 (1994), 74–104, at pp. 74 and 93.

11. McCoy, "Historic and Biographic," p. 48.

12. *Daily Herald,* San Antonio, Texas, June 28, 1867, as quoted in William L. Urban, "The Juvenal Cattle Drive of 1870," *Kansas Historical Quarterly* 39 (1973), pp. 200–205, at p. 200. Juvenal evidently liked the route to Baxter Springs, since he and his brothers continued to follow the Shawnee Trail as far as Baxter Springs and then headed for Abilene by following the Neosho River into central Kansas in 1869 and 1870, even after the Chisholm Trail had become the established route to the railhead; Ibid., pp. 201–203.

13. McCoy, "Historic and Biographic," p. 49.

14. Gard, *Chisholm Trail,* pp. 60–67; Joseph G. McCoy, *Historic Sketches of the Cattle Trade of the West and Southwest* (Kansas City, Missouri: Ramsey, Millett & Hudson, 1874), pp. 39–40, 74–5. Although McCoy does not date his encounter with Myers in his book, he evidently gave an interview to the New York *Daily Tribune* for an article that appeared 5 September 1867. The newspaper account dates the reconnaissance trip to June 1867; "The Chisholm Trail," *Kansas Historical Quarterly* 33 (1967), 129–37, at p. 134.

15. McCoy, *Historic Sketches of the Cattle Trade,* pp. 40–4.

16. Unfortunately, the plant failed after the 1868 season; McCoy, *Historic Sketches of the Cattle Trade,* pp. 60, 106.

17. McCoy, *Historic Sketches of the Cattle Trade,* p. 41.

18. C.F. Gross to J.B. Edwards, 13 April 1922, p. 5, Papers of J.B. Edwards: Letters Received, Series A, J.B. Edwards collection, Manuscripts Department, Kansas State Historical Society. *NB* that spelling and punctuation have been regularized in all quotations from this source.

19. McCoy, *Historic Sketches of the Cattle Trade,* p. 50.

20. John Rossel, "The Chisholm Trail," *Kansas Historical Quarterly* 5 (1936), 3–14, at p. 4.

21. McCoy, *Historic Sketches of the Cattle Trade,* pp. 50–1; McCoy, "Historic and Biographic," p. 52.

22. McCoy, *Historic Sketches of the Cattle Trade,* p. 65.

23. McCoy, *Historic Sketches of the Cattle Trade,* pp. 51, 53, 106–7. Dary, *Cowboy Culture,* p. 210.

24. "The Chisholm Trail," *Kansas Historical Quarterly* 33 (1967), 129–37, at pp. 136–7.

25. McCoy, *Historic Sketches of the Cattle Trade,* p. 102.

26. Edwards, *Early Days,* p. 2.

27. C.F. Gross to J.B. Edwards, 13 April 1922, p. 4. *NB* that spelling has been regularized in all quotations from this source. Evidently Gross was confused on the timing (as he readily admitted throughout his papers he was prone to do), for he dates this expedition to the winter of 1866, far too early to be accurate. Gross's recollection must be of the second trip south, which McCoy records as occurring during the winter of 1867–68; *Historical Sketches,* p. 115.

28. Dary, *Cowboy Culture,* pp. 187–8.

29. McCoy, *Historical Sketches of the Cattle Trade,* p. 116.

30. C.F. Gross to J.B. Edwards, 13 April 1922, pp. 4, 13.

31. McCoy, *Historic Sketches of the Cattle Trade,* p. 150; C.F. Gross to J.B. Edwards, 13 April 1922, p. 9; Robert Dykstra, "The Last Days of 'Texan' Abilene: A Study in Community Conflict on the Farmer's Frontier," *Agricultural History* 34 (1960), 107–119, at p. 107.

32. McCoy, "Historic and Biographic," p. 53; Gard, *Chisholm Trail,* p. 149; Dykstra, "Last Days," pp. 108, 111; Edwards, *Early Days,* p. 2.

33. Lyrics from the cowboy ballad, the "Old Chisholm Trail."

34. The earliest known references to the Chisholm Trail in print were in the *Kansas Daily Commonwealth* of May 27 and October 11, 1870.

35. Rossel, "Chisholm Trail," pp. 4–7.

36. George W. Saunders, president of the Old Trail Drivers' Association, thought it more technically correct to label the Texas portion of the route to Abilene as the "Eastern Trail" (in distinction from the later "Western Trail" to Dodge City) and to call it the Chisholm Trail only once the path had crossed into Indian Territory and then on to Kansas. For a very helpful review of the ongoing debate, see Gary and Margaret Kraisinger, *The Western: The Greatest Texas Cattle Trail, 1874–1886* (Newton, Kansas: Mennonite Press, 2010), pp. 3–20.

37. Wayne Gard, "Retracing the Chisholm Trail," *Southwestern Historical Quarterly* 60 (1956), 53–68; Gard, *Chisholm Trail,* pp. 76–82.

38. The following account of cowboy life on the trail has been compiled from Dary, *Cowboy Culture,* pp. 190–97; Gard, *Chisholm Trail,* pp. 106–39; Everett Dick, "The Long Drive," *Collections of the Kansas State Historical Society* 17 (1926-28), 27–97.

39. "The Chisholm Trail," *Kansas Historical Quarterly* 33 (1967), 129–37, at p. 135.

40. Gard, *Chisholm Trail,* p. 109.

41. See Dary's very helpful drawing of a herd formation; *Cowboy Culture,* p. 191.

42. Henry King, "Over Sunday in New Sharon," *Scribner's Monthly* 19 (1880), 768–775, at p. 770.

43. George L. Cushman, "Abilene, First of the Kansas Cow Towns," *Kansas Historical Quarterly* 9 (1940), 240–58; C.F. Gross to J.B. Edwards, 29 May 1922, p. 3; Henry B. James, *Heroes by the Dozen: Abilene—Cattle Days to President Ike* (Abilene, Kansas: Shadinger-Wilson Printers, 1961), pp. 23–30, *NB* the very helpful map of early Abilene on p. 20; Edwards, *Early Days,* pp. 8–10.

44. Dary, *Cowboy Culture,* pp. 211–13.

45. J. Marvin Hunter and George W. Saunders, *The Trail Drivers of Texas* (Austin: University of Texas Press, 1985), pp. 486–7.

46. Edwards, *Early Days,* p. 3.

47. C.F. Gross to J.B. Edwards, 13 April 1922, p. 13.

48. T.C. Henry, "Thomas James Smith, of Abilene," *Kansas Historical Collections* 9 (1905–6), 526–32.

49. "Wild Bill," *Harper's New Monthly Magazine* 34 (1867), 273–85.

50. Frank Wilkeson, "Cattle-raising on the Plains," *Harper's New Monthly Magazine* 72 (1886), 788–795, at p. 789.

51. Gard, *Chisholm Trail,* pp. 173–9.

52. Ibid., pp. 180–1. Henry, "Thomas Smith," p. 532.

53. Dykstra, *Cattle Towns,* pp. 30–41; Gard, *Chisholm Trail,* pp. 187–91; F.B. Streeter, "Ellsworth as a Texas Cattle Market," *Kansas Historical Quarterly* 4 (1935), pp. 388–98.

54. Sara R. Massey, ed., *Texas Women on the Cattle Trails* (College Station, Texas: Texas A&M University Press, 2006), pp. 48–5.

55. Jelinek, *90 Years,* p. 40.

56. Ibid., pp. 48–9.

57. Gary L. Cunningham, "Moral Corruption and the American West: A Socio-Historical Investigation of the Origins and Nature of Prostitution and Gambling in the Primary Cattle Towns of Kansas, 1867–1886" (Ph.D. Thesis, University of California at Santa Barbara, 1980), pp. 181–5.

58. Gard, *Chisholm Trail,* pp. 181, 188.

59. Ibid., pp. 158–9, 183.

60. Ibid., pp. 182–6, 212.

61. Ibid., pp. 198–205.

62. Stan Hoig, *Cowtown Wichita and the Wild, Wicked West* (Albuquerque, New Mexico: University of New Mexico Press, 2007), p. 114–17.

63. Gard, *Chisholm Trail,* p. 203.

64. Massey, *Texas Women,* pp. 89–118.

65. Gard, *Chisholm Trail,* pp. 211–5.

66. Dykstra, *Cattle Towns,* pp. 307–18.

67. Ibid., pp. 319–28.

68. James R. Shortridge, *Cities on the Plain: The Evolution of Urban Kansas* (Lawrence, Kansas: University Press of Kansas, 2004), p.147.

69. Kraisinger, *The Western,* pp. 30–3.

70. Shortridge, *Cities on the Plain,* p.147.

71. Kraisinger, *The Western,* pp. 33–7.

72. Ibid., pp. 49–67.

73. Harry E. Chrisman, *Lost Trails of the Cimarron* (Norman, Oklahoma: University of Oklahoma Press, 1998), p. 67.

74. Rossel, "Chisholm Trail," p. 13.

75. Robert M. Wright, *Dodge City: Cowboy Capital and the Great Southwest* (Dodge City, Kansas: n.p., 1913), p. 149.

76. Gary L. Cunningham, "Gambling in the Kansas Cattle Towns: A Prominent and Somewhat Honorable Profession," *Kansas History: A Journal of the Central Plains* 5 (1982), 2–22, at p. 9.

77. Ibid., p. 10.

78. Ibid.

79. Joanne S. Liu, *Barbed-Wire: The Fence that Changed the West* (Missoula, Montana: Mountain Press, 2009), p. 50.

80. Gard, *Chisholm Trail,* pp. 256–7.

81. G.D. Freeman, *Midnight and Noonday,* ed. By Richard L. Lane (Norman, Oklahoma: University of Oklahoma Press, 1984), pp. 216–28.

82. C.W. McCampbell, "W.E. Campbell, Pioneer Kansas Livestockman," *Kansas Historical Quarterly* 16 (1948), pp. 245–73.

83. His land was the first purchase made toward today's CK Ranch; Bud Snidow, *50 Years of Kansas Ranching: CK Ranch* (Brookville, Kansas: n.p., 1984), p. 5.

84. Urban,"Juvenal," pp. 204–5.

85. Charles F. Colcord, "Reminiscences," *Chronicles of Oklahoma* 12 (1934), 5–18, at pp. 7–8.

86. Hutson, "Texas Fever," pp. 80–8.

87. *The Breeder's Gazette,* V (April 17, 1884), p. 584.

88. J.N. Bradley, "Cattle Interest West of the Mississippi," *Second Annual Report of the U.S. Bureau of Animal Industry for the year 1885* (Washington, DC: GPO, 1886), p. 425.

89. *First Biannual Report of the Live-Stock Sanitary Commission of the State of Kansas, 1887–88,* p. 3.

Chapter Two

1. For more on *Beef Empire Days,* typically held in June, visit http://www.beefempiredays.com/.

2. Julia Kathryn Garrett, *Fort Worth: A Frontier Triumph* (Austin: The Encino Press, 1972), p. 267.

3. For more on George Grant and the role played by hopeful investors, see James L. Forsythe, "George Grant of Victoria: Man and Myth," *Kansas History* 9, no. 3 (1986).

4. One economic historian reports that in 1784 woolens, hardware, and other manufactures rushed into the United States as Americans longed to "restore depleted stocks of once-familiar goods." Meanwhile, American exports to Britain, still mostly agricultural products, declined for a few years after 1776 as Britain turned to other sources of tobacco and rice. Nevertheless, Great Britain was easily America's most important export market. As Campbell notes, "More accurate statistics are available beginning in 1790 when United States exports to Britain amounted to $7,000,000, over a third of the total exports of $20,000,000." Charles S. Campbell, *From Revolution to Rapprochement: The United States and Great Britain, 1783–1900,* ed. Robert A. Divine, America and the World (New York: John Wiley & Sons, Inc., 1974), pp. 1, 3–5.

5. Walter L. Arnstein, *Britain Yesterday and Today: 1830 to the Present,* ed. Lacey Baldwin Smith, A History of England (Boston: D.C. Heath and Company, 1966), pp. 37–39.

6. In his free trade budget of 1842, Prime Minister Sir Robert Peel allowed livestock imports for the first time. However, they were subject to tariffs until 1846. Richard Perren, *The Meat Trade in Britain 1840–1914,* ed. F.M.L. Thompson, Studies in Economic History (London: Routledge & Kegan Paul Ltd., 1978), pp. 21, 74.

7. Christopher Harvie, "Revolution and the Rule of Law (1789–1851)," in The Oxford History of Britain, ed. Kenneth O. Morgan (Oxford and New York: Oxford University Press, 1988), p. 510. Perren, *The Meat Trade in Britain,* p. 74.

8. However, those involved in the preserved meat trade did, almost immediately, appreciate the removal of tariffs on meat imports. Traders in Liverpool, for example, had decried in 1844 the duty on American beef and pork packed in barrels. This duty was removed in 1846. R. Thompson, "The Liverpool Provision Trade Association Ltd.: Commercial History," in *Provision Trade Association Archives* (Liverpool: Mersey Docks and Harbour Board | Merseyside Maritime Museum Archives, 1980), sheet 1.

9. Michael Tracy, *Government and Agriculture in Western Europe 1880–1988,* 3rd ed. (New York: New York University Press, 1989), p. 41. John P. Huttman, "British Meat Imports in the Free Trade Era," *Agricultural History* 52, no. 2 (1978), pp. 249–250, 253–254. Harvie, "Revolution and the Rule of Law", p. 514.

10. Several of these feuds were manifestations of British concern over American expansion westward. While Britain was most eager to protect its North American possessions, it was also interested in checking U.S. expansion to the southwest. Evidence of this can be seen in Britain propping up the Republic of Texas as a sovereign nation-state. To some Britons, an independent Texas would not only check U.S. expansion but also give British manufacturers an alternative source of supply for cotton and possibly new low-tariff markets. However, if Washington would acquiesce to Texan pleas for annexation, such hopes would end. Eventually, in 1845, the U.S. annexed Texas. *Campbell, From Revolution to Rapprochement,* pp. 50–69. "Recognition of Texas," *The Times,* 23 December 1840. In London, one can dine in a monument to Britain's diplomatic efforts during this era. Just off Trafalgar Square, on Cockspur Street, tourists will find the Texas Embassy Cantina—a restaurant and an allusion to the actual Texas Embassy that existed on St. James Street. In addition to being an historical marker, the Texas Embassy offers sizeable entrees and free drink refills.

11. Many of the 1860s treaties espoused the Most Favored Nation (MFN) principle, today the cornerstone of the *General Agreement on Tariffs and Trade* and the World Trade Organization. M. Trebilcock and R. Howse, *The Regulation of International Trade* (London: Routledge, 1999), p. 18 and Herbert Heaton, *Economic History of Europe,* ed. Guy Stanton Ford, Revised Edition ed., Harper's Historical Series (New York, New York: Harper & Row, 1948), pp. 641–43.

12. Perren, *The Meat Trade in Britain,* pp. 3–4.

13. Ibid., p. 131. Huttman, "British Meat Imports in the Free Trade Era," pp. 253–4.

14. Speech of Richard Cobden on October 25, 1862, as quoted in *Campbell, From Revolution to Rapprochement,* p. 106. Cobden was indeed popular; years before, he was one of the most celebrated British politicians in Europe, "fêted in capital after capital." Harvie, "Revolution and the Rule of Law," p. 513. Cobden had been one of the most renowned members of the free-trade oriented Anti-Corn Law League. James Foreman-Peck, *A History of the World Economy: International Economic Relations since 1850* (Totawa, New Jersey: Barnes & Noble Books, 1983), p. 30 (fn10).

15. Campbell, *From Revolution to Rapprochement,* p. 106.

16. "Proceedings of the Society: Food Committee," *Journal of the Society of Arts* XV, no. 756 (1867), p. 414.

17. Ibid., pp. 414–16.

18. The Liverpool Provision Trade Association was formed in 1874 to manage the trade in packed meat, but it later expanded to include dairy products, eggs, and canned goods. Gordon Read and Michael Stammers, *Guide to the Records of Merseyside Maritime Museum* (St. John's, Newfoundland: Trustees of the National Museums and Galleries on Merseyside & International Maritime Economic History Association, 1995), p. 78. John Burnett, *Plenty and Want, a Social History of Diet in England from 1815 to the Present Day* (London: Thomas Nelson and Sons Ltd., 1966), pp. 100–101.

19. "Proceedings of the Society: Food Committee," p. 414. Britain's preference for mild-salted pork would persist. In 1888, the Liverpool Provision Trade Association reminded the U.S. consul at Liverpool of this, as evidenced in a letter dated March 29, 1888, which is available in Liverpool's Merseyside Maritime Museum archive. *Liverpool Provision Trade Association: Minute Book Volume 1* (Liverpool: Merseyside Maritime Museum Archives, 1874–1889), pp. 374–375.

20. Pioneers were John Bell & Sons (Glasgow), Swan & Sons (Edinburgh), and Nelson Morris (Chicago). John Dyke conducted shipping experiments involving Canadian cattle in 1873. W.D. Zimmerman, "Live Cattle Export Trade between United States and Great Britain, 1868–1885," *Agricultural History* 36 (1962), p. 46. Fred Wilbur Powell, *The Bureau of Animal Industry: Its History, Activities and Organization,* ed. Institute for Government Research, Service Monographs of the United States Government (Baltimore: The Johns Hopkins Press, 1927), p. 4. "Cattle Diseases and the Meat Supply," *The Times,* 10 February 1890.

21. Eastman was the largest U.S. exporter of live cattle between 1877 and 1885. Zimmerman, "Live Cattle Export Trade between United States and Great Britain, 1868–1885," p. 47. See also James MacDonald, *Food from the Far West: American Agriculture with Special Reference to the Beef Production and Importation of Dead Meat from America to Great Britain* (London: William P. Nimmo, 1878), pp. 143–147.

22. Perren, *The Meat Trade in Britain,* pp. 114, 126. James Troubridge Critchell and Joseph Raymond, *A History of the Frozen Meat Trade* (London: Constable & Company Ltd., 1912), p. 191. Zimmerman, "Live Cattle Export Trade between United States and Great Britain, 1868–1885," p. 48.

23. May 1877 issue of *The National Livestock Journal,* quoted in Zimmerman, "Live Cattle Export Trade between United States and Great Britain, 1868–1885," p. 48.

24. W. W. Rostow, *British Economy of the Nineteenth Century* (Oxford: Oxford University Press, 1948; reprint, 1952), p. 231. *P.L. Cottrell, British Overseas Investment in the Nineteenth Century,* Studies in Economic and Social History (London: The MacMillan Press Ltd., 1975), p. 22.

25. Foreman-Peck, *A History of the World Economy,* pp. 16, 21.

26. Heaton, *Economic History of Europe,* p. 575.

27. Cottrell, *British Overseas Investment in the Nineteenth Century,* pp. 34, 36.

28. Heaton, *Economic History of Europe,* p. 578.

29. Using the U.S. Treasury Department statistics for 1879, less than 20 percent of American exports were carried by American ships; most of the rest were British bottoms. US Bureau of Statistics, *Annual Report on the Foreign Commerce of the United States* (Washington: Government Printing Office, 1879), p. xxix.

30. See Cottrell, *British Overseas Investment in the Nineteenth Century,* pp. 22, 40.

31. MacDonald, *Food from the Far West.* Jimmy M. Skaggs, *Prime Cut: Livestock Raising and Meatpacking in the United States, 1607–1983* (College Station: Texas A&M University Press, 1986), p. 59.

32. Rendition of opening page of Chapter X, in MacDonald, *Food from the Far West,* p. 70.

33. Rendition of opening page of Chapter XI, in Ibid., p. 79.

34. Forsythe, "George Grant of Victoria: Man and Myth," p. 103.

35. *English Enterprise in America: Notes Addressed to Investors and Settlers Concerning the Estate of Victoria (Ellis County, Kansas, U.S.), the Property of George Grant (Late of the Firm Grant and Gask, Now Gask and Gask, Oxford Street, London),* (London: T. Pettitt & Co., 1876), p. 7. The author is grateful to Paul and Lucy Baier, owners of the George Grant Villa, for allowing us to scan this important piece of Kansas cattle history.

36. *150 Years of Kansas Beef* research assistant Blair Tenhouse and her husband Doug Tenhouse attended this event, and the 150 Years of Kansas Beef podcast series (at www.beefcattleinstitute.org) features a summary of the event.

37. On October 8, 1879, an editorial in *The Times,* a London newspaper, described how much more efficient American farms and ports were compared to their British counterparts. Americans, *The Times*' correspondent noted, have an "instinctive aversion to the pursuit of an unprofitable calling." For the article, which was cited in diplomatic correspondence, see p. 467 of "Diplomatic Correspondence Paper No. 209: Mr. Hoppin to Mr. Evarts, October 11, 1879," *Papers Relating to the Foreign Relations of the United States* (1879). Agricultural historian Marchildon confirms that nineteenth-century North America was rapidly becoming the world expert in handling and transporting food. Gregory P. Marchildon, "Canadian-American Agricultural Trade Relations: A Brief History," *American Review of Canadian Studies* 28, no. 3 (1998), p. 237.

38. J.H.D., "Texas Live Stock: An Englishman's Opinion of Its Value for Exportation," *New York Times,* 14 August 1878.

39. W.M. Pearce, *The Matador Land and Cattle Company* (Norman: University of Oklahoma Press, 1964), pp. vii, 10; Skaggs, *Prime Cut,* p. 59; For more on the Parliamentary commission, see Zimmerman, "Live Cattle Export Trade between United States and Great Britain, 1868–1885," pp. 47–48.

40. Harmon Ross Mothershead, *The Swan Land and Cattle Company, Ltd.* (Norman: University of Oklahoma Press, 1971), pp. 16–17.

41. Skaggs, *Prime Cut,* p. 59.

42. Louise Barry, "A British Bride in Manhattan, 1890–1891: The Journal of Mrs. Stuart James Hogg," *Kansas Historical Quarterly* 19 (1951).

43. Resentment was strong enough to spawn by the late 1880s anti-foreign investment legislation. Congress and several states passed such laws. Nevertheless, British investment continued unabated; often the laws were not enforced or easily circumvented by hiring local representatives. See Skaggs, *Prime Cut,* pp. 63–65, and Roger V. Clements, "British-Controlled Enterprise in the West between 1870 and 1900, and Some Agrarian Reactions," *Agricultural History* 27 (1953).

44. Marchildon, "Canadian-American Agricultural Trade Relations," p. 237.

45. Eastman exported cattle raised in Paris, Kentucky, and Macon County, Illinois. John Gillet, the Macon County farmer responsible for Eastman's inaugural beef shipments, was visited by James MacDonald during his North American tour. MacDonald, *Food from the Far West,* pp. 143–47. Zimmerman, "Live Cattle Export Trade between United States and Great Britain, 1868–1885," p. 47.

46. Zimmerman, "Live Cattle Export Trade between United States and Great Britain, 1868–1885."

Chapter Three

1. Personal communication with Bill, Judy, Jerry, and Diane Paige, 3 October 2008.

2. Personal communication with Leon Mosteller, 27 November 2010.

3. Personal communication with Howard Blender, 26 November 2010.

4. A good example of this may be found in popular culture—for example, in electronic encyclopedic entries. See Encyclopedia.com, "Kansas-Nebraska Act (1854)," in *Gale Encyclopedia of U.S. Economic History* (1999).

5. Personal communication with Norm Wilson, 28 November 2010.

6. Leola Howard Blanchard, *Conquest of Southwest Kansas,* 2nd ed. (Wichita: Wichita Eagle Press, 1931), p. 55.

7. Jim Hoy, *Flint Hills Cowboys: Tales from the Tallgrass Prairie* (Lawrence, Kansas: University Press of Kansas, 2006), Chapter 13 "Good Fences," p. 135.

8. Ibid., p. 137.

9. For more on the scenic Kansas byways, visit http://www.ksbyways.org/index.html.

10. Personal communication with Leon Mosteller, 27 November 2010.

11. Personal communication with Norm Wilson, 28 November 2010.

12. Ibid.

13. Personal communication with Hal Luthi, December 2007.

14. U.S. Department of Agriculture National Agricultural Statistics Service, "Kansas Farm Facts," (2010), p. 60.

15. Personal communication with Hal Luthi, December 2007.

16. Ibid.

17. Ibid.

18. Personal communication with Howard Blender, 26 November 2010.

19. Ibid.

20. State of Kansas Flint Hills Smoke Management Plan (in draft form as of December 2010).

21. Personal communication with Howard Blender, 26 November 2010.

22. Ibid.

23. "Family of Dr. O. M. and Anne Ohisen Franklin", Eloise Lane Articles 161–180, no. 174 (n.d.).

24. USDA APHIS Infosheet, "Treatment of Respiratory Disease in U.S. Feedlots," (October 2001).

25. A.P. Raun and R.L. Preston, "History of Diethylstilbestrol Use in Cattle (American Society of Animal Science)," (2002).

26. Personal communication with Howard Blender, 26 November 2010.

27. For an online demonstration, see http://www.bqa.org/bqainjectionsitedemo.aspx.

28. Ibid.

29. Personal communication with Bill McIntire, 14 November 2010.

30. Angus E-List, 21 February 2008, www.anguselist.com.

31. Personal communication with Bill McIntire, 14 November 2010.

32. Personal communication with Arden Oleen, owner, Oleen Brothers, 4 November 2010.

33. Ibid.

34. Ibid.

35. Personal communication with Greg McCurry, 24 November 2010.

36. Personal communication with Arden Oleen, owner, Oleen Brothers, 4 November 2010.

37. Personal communication with Norm Wilson, 28 November 2010.

38. Personal communication with Howard Blender, 26 November 2010.

39. Ibid.

40. Jack C. Whittier, James E. Ross, and University of Missouri–Extension Division of Animal Sciences, "Freeze Branding Cattle," (February 2010).

41. Personal communication with Bill McIntire, 14 November 2010.

42. Personal communication with Bill, Judy, Jerry, and Diane Paige, 3 October 2008.

43. Ibid.

44. Ibid.

45. Personal communication with Arden Oleen, owner, Oleen Brothers, 4 November 2010.

46. Ibid.

47. Personal communication with Warren Weibert, owner, Decatur County Feed Yard, 24 November 2010.

48. "Fed Cattle Sold 2007," *Beef Magazine* 2007, http://beefmagazine.com/maps/BEEF_FedCattleMap_2007.pdf.

49. U.S. Department of Agriculture National Agricultural Statistics Service, "Kansas Farm Facts," p. 45.

50. Personal communication with Warren Weibert, owner, Decatur County Feed Yard, 24 November 2010.

51. Ibid.

52. Ibid.

53. Ibid.

54. For more on the Animal Health Corridor, see http://www.kcanimalhealth.com/.

55. Personal communication with Warren Weibert, owner, Decatur County Feed Yard, 24 November 2010.

56. Ibid.

57. Personal communication with John Atwater, 27 November 2010.

58. Ibid.

59. Personal communication with Warren Weibert, owner, Decatur County Feed Yard, 24 November 2010.

Chapter Four

1. O.J. Hazlett, "Regulation in the Livestock Trade: The Origins and Operations of the Kansas City Livestock Exchange 1886–1921" (Ph.D. Dissertation, Oklahoma State University, 1987).

2. E.R. Johnson, "American Produce Exchange Markets," *The Annals of the American Academy of Political and Social Sciences* 38 (1911), pp. 319–320.

3. O.J. Hazlett, "Cattle Marketing in the American Southwest," *Kansas History* 18, no. 2 (1995), pp. 100–115.

4. W.H. Thompson, "Livestock Exchanges," in *Proceedings of the National Live Stock Association* (Denver, Colorado: Smith-Brooks Printing Co., 1900), pp. 232–236.

5. Hazlett, "Cattle Marketing in the American Southwest."

6. C.E. Ball, *The Finishing Touch: A History of the Texas Cattle Feeders Association and Cattle Feeding in the Southwest* (Amarillo, Texas: Texas Cattle Feeders Association, 1992).

7. Hazlett, "Cattle Marketing in the American Southwest."

8. D. Galenson, "The Profitability of the Long Drive," *Agricultural History* 51 (1977), pp. 737–758.

9. Hazlett, "Cattle Marketing in the American Southwest."

10. Ibid.

11. Ibid.

12. E. Snyder, "Livestock Markets of Kansas," in *Kansas State Board of Agriculture, Quarterly Report* (Topeka, Kansas: Hamilton Printing Co., March 1893).

13. G.K. Renner, "The Kansas City Meat Packing Industry before 1900," *Missouri Historical Review* 55 (1960), pp. 18–29.

14. C.H. Taylor, *History of the Chicago Board of Trade,* Vol. 1 (Chicago, Illinois: Robert O. Law Co., 1917).

15. Hazlett, "Cattle Marketing in the American Southwest."

16. Renner, "The Kansas City Meat Packing Industry before 1900."

17. Snyder, "Livestock Markets of Kansas."

18. Hazlett, "Cattle Marketing in the American Southwest."

19. ———, "Regulation in the Livestock Trade: The Origins and Operations of the Kansas City Livestock Exchange 1886–1921."

20. ———, "Cattle Marketing in the American Southwest."

21. Ibid.

22. Hazlett, "Regulation in the Livestock Trade: The Origins and Operations of the Kansas City Livestock Exchange 1886–1921."

23. E.L. Atkinson, "Kansas City's Livestock Trade and Packing Industry, 1870–1914" (Ph.D. dissertation, University of Kansas, 1971).

24. Hazlett, "Cattle Marketing in the American Southwest."

25. ———, "Regulation in the Livestock Trade: The Origins and Operations of the Kansas City Livestock Exchange 1886–1921."

26. ———, "Cattle Marketing in the American Southwest."

27. O.J. Hazlett, "Chaos and Conspiracy: The Kansas City Livestock Trade," *Kansas History* 15, no. 2 (1992), pp. 126–144.

28. J.F. Smithcors, *The American Veterinary Profession: Its Background and Development* (Ames, Iowa: Iowa State University Press, 1963).

29. Ibid.

30. Hazlett, "Chaos and Conspiracy: The Kansas City Livestock Trade."

31. ———, "Regulation in the Livestock Trade: The Origins and Operations of the Kansas City Livestock Exchange 1886–1921."

32. Ibid.

33. Hazlett, "Chaos and Conspiracy: The Kansas City Livestock Trade."

34. ———, "Regulation in the Livestock Trade: The Origins and Operations of the Kansas City Livestock Exchange 1886-1921."

35. ———, "Cattle Marketing in the American Southwest."

36. ———, "Regulation in the Livestock Trade: The Origins and Operations of the Kansas City Livestock Exchange 1886–1921."

37. Ibid.

38. Hazlett, "Chaos and Conspiracy: The Kansas City Livestock Trade."

39. ———, "Regulation in the Livestock Trade: The Origins and Operations of the Kansas City Livestock Exchange 1886–1921."

40. Ibid.

41. Hazlett, "Chaos and Conspiracy: The Kansas City Livestock Trade."

42. ———, "Regulation in the Livestock Trade: The Origins and Operations of the Kansas City Livestock Exchange 1886–1921."

43. Ibid.

44. Ibid.

45. Ibid.

46. Ibid.

47. Ibid.

48. E. Snyder, "Livestock Markets of Kansas," in *Kansas State Board of Agriculture, Quarterly Report* (Topeka, Kansas: Hamilton Printing Co., March 1892).

49. J.W. Hurst, *Law and the Conditions of Freedom in the Nineteenth Century United States* (Madison, Wisconsin: University of Wisconsin Press, 1956).

50. J. Lurie, *The Chicago Board of Trade, 1859–1905: The Dynamics of Self Regulation* (Urbana, Illinois: University of Illinois Press, 1979).

51. Hazlett, "Regulation in the Livestock Trade: The Origins and Operations of the Kansas City Livestock Exchange 1886–1921."

52. ———, "Chaos and Conspiracy: The Kansas City Livestock Trade."

53. ———, "Regulation in the Livestock Trade: The Origins and Operations of the Kansas City Livestock Exchange 1886–1921."

54. Ibid.

55. Ibid.

56. Hazlett, "Cattle Marketing in the American Southwest."

57. ———, "Regulation in the Livestock Trade: The Origins and Operations of the Kansas City Livestock Exchange 1886–1921."

58. Ibid.; ———, "Chaos and Conspiracy: The Kansas City Livestock Trade."

59. Hazlett, "Regulation in the Livestock Trade: The Origins and Operations of the Kansas City Livestock Exchange 1886–1921."

60. Ibid.

61. Hazlett, "Cattle Marketing in the American Southwest."

62. ———, "Chaos and Conspiracy: The Kansas City Livestock Trade."

63. Ibid.

Chapter Five

1. Derek Brewer, *Chaucer in His Time* (London: Thomas Nelson and Sons Ltd., 1963), p. 63.

2. Personal communication with Mark Gardiner, 20 November 2010.

3. Ibid.

4. Ibid.

5. Ibid.

6. Personal communication with Tom, Carolyn, Matt, and Amy Perrier, 20 October 2010.

7. Ibid.

8. Ibid. The mission statement is also available on the Dalebanks Angus website: http://dalebanks.com/history.html.

9. Personal communication with Tom, Carolyn, Matt, and Amy Perrier, 20 October 2010.

Chapter Six

1. College circular, as quoted on p. 13 of Miles McKee, *Building the Legacy: A History of the Kansas State University Department of Animal Sciences & Industry* (Manhattan, Kansas).

2. Ibid., p. 14.

3. Previous accounts, cited in Ibid., p. 18.

4. Ibid., p. 21.

5. George C. Wheeler, in the December 18, 1905, edition of *The Industrialist,* as quoted on p. 25 of Ibid.

Chapter Seven

1. For more on the 2011 Sesquicentennial, visit http://ks150.kansas.gov/.

2. For more on the Trade Development Division's activities during 2007, see Kansas Department of Commerce, *2008 Annual Report* (Topeka, Kansas: State of Kansas, 2008), pp. 15–16.

3. The trade mission featured a Kansas-Mexico Food Safety Forum that promoted cross-border cooperation to facilitate international trade.

Bibliography

Arnstein, Walter L. *Britain Yesterday and Today: 1830 to the Present*. Edited by Lacey Baldwin Smith, A History of England. Boston: D.C. Heath and Company, 1966.

Atkinson, E.L. "Kansas City's Livestock Trade and Packing Industry, 1870–1914." Ph.D. dissertation, University of Kansas, 1971.

Ball, C.E. *The Finishing Touch: A History of the Texas Cattle Feeders Association and Cattle Feeding in the Southwest*. Amarillo, Texas: Texas Cattle Feeders Association, 1992.

Barry, Louise. "A British Bride in Manhattan, 1890–1891: The Journal of Mrs. Stuart James Hogg." *Kansas Historical Quarterly* 19 (1951): 269–86.

Blanchard, Leola Howard. *Conquest of Southwest Kansas*. 2nd ed. Wichita: Wichita Eagle Press, 1931.

Brewer, Derek. *Chaucer in His Time*. London: Thomas Nelson and Sons Ltd., 1963.

Burnett, John. *Plenty and Want, a Social History of Diet in England from 1815 to the Present Day*. London: Thomas Nelson and Sons Ltd., 1966.

Campbell, Charles S. *From Revolution to Rapprochement: The United States and Great Britain, 1783–1900*. Edited by Robert A. Divine, America and the World. New York: John Wiley & Sons, Inc., 1974.

"Cattle Diseases and the Meat Supply." *The Times*, 10 February 1890, 4, column b.

Clements, Roger V. "British-Controlled Enterprise in the West between 1870 and 1900, and Some Agrarian Reactions." *Agricultural History* 27 (1953): 132–41.

Cottrell, P.L. *British Overseas Investment in the Nineteenth Century, Studies in Economic and Social History*. London: The MacMillan Press Ltd, 1975.

Critchell, James Troubridge, and Joseph Raymond. *A History of the Frozen Meat Trade*. London: Constable & Company Ltd, 1912.

"Diplomatic Correspondence Paper No. 209: Mr. Hoppin to Mr. Evarts, October 11, 1879." *Papers Relating to the Foreign Relations of the United States* (1879): 466–67.

Encyclopedia.com. "Kansas-Nebraska Act (1854)." In *Gale Encyclopedia of U.S. Economic History*, 1999.

English Enterprise in America: Notes Addressed to Investors and Settlers Concerning the Estate of Victoria (Ellis County, Kansas, U.S.), the Property of George Grant (Late of the Firm Grant and Gask, Now Gask and Gask, Oxford Street, London). London: T. Pettitt & Co., 1876.

"Family of Dr. O. M. and Anne Ohisen Franklin." *Eloise Lane Articles* 161–180, no. 174 (n.d.).

"Fed Cattle Sold 2007." *Beef Magazine*, 2007, http://beefmagazine.com/maps/BEEF_FedCattleMap_2007.pdf.

Foreman-Peck, James. *A History of the World Economy: International Economic Relations since 1850*. Totawa, New Jersey: Barnes & Noble Books, 1983.

Forsythe, James L. "George Grant of Victoria: Man and Myth." *Kansas History* 9, no. 3 (1986): 102–14.

Galenson, D. "The Profitability of the Long Drive." *Agricultural History* 51 (1977): 737–58.

Garrett, Julia Kathryn. *Fort Worth: A Frontier Triumph*. Austin: The Encino Press, 1972.

Harvie, Christopher. "Revolution and the Rule of Law (1789–1851)." In *The Oxford History of Britain*, edited by Kenneth O. Morgan. Oxford and New York: Oxford University Press, 1988.

Hazlett, O.J. "Cattle Marketing in the American Southwest." *Kansas History* 18, no. 2 (1995): 100–15.

———. "Chaos and Conspiracy: The Kansas City Livestock Trade." *Kansas History* 15, no. 2 (1992): 126–44.

———. "Regulation in the Livestock Trade: The Origins and Operations of the Kansas City Livestock Exchange 1886–1921." Ph.D. dissertation, Oklahoma State University, 1987.

Heaton, Herbert. *Economic History of Europe*. Edited by Guy Stanton Ford. Revised Edition ed, Harper's Historical Series. New York: Harper & Row, 1948.

Hoy, Jim. *Flint Hills Cowboys: Tales from the Tallgrass Prairie*. Lawrence, Kansas: University Press of Kansas, 2006.

Hurst, J.W. *Law and the Conditions of Freedom in the Nineteenth Century United States*. Madison, Wisconsin: University of Wisconsin Press, 1956.

Hurt, Douglas. *American Agriculture: A Brief History*. Ames, Iowa: Iowa State University Press, 1994.

Huttman, John P. "British Meat Imports in the Free Trade Era." *Agricultural History* 52, no. 2 (1978): 247–62.

J.H.D. "Texas Live Stock: An Englishman's Opinion of Its Value for Exportation." *New York Times*, 14 August 1878, 3, column 3.

Johnson, E.R. "American Produce Exchange Markets." *The Annals of the American Academy of Political and Social Sciences* 38 (1911): 319–20.

Kansas Department of Commerce. *2008 Annual Report*. Topeka, Kansas: State of Kansas, 2008.

Liverpool Provision Trade Association: Minute Book Volume 1. Liverpool: Merseyside Maritime Museum Archives, 1874–1889.

Lurie, J. *The Chicago Board of Trade, 1859–1905: The Dynamics of Self Regulation*. Urbana, Illinois: University of Illinois Press, 1979.

MacDonald, James. *Food from the Far West: American Agriculture with Special Reference to the Beef Production and Importation of Dead Meat from America to Great Britain*. London: William P. Nimmo, 1878.

Marchildon, Gregory P. "Canadian-American Agricultural Trade Relations: A Brief History." *American Review of Canadian Studies* 28, no. 3 (1998): 233–52.

McKee, Miles. *Building a Legacy: A History of the Kansas State University Department of Animal Sciences & Industry*. Manhattan, Kansas.

Mothershead, Harmon Ross. *The Swan Land and Cattle Company, Ltd*. Norman: University of Oklahoma Press, 1971.

"An Old Dream in Trouble." *The Economist*, 2 June 2001, 29–30.

Pearce, W.M. *The Matador Land and Cattle Company*. Norman: University of Oklahoma Press, 1964.

Perren, Richard. *The Meat Trade in Britain 1840–1914*. Edited by F.M.L. Thompson, Studies in Economic History. London: Routledge & Kegan Paul Ltd., 1978.

Powell, Fred Wilbur. *The Bureau of Animal Industry: Its History, Activities and Organization*. Edited by Institute for Government Research, Service Monographs of the United States Government. Baltimore: The Johns Hopkins Press, 1927.

"Proceedings of the Society: Food Committee." *Journal of the Society of Arts* XV, no. 756 (1867): 414–17.

Raun, A.P., and R.L. Preston. "History of Diethylstilbestrol Use in Cattle (American Society of Animal Science)." (2002).

Read, Gordon, and Michael Stammers. *Guide to the Records of Merseyside Maritime Museum*. St. John's, Newfoundland: Trustees of the National Museums and Galleries on Merseyside & International Maritime Economic History Association, 1995.

"Recognition of Texas." *The Times*, 23 December 1840, 3, column d.

Renner, G.K. "The Kansas City Meat Packing Industry before 1900." *Missouri Historical Review* 55 (1960): 18–29.

Rostow, W.W. *British Economy of the Nineteenth Century*. Oxford: Oxford University Press, 1948. Reprint, 1952.

Skaggs, Jimmy M. *Prime Cut: Livestock Raising and Meatpacking in the United States, 1607–1983*. College Station: Texas A&M University Press, 1986.

———. *The Cattle Trailing Industry: Between Supply and Demand, 1866–1890*. Lawrence, Kansas: The University of Kansas Press, 1973.

Smithcors, J.F. *The American Veterinary Profession: Its Background and Development*. Ames, Iowa: Iowa State University Press, 1963.

Snyder, E. "Livestock Markets of Kansas." In *Kansas State Board of Agriculture, Quarterly Report*. Topeka, Kansas: Hamilton Printing Co., March 1892.

———. "Livestock Markets of Kansas." In *Kansas State Board of Agriculture, Quarterly Report*. Topeka, Kansas: Hamilton Printing Co., March 1893.

Taylor, C.H. *History of the Chicago Board of Trade*. Vol. 1. Chicago, Illinois: Robert O. Law Co., 1917.

Thompson, R. "The Liverpool Provision Trade Association Ltd.: Commercial History." In *Provision Trade Association Archives*. Liverpool: Mersey Docks and Harbour Board | Merseyside Maritime Museum Archives, 1980.

Thompson, W.H. "Livestock Exchanges." In *Proceedings of the National Live Stock Association*, 232–36. Denver, Colorado: Smith-Brooks Printing Co., 1900.

Tracy, Michael. *Government and Agriculture in Western Europe 1880–1988*. 3rd ed. New York: New York University Press, 1989.

Trebilcock, M., and R. Howse. *The Regulation of International Trade*. London: Routledge, 1999.

U.S. Bureau of Statistics. *Annual Report on the Foreign Commerce of the United States*. Washington: Government Printing Office, 1879.

U.S. Department of Agriculture National Agricultural Statistics Service. "Kansas Farm Facts." (2010).

USDA APHIS Infosheet. "Treatment of Respiratory Disease in U.S. Feedlots." (October 2001).

Whittier, Jack C., James E. Ross, and University of Missouri-Extension Division of Animal Sciences. "Freeze Branding Cattle." (February 2010).

Zimmerman, W.D. "Live Cattle Export Trade between United States and Great Britain, 1868–1885." *Agricultural History* 36 (1962): 46–52.

About the Authors and Photographers

Justin Kastner

Justin Kastner, Ph.D., is an assistant professor in the College of Veterinary Medicine at Kansas State University. A Fulbright and Truman Scholar who graduated from Kansas State University, Dr. Kastner holds graduate degrees from the University of Guelph, London South Bank University, and the University of Edinburgh. At K-State, he provides educational leadership as co-director of the Frontier program for the historical studies of border security, food security, and trade policy (http://frontier.k-state.edu). Justin is married to Susie (Viterise) Kastner, a native of Garden City, Kansas, and they have two children, Ian and Sally. Having thoroughly enjoyed the collegiality of a skilled project team, it has been his privilege to serve as director for *150 Years of Kansas Beef.*

Blair Tenhouse

Blair (Bryant) Tenhouse is both a research assistant and the programs and curricula coordinator for the Frontier program in the College of Veterinary Medicine at Kansas State University. In 2008, she graduated with honors from Kansas State University, where she earned a B.S. in food science and industry with a minor in agribusiness. While growing up in south-central Kansas, Blair raised and exhibited registered Chianina and Simmental cattle on the local, state, and national levels. Blair and her husband, Doug, a native of Liberty, Illinois, reside in the Manhattan area. With her passion for agriculture and the beef industry, Blair could not have experienced more fulfilling work than she has as the principal program coordinator and research assistant for the *150 Years of Kansas Beef* project.

Barbara F. Bragg

Barbara F. Bragg is a great-granddaughter of Thomas A. Bragg and the daughter of Joseph Bragg. Barbara holds a degree in elementary education, a master in music, and a B.S. in music with a second major in Italian studies from Indiana University. After she became a master's candidate at Indiana University, she continued her studies of the Italian language and performance in voice in Perugia, Italy. She is now a retired teacher and resides in Topeka, Kansas.

Lewis Browder

Lewis Browder is a confirmed Kansan of fifty-three years by choice. He is a K-State graduate and a retired plant pathologist who enjoys family, family history, Kansas history, gardening, prairie grasslands, and Kansas cattle. For more of his photographs of Kansas subjects, please visit http://www.flickr.com/zadalew.

H. Hurst Coffman

H. Hurst Coffman is a great-grandson of Thomas A. Bragg and the son of Floyd H. Coffman. Hurst has a B.A. from Ottawa University and a J.D. from the Washington College of Law at the American University in Washington, D.C. Since 1976, he has practiced law in Topeka. Hurst and his sister, Gerry Coffman, raise beef cattle at their farm in Douglas County, and they still retain ownership of a small parcel from the Bragg Ranch in Ford County.

Curtis Kastner

Dr. Curtis L. Kastner is director of the Food Science Institute at Kansas State University. He has served as a member of the U.S. Department of Agriculture (USDA) National Advisory Committee on Meat and Poultry Inspection. He is a founding co-leader of the longstanding USDA-funded Food Safety Consortium, which is tasked with developing hazard detection and elimination techniques to enhance meat safety. Dr. Kastner also is the education theme leader for the National Center for Food Protection and Defense, a Center of Excellence for the U.S. Department of Homeland Security.

Holli Kepner Spielbusch

Holli Kepner Spielbusch was born in St. Joseph, Missouri, and graduated from Missouri Western State University with a B.A. in English and journalism in 2003. She currently lives in Kansas City with her husband, John, and works at FOX4-TV in sales and marketing. In her spare time, she serves as a freelance photographer. Some of her recent freelance work is on display at the Kansas City International Airport.

Miles McKee

A native of Cottonwood Falls, Kansas, Miles McKee received degrees from Kansas State University in agriculture (B.S., 1951) and animal husbandry (M.S., 1963). He was awarded a Ph.D. (1968) in animal sciences from the University of Kentucky. In addition to working for his family's livestock operation in Chase County, he worked for Titus and Stout Hereford Ranch, Moxley Hall Hereford Ranch, Luckhardt Farms, and L&J Crusoe Ranch. In 1959, he was appointed to the faculty of K-State's Department of Animal Husbandry, which is now the Department of Animal Sciences and Industry. Dr. McKee retired in 2005 but continues to help with student advising and with chronicling the department's history. Miles and his wife, Marjorie, reside in Manhattan. They have four children and nine grandchildren.

Rhonda McCurry

Rhonda Nida-McCurry has written about agriculture and the beef industry for ten years. A K-State graduate of agricultural journalism and an Angus cattle producer, she has an appreciation for Kansas ranchers and their efforts to be productive and profitable in the country's greatest profession.

Ashley Null

The great-grandson of a Kansas homesteader and reared in Salina, the Reverend John Ashley Null, Ph.D., is the canon theologian of the Episcopal Diocese of Western Kansas. The recipient of Fulbright, National Endowment for the Humanities, and Guggenheim fellowships for his work on the English reformation, Dr. Null has also been elected as a fellow of two British learned societies—the Society of Antiquaries and the Royal Historical Society. He is currently editing the theological research notes of Archbishop Thomas Cranmer—the author of the first Anglican prayer books—for Oxford University Press. As part of this five-volume project, he is a visiting fellow of both the theological faculty of Humboldt University of Berlin and the faculty of divinity at the University of Cambridge. He warmly welcomed the opportunity to contribute to the *150 Years of Kansas Beef* project so he could share with a wider audience the rich Kansas heritage that shaped and continues to inspire him in his work today.

K.C. Olson

Dr. K.C. Olson is a faculty member in the Department of Animal Sciences and Industry at Kansas State University. He takes great pride in helping to train the next generation of Kansas ranchers. K.C., his wife Karli, and their sons Charles and Theodore live on a beautiful Flint Hills ranch in northwestern Lyon County, where they custom-graze stocker cattle.

Corinne Patterson

Corinne Patterson grew up on her family's fourth-generation stocker ranch in the heart of the Flint Hills. After earning bachelor degrees in agricultural journalism and animal sciences and industry, she became an associate editor for the *Angus Journal* and the *Angus Beef Bulletin.* When Corinne and her husband, Tom, had the opportunity to return to her parents' ranch in Chase County, she began freelance writing and is a contributing editor to *Working Ranch* magazine. Corinne educates the next generation as a 4-H youth development extension agent for K-State Research and Extension.

Susan S. Russell

Susan Schlickau Russell was active in her family's registered Hereford ranch and breed activities during her youth. She is a K-State journalism graduate, who then did master's work in journalism ethics at the University of Nebraska at Lincoln. She was a newspaper and magazine editor for publications in Iowa, Nebraska, and Colorado. She, her husband, and their sons run a registered cow-calf and yearling operation and a feed store in Colorado. Susan currently serves on regional and state cattle boards, aids with the National Western Stock Show, and is a western trustee on the American Simmental Association board.

John L. Schlageck

John Schlageck is a leading commentator on agriculture and rural Kansas. Born and raised on a diversified farm in northwestern Kansas, his writing and photography reflect a lifetime of experience, knowledge, and passion. John pens a weekly column, "Insight," that is distributed across Kansas to radio, television, and newspaper outlets. He also blogs on behalf of the farmers and ranchers of Kansas Farm Bureau. He pioneered *Kansas Living* magazine and continues to write and photograph for this publication while serving on the publication's creative team.

Lois R. Schlickau

Lois Schlickau is an agricultural advocate, an active partner in Schlickau Herefords, and a volunteer on numerous local, county, state, and national civic, religious, agricultural, and livestock boards. She was the first woman president of the Kansas State Fair Board and Kansas State Board of Agriculture. Lois and her husband, George, have four children. Her hobbies include reading, baking, and spending time with their nine grandchildren.

Chris Stephens

Chris Stephens grew up in the registered Hereford cattle business and has a vast working knowledge of the complex organizational and management needs associated with running multiple ranching operations. Chris is currently director of operations for EE Ranches, Inc., in Dallas, Texas. In addition to the cutting horse stallion station in Whitesboro, Texas, EE Ranches, Inc., has a registered Hereford and Angus cattle operations in Kansas and Mississippi, a cutting horse training division in Pilot Point, Texas, and an alfalfa hay operation in Wheatland, Wyoming. Chris graduated from Oklahoma State University in 2002 and has served as director of both the Hereford Youth Foundation and the American Hereford Association's Youth Activities program. Chris served as the College of Veterinary Medicine's development officer for the Kansas State University Foundation until April 2010.

FREE!

FINE AMERICAN JEAN! *Beautiful Ornamental Stitching,* Perfect Fit.

Give your Corset measure

Usual Price, $1.25.

Five hundred thousand to be given away in six months!!

The Mme. Demorest Illustrated Monthly Fashion Journal

Contains **36 pages** on the finest paper, and is the most BEAUTIFULLY ILLUSTRATED PUBLICATION in the world. It covers every possible field of Fashions, Fancy Work, House Decoration, Cooking, etc., etc. Subscription price only **50 cents per Year.** Mention this paper, and send **50 cents** for subscription and **25 cents** additional to pay postage and packing, **75 cents** in all, and we will mail you one of these handsome Corsets free. Address

DEMOREST FASHION AND SEWING MACHINE CO.,
17 East 14th Street, New York City.

What is CASTORIA

Castoria is Dr. Samuel Pitcher's prescription for Infants and Children. It contains neither Opium, Morphine, nor other Narcotic substance. It is a harmless substitute for Paregoric, Drops, Infant Syrups, and Castor Oil. It is Pleasant. Its guarantee is thirty years' use by Millions of Mothers. Castoria kills Worms. Castoria is the Children's Panacea—the Mother's Friend.

Castoria.

Castoria cures Colic, Constipation, Sour Stomach, Diarrhœa, Eructation, Gives healthy sleep and promotes digestion, Without injurious medication.

Castoria.

"Castoria is so well adapted to children that I recommend it as superior to any prescription known to me." H. A. ARCHER, M. D., 111 So. Oxford St., Brooklyn, N. Y.

The Centaur Company, 77 Murray Street, N. Y.

People Differ.

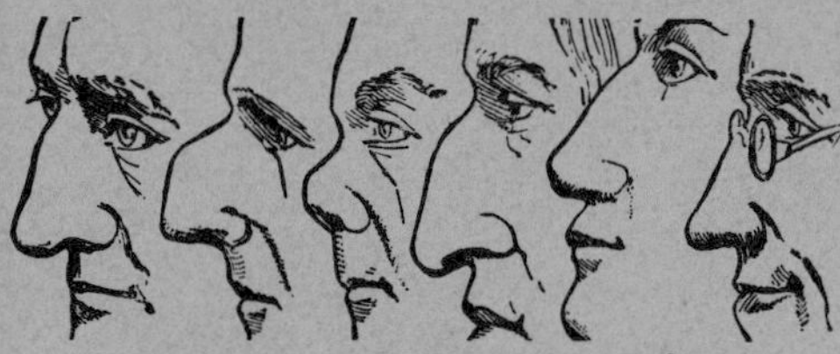

IF YOU WANT SOMETHING

that will interest you more than anything you have ever read and enable you to understand all the differences at a glance, by the SIGNS OF CHARACTER, send for a copy of

HEADS and FACES;

HOW TO STUDY THEM.

A new manual of Character Reading for the people. A knowledge of Human Nature would save many disappointments in social and business life.

This is the most comprehensive and popular work ever published for the price, 75,000 copies having been sold, containing 200 large octavo pages and 250 portraits. We will send it carefully, by mail, postpaid, on receipt of price, 40 cents, in paper, or $1, in extra cloth binding. If not satisfied, money will be returned. Address,

FOWLER & WELLS CO., Publishers,

785 BROADWAY, NEW YORK.

N. B.—Sent Free, a sample copy of the PHRENOLOGICAL JOURNAL, an illustrated Magazine of Human Nature, monthly, $1.50 a year, 15c. a number, with list of Books on Phrenology, Physiognomy and all phases of Human Nature, to all who mention THE CHAUTAUQUAN.

Estab'd | HENRY HART, Jeweler. | 1880.

Tenth year of advertising in THE CHAUTAUQUAN

This "No. O" size Elgin Ladies' Watch, eleven jewel movement, stem wind and set, filled cases (twenty year guarantee) $18.75. Solid 14 K Case, $30; Diamond Set 14 K Case, $45.

No. "6" (ordinary) size same prices with seven jewel movement. Ladies SILVER STEM-WIND AND SET SWISS, $6.50. Same in nickel, $4.50.

GENTLEMEN'S WATCHES.

KK filled Cases (Elgin), $21; 1214 (Elgin), $18.75.

4 oz. COIN SILVER Hunting-Case ELGIN WATCH, $13.75. Same in nickel case, $7.25.

A HANDSOME and DURABLE CHAIN given with each watch and sent express or post-paid if CASH ACCOMPANIES ORDER; or we send C. O. D. with privilege of examination on receipt of $1 to guarantee charges. INSCRIPTION or MONOGRAM engraved FREE. Enclose stamp for price-list of solid gold and silver watches and testimonials of our patrons.

I WANT YOUR OLD WATCH. Either Solid Gold or Silver, Filled Gold or Nickel, in part payment for new, or will pay *cash* for *solid only*. If "only a case" send for an estimate (it is IMPOSSIBLE TO ESTIMATE from a WRITTEN DESCRIPTION) when you can decide whether to sell or exchange.

OLD GOLD or silver jewelry, etc., also wanted for cash or new goods. *Send all to us for estimate* or check by return mail as desired.

College, School, Society, and Club Badges, Rings, and Medals.

(Name school, etc., desiring, state number wanted, and price intended, and we will send special designs.) I can furnish as fine designs and workmanship at as reasonable prices as any house in the U. S. New Illust. Catalogues of Medals and Badges, 4 cents

Solid gold '90 dates ready for attaching to class badges, 75c. Stamps taken. Also dates '87, '88, '89, '91, '92, same price. If registry of anything in this advertisement desired, enclose fee Remit New York Draft, Money Order, or Registered Letter. Address,

'91

HENRY HART, Box 183 Rochester, N. Y.

How to Cure Skin & Scalp DISEASES with the CUTICURA REMEDIES.

THE MOST DISTRESSING FORMS OF SKIN AND scalp diseases, with loss of hair, from infancy to old age, are speedily, economically, and permanently cured by the CUTICURA REMEDIES, when all other remedies and methods fail.

CUTICURA, the great Skin Cure, and CUTICURA SOAP, an exquisite Skin Beautifier, prepared from it, externally, and CUTICURA RESOLVENT, the new Blood Purifier, internally, cure every form of skin and blood diseases, from pimples to scrofula.

Sold everywhere. Price, CUTICURA, 50c.; SOAP, 25c.; RESOLVENT, $1. Prepared by the POTTER DRUG AND CHEMICAL CORPORATION, BOSTON, MASS.

Send for "How to Cure Skin Diseases."

Pimples, blackheads, chapped and oily skin prevented by CUTICURA SOAP.

Relief in one minute, for all pains and weaknesses, in CUTICURA ANTI-PAIN PLASTER, the only pain-killing plaster. 25c.

DIXON'S AMERICAN GRAPHITE PENCILS are unequaled for smooth and tough leads. If your stationer does not keep them, mention THE CHAUTAUQUAN and send 16 cents for samples worth double the money

JOSEPH DIXON CRUCIBLE CO.,
JERSEY CITY, N. J.

COWDREY'S SOUPS

Delicious, Appetizing, Nourishing.

Tomato, Mock Turtle, Ox Tail, Consomme, Julienne, Chicken, Vegetable, Mutton, Printanier, Green Turtle, Terrapin, Macaroni, Beef, Pea, Okra, Vermicelli, Soup & Bouilli, Clam Broth, Mulligatawny, Purée of Game.

Cowdrey's TOMATO SOUP MANUFACTURED BY THE E.T. COWDREY CO. BOSTON, U.S.A.

Sample mailed on receipt of 12c. to pay postage.

E. T. COWDREY CO., Boston, U. S. A.

Ideal Felt Tooth Polisher.

ENDORSED BY LEADING DENTISTS.

NON-IRRITATING TO GUMS OR ENAMEL

FOR SALE BY ALL DRUGGISTS

BATH CABINET. Affording a refreshing Turkish Bath at home.

ROLLING CHAIR. A Priceless Boon to those who are unable to walk.

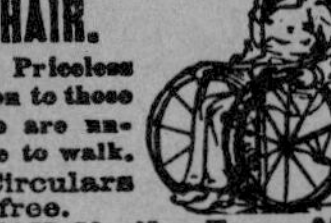

Descriptive Circulars of both mailed free.

NEW HAVEN CHAIR CO., New Haven, Ct.

RECOLLECTIONS OF THE LYCEUM & CHAUTAUQUA CIRCUITS

Recollections of the Lyceum & Chautauqua Circuits

By Irene Briggs & Raymond F. DaBoll

Part I—*As Told in a Person-to-Person Interview*
Part II—*Notes Historical, Personal & Musical*
Part III—*On Calligraphy & Scribal Writing With a Few Musical Overtones: Designed & Written Out in the Scribal Hand of R.F.D*

The Bond Wheelwright Company *Publishers* · Porter's Landing, Freeport, Maine

ACKNOWLEDGMENTS

The various items, including pictures, letters, and excerpts from letters, reproduced in this 3-part book have come from so many different sources that it would be impractical to list them all on this page. We have identified the source of most of them on the pages where they appear. We wish to express here our thanks to all the donors for their gracious permission to reproduce their excellent material. In addition, we are grateful to:—

MISS DOROTHY ABBE, Hingham, Massachusetts, for permission to reproduce W.A. Dwiggins's calligraphic rendering of the quotation from Hokusai's "Preface".

BROWN & BIGELOW, St. Paul, Minnesota, for permission to use the "Disciplined Freedom" broadside.

COLLINS, MILLER & HUTCHINGS, INC., Chicago, Illinois, for the 4-color negatives they supplied for the landscape by Robert Briggs, and our "40".

MISS MARGARET COPELAND, Librarian, Smith Memorial Library, and the New York Chautauqua, for her assistance and their permission to reproduce pictures from A READING JOURNEY THROUGH CHAUTAUQUA, as well as from photographs from the Library files and from issues of the CHAUTAUQUAN.

CHARLES M. SCHULZ, for permission to reproduce the PEANUTS cartoon strip, copyright 1961, United Feature Syndicate.

MRS CHARLOTTE DOBSON, Secretary, The Society for Italic Handwriting, QUARTERLY JOURNAL London, England, for permission to reproduce the calligraphy of Beryl Lindop.

SPENCER MUSIC CORPORATION for permission to reproduce the refrain of "Good Night, Irene", copyright 1950.

YALE UNIVERSITY PRESS, New Haven, Conn., for permission to reprint the facsimile of a page and its translation from THE FIRST WRITING BOOK—ARRIGHI'S OPERINA, copyright 1954, by John Howard Benson.

WE are particularly grateful—and fortunate—to have had the editorial help of Thea Wheelwright, especially with regard to Part II, for her indefatigable efforts in the researching of illustrations and digging up facts about Lyceum & Chautauqua. This part of the book is largely her own writing.

IF THERE ARE any others whom we have failed to identify, we trust that the oversight will be forgiven.

Irene Briggs Da Boll
Raymond F. Da Boll

The Bond Wheelwright Company, Freeport, Maine 04032
Standard Book Number 87027-107-5 Library of Congress Card Number 66-19773
Printed in the United States of America

In Canada, Abelard-Schuman, Canada, Ltd.

To Revive Memories
of the old Lyceum & Chautauqua days
and to show that the Art of Scribal
Writing still Flourishes
this Book is Written

CONTENTS

LIST OF ILLUSTRATIONS

(calligraphic & otherwise)

PART III:———

PART I
Dialogue

IRENE GERTRUDE BRIGGS (AGE 13) SOPRANO. (COMPASS 3 OCTAVES.)

FOREWORD

WITH THE ENCOURAGEMENT of favorable press reviews, it was decided to publish the following "person-to-person" interview, first spoken at a meeting of The Musical Arts Club of Batesville, Ark., April 7, 1959. The speakers realize that Readers-at-large, be they music buffs or Calligraphiles, may not be as interested as one's own li'l ole music club whose program chairman, in search of data on Lyceum & Chautauqua, had persuaded an "old trouper" to talk about those bygone forms of educational entertainment. Dialogue was chosen as the most concise way to present the subject. It went over well and has been given before other musical groups with equal success. Now we trust it will prove to be of value to others Researching those earlier forms of show business.

Lyceum is from the Greek ΛΥΚΕΙΟΝ — a lecture hall.

BE IT KNOWN — Your Querying scribe knew not the sweet singer until after her services for Redpath were over — therefore — the Questions as well as the answers came from her own experiences "on the Road" and were propounded mainly by her to keep the story moving. It may help the Reader to know that when this little mark ✻ precedes a Question, it indicates a quick change of time &/or place, or that the Questioner was cued with only a word or two. A few words of historical substance are also in order here — as follows: —

There may be a few still in their 50's who will recollect that lyceum & Chautauqua were associated for many years in the promotion of better education for Americans IF ONLY to point out that the present emphasis on that subject is NOT unique. Launched in 1826 by Josiah Holbrook at Millbury, Mass., Lyceum became a wide-spread movement carried on with lecture courses in Science, Religion, literature & philosophy. Suffering a 90% decline during the Civil War because of dissenting social & political opinion, it came back strong with such eminent lecturers as Ralph Waldo Emerson, Henry Ward Beecher and Mark Twain.

The name Chautauqua is derived from the Indian JADAQUA — pertaining to water. The original CHAUTAUQUA INSTITUTION is located at Chautauqua Lake in the S.W. corner of New York State. Its purpose is purely educational & not for profit.

WHEN THE DEMAND for good speakers could not be met, musicians, singers & dramatic Readers were brought in to fill out the programs. The support of the Chautauqua Societies, initiated by Bishop John Vincent & Lewis Miller at Fair Point, N.Y., in 1874 intensified the life of the Lyceum movement most remarkably. Then University Extension work with its many Ramifications sprang from the Chautauquas. By the turn of the century, over 100 bureaus for the placement of talent had come into the field, supplying thousands of entertainers of all kinds—from Phrenologists and Grand Opera singers to Chalk-talkers and Swiss Bell Ringers.

PHINEAS T. BARNUM was one of the later Lecturers. Perhaps his earlier highly successful importation in 1850 of Jenny Lind* "The Swedish Nightingale" and greatest attraction ever offered by that famous entrepreneur, may have suggested the power of music to inspire people and to MOVE them—into the auditoriums, pavilions and tents of the Lyceum and Chautauqua Circuits that covered the nation. According to The Encyclopedia Americana such performances played an important part in the cultural development & life of our people in those now almost forgotten times.

*At $1,000 a performance—a fee hitherto unheard of for a singer

BY 1911 the Cinema & phonograph had progressed from the Nickelodeon & player-piano stage to a point indicating that a new era with a different culture was taking over and with Hollywood setting the pace—Radio, Television, Video, in quick succession have stepped it up—and now 'tis said we have a flood of entertainment with an ADmixture of commercials and a SUBnormal culture dominated by Westerns, Violence & Sex.

THAT MAY BE too harsh—considering that we do enjoy the privilege of listening to those splendid broadcasts of Metropolitan Grand Opera sponsored by Texaco, and of witnessing as well as hearing such fine educational entertainment as that provided by artists & budding artists on The Bell Telephone Hour. Another example of generous and conscientious sponsorship is the regular broadcasting of NET—National Educational Television—with no commercials at all.

HAPPILY, TOO, TV as an aid to class-room study, and relief for over-worked teachers, is increasing. The USA is leading in this field, with Japan a close second. People hungry for Peace (and who isn't?) can have it by listening to Pablo Casals—and HEEDING the all-embracing message of his great Oratorio—El Pessebre—a Spiritual feast for the nourishment of World Peace. We need more programs of that kind—not the MUZAK mush piped into restaurants, elevators & airplanes.

This is Important

WE WOULD NOT go back to horse & buggy days—good only in retrospect—but we do long for more bel canto & less hullabaloo in a peaceful world where everything (including US) can end properly resolved on a good healthy TONIC chord.

MEANWHILE, we marvel, mystified by messages bounced from satellites orbited in Outer Space! And above all, we shall remain eternally grateful to the late EDWARD R. MURROW for his memorable "person-to-person" interviews that so clearly pointed the way to the presentation of our own in 1959. It was given by pure coincidence on the eve of our 42nd Wedding Anniversary at the afore mentioned meeting of our Musical Arts Club—a chapter of the National Federation of Music Clubs, which has an adult & junior membership of over 520,000 with educational aims closer to those of the old Lyceums and Chautauquas than any other musical organization in America

His way on TV was inimitable—but, at least, we tried.

AS TO THE UNUSUAL Calligraphic Style of this manuscript, here again we look backward searchingly in order to go forward intelligently, guided by the master hand of the 16th century scribe—ARRIGHI and others of his day when scribal writing attained its vertex and movable type was invented—only 5 centuries ago.

TYPOGRAPHY—the craft of setting type—has been aptly called "The art preservative of the arts", also "The art that was born perfect". Why? (if you ask) Because its letter forms and principles were evolved & perfected by scribes during a period of more than two thousand years — and are a priceless heritage.

WE TRUST this message of Music & Calligraphy will be construed as the marriage of two timeless & noble arts, and that those who read it will be with us in spirit.

Irene + Ray DaBoll

pendulum swings with measured pace—as words and music soar through space—some bearing marks of ancient date—all potent with power to educate—

Irene in the Redpath Era

PERSON TO PERSON

Good evening, Mrs. DaBoll.	Good evening, Raymond. It's so long since I've seen you.
I wish you would call me Ray; may I call you Irene ?	Please do.
It is indeed kind of you to grant this interview in such pleasant surroundings. We do greatly appreciate it.	Isn't it lovely—and wasn't that a delicious dinner? . . . Shall we dance?
Thank you, Irene, I'm not a dancer. Let's discuss your career as a singer. When did you start singing on the Redpath Circuits ?	I must have been about twenty.
I mean what year ?	If I tell you, folks will start counting on their fingers.
In that case- let's skip it.	OKay.
Was this high class entertainment that you were in ?	Oh yes indeed.
**PHEW is the cue word here. . . .*	I was a member of THE JESS PUGH CONCERT COMPANY.
Oh! I see! It's spelled P-U-G-H	That's right.
How many were in your group ?	Only three, but each was required to do at least two things. The personnel was kept small because it was necessary to hold expenses down.
What did Mr. Pugh do ?	He was a raconteur, sang baritone songs and acted as manager of the company.
And the other one ?	Was a violinist. In fact, there were four violin replace-ments—all women—during the time I was with Redpath. It seems as though they were all just marking time until they could get married.
And you ?	I sang love songs and let it go at that, confident that SOME DAY MY PRINCE WOULD COME.

Maybe it's just as well you waited. Yes — I'm sure it was.

What did the Violinists do — aside from fiddling and writing Love Letters? They had to play a good piano accompaniment for my songs. I, in turn, accompanied their selections and Mr. Pugh's songs and some of his readings, for instance, the beautiful poem ANNABEL LEE by Edgar Allan Poe. Now-a-days we seldom hear words spoken to music. I loved it and think it should be done more often.

So do I. What were some of the unaccompanied readings? Such things as Kipling's THE LIGHT THAT FAILED, THE GADSBYS and, of course, GUNGA DIN. Pugh was from Indiana and naturally devoted to the Hoosier poet, James Whitcomb Riley. I recall with what intensity of feeling he rendered LET'S GO BACK TO GRIGGSBY'S STATION WHERE WE USTER BE SO HAPPY AND SO PORE.

Jess Pugh could draw tears from a turnip.

What songs did he sing? They were mostly on the light side — Scottish dialect songs that Harry Lauder had made popular — like ROAMIN' IN TH' GLOAMIN' — IT'S A BRAUGH, BRICHT MOONLICHT NICHT TONICHT — I'VE LOVED HER EVER SINCE SHE WAS A BABY — and THAT'S TH' REASON NOO I WEAR A KILT (*singing*)

Fer th' wife, she used to ramble through
M' britches when I was fast asleep
Beneath th' quilt.
In th' marnin' when I woke,
I was always stony broke,
An' that's th' reason noo I wear a kilt.

Nothing heavier than that? Oh yes — there was always INVICTUS which he seldom failed to sing. At first I thought it was magnificent but after hearing it and playing the accompaniment for two solid years I couldn't bear the sound of it.

I don't blame you. Were there any concerted numbers? We did several duets — LA CI DAREM LÀ MANO from DON GIOVANNI — ON WINGS OF MUSIC by Mendelssohn with violin obbligato was a great favorite. Or something funny, like TH'-TH'-TH'-S-S-S-TUT-TUT-TUT-ERING L-L-L-LOVERS. Variety was important.

Were your programs varied? We had a standard program but our repertoire was large enough so we could change it at will with an announcement by Mr. Pugh. We were quite sensitive to audience reaction and anxious to please so the home-office would get a good report on our performance.

What were some of the Violin selections?

Mostly standard classics such as MEDITATION FROM THAÏS, INTERMEZZO, AIRS RUSSES, parts of THE MENDELSSOHN CONCERTO, CAPRICE VIENNOIS. I remember ZEGOINERVOISSEN particularly because I found it tricky to accompany.

In those days, of course, no program was complete without HUMORESQUE; and SOUVENIR was a great favorite, too, popularized by one of Maude Powell's recordings for Victor.

She was indeed an Excellent Artist, as no doubt many of our audience will remember. Was she on your Circuit?

Yes—sometimes only one jump ahead of us—making it a bit tough for our fiddler to follow.

Do you recall any other well-known personalities?

On Chautauqua we often shared programs with such lecturers as STRICKLAND GILLILAN, OPIE READ and WM RAINEY BENNETT. I also remember the Spanish-AMERICAN WAR hero, RICHMOND P. HOBSON.

That must have been right after he almost "bottled up" the Spanish fleet in Santiago Harbor!

No—and don't you try to date me like that!

I'm sorry!

This was years after that naval exploit; meanwhile he had become a rabid prohibitionist—lecturing on THE GREAT DESTROYER—ALCOHOL.

What songs did you sing?

Your question reminds me of the time we were filling a Chautauqua date in Wassau, Wisconsin, and sharing the program with a lecturer. Just before I went on stage, he asked me what I was going to sing. I told him: SONGS MY MOTHER TAUGHT ME. They are always nice, he agreed, not knowing that was just the name of the song by DVOŘÁK. I didn't tell him any different—thought I'd let him find out for himself.

MOTHER

In that connection, my mother was a great help

to me musically. She had a keen appreciation of music and knew dozens of German and American folk songs that she sang from memory to her own accompaniment on the guitar, and was never happier than when teaching me the German lieder.

Let's get back to that lyceum program. What did you sing?

Such arias as UNA VOCE POCO FA from THE BARBER OF SEVILLE—VOI CHE SAPETE from THE MARRIAGE OF FIGARO—VOI LO SAPETE, O MAMMA from CAVALLERIA RUSTICANA. I am sure that of all the arias I sang, I put the most genuine fervor into the nostalgic KENST DU DAS LAND? sung by the kidnapped and homesick girl-MIGNON: KNOWEST THOU THE LAND? (*singing*)

> *'Tis there that my heart desires to be,*
> *To live, to love, and to die . . . 'tis there, ah 'tis there.*

Did you keep in touch with the old folks at home?

I wrote home every day and daily received a letter from some member of the family. I always waited to read it in the privacy of my room because my tears were apt to flow. And speaking of tears, I loved to sing HEAVEN HATH SHED A TEAR by Kücken to my own piano accompaniment and with the lovely violin obbligato.

And as to *floods*, OCEAN, THOU MIGHTY MONSTER from OBERON was one of my favorites. SCENA and PRAYER from DER FREISCHÜTZ was another. The SCENA is very dramatic while the PRAYER is simple and light for contrast. Such a lovely melody.

Please play a few measures of it?

What do you know! I happen to have the music here!

Any songs by American composers? Many—here are several that were very well received—CADMAN'S AT DAWNING—a favorite with the lovelorn. MACDERMID'S IF I KNEW YOU AND YOU KNEW ME. (By the way, that's a good ice breaker if you have a frigid audience.) It goes in part like this (*singing*)

I'm sure that we would differ less
And clasp our hands in friendliness;
Our thoughts would pleasantly agree
If I knew you and you knew me.

Also by MACDERMID—IF YOU WOULD LOVE ME. And there was MRS. H. A. BEACH'S THE YEAR'S AT THE SPRING—always good as an encore. I loved THE SPIRIT FLOWER by CAMPBELL TIPTON, and THOU ART SO LIKE A FLOWER by GEORGE W. CHADWICK was equal in my opinion to the settings of the same words by SCHUMANN, LISZT, and RUBENSTEIN (*Anton, not Arthur*).

I reckon each one was your favorite when you were singing it. Yes, and I loved them all. ARTHUR DUNHAM'S I MADE A PILGRIMAGE was unusual but good as a religious number; MAY THE MAIDEN by JOHN ALDEN CARPENTER was fine—so were several that CARRIE JACOBS BOND had given me personally. I added a second part to her well-known THE END OF A PERFECT DAY. It turned out to be good ensemble number for ending the program with the audience joining in as I warbled a high obbligato. It was popular then and everybody was singing it.

With that list, you surely have succeeded in dating yourself. That's all right with me.

Me too! Maybe we are just a couple of old squares . . . .

Did you find that some songs went over better in one locality than in an other? I remember that BY THE WATERS OF MINNETONKA, THE INDIAN LOVE CALL, and FROM THE LAND OF THE SKY-BLUE WATER were well liked in The Land O'Lakes Country—Minnesota, Wisconsin & Michigan, if that has any particular significance. All three are by American composers.

Who were your Violinists? The first was Susan Shelley of Poughkeepsie, New York—an excellent musician in a constant state of irritation, partly because she was buying an expensive old instrument on time, always seemed to be short of money and, on that account, short of temper. Some might have called it temperament but Pugh & I knew better. She'd throw a tantrum if a date had to be canceled; she needed every cent she could make. We were truly sorry for her. I could write a book about Shelley.

If you do, watch out for a libel suit. What happened to her?

She left to be married. And speaking of suits, we heard that her husband sued for a divorce within a year. Pugh said she left Redpath to go on the warpath.

The way I've heard it, Shelley was always on the warpath. Who replaced her—I mean as a violinist—not as a wife?

A sweet & accomplished British girl—Miss Austin, I can't recall her first name. We were known by our last names and usually addressed each other that way. I was impressed by Austin's fine diction and a bit shocked at first because she smoked cigarettes—*even in public!* Our nice American girls didn't do that in those days. How our customs have changed! Austin told me her father had given her a large supply of cigarettes when she left for America. Most English women smoked, she said.

STATION BREAK

(lighting Cigarette & starting Commercial for Faulty-Filters)

Puff Puff Faulty-Filter Cigarettes

are fine for your daughter, and they taste good (cough-cough) like a Cigarette oughter

(Violent fit of coughing)

Please "filter out" something more (cough-cough) about Austin.

She was the first young lady I had seen with bobbed hair. This gave her a very girlish appearance and kept Pugh and me, as well as our audience, guessing as to her age. She never divulged it.

Cough / Cough

Once, following a performance, a lady came back stage ostensibly to compliment her on her fine playing, then asked:—and just how old are you, my dear? Austin smiled sweetly and replied:—

I am just as old as my tongue and a little bit older than my teeth. So Pugh & I, as well as the lady, were foiled again.

If that lady was on the Music Committee, I doubt the home office received a favorable report on Austin. Did she leave to be married?

No, she was homesick—as I was. She might have been lovesick too; she didn't tell us. Anyway she went back to England and I was sad to see her go.

Who took her place?

Clarissa Max, better known as "Maxie" of Yankton, South Dakota. Musical talent is wherever you can find it—even if it originates in a whistle stop . . . you might be surprised. *Incidentally—Yankton is not a Whistle Stop* *(Population as of 1962 is 9,279)*

WESTERN EPISODE—how come?

Making a train connection one day, we stopped off at Maxie's home town and were having lunch at the station restaurant when—

Ah reckon you mean bah, don'tcha, podnah?

Well—snack bah. Pugh, in a kidding way, said to Maxie: So—this is where you come from, YANKTON, SOME TOWN! A tough hombre who had been sitting at a table back of us and was evidently full of beer & civic pride, stalked over to our table and growled menacingly YEH—THIS IS YANKTON—DO YUH WANNA MAKE SUMPTHIN' OUT OF IT?

Gosh! Were you all skeered?

We certainly were frightened—especially Pugh who didn't want his nose broken.

Was there a Fight?

Luckily not. Pugh stepped aside so we girls would be out of the line of fire—or of flying fists—and in his most charming manner, managed to wiggle out of a most unpleasant situation.

What happened to Maxie?

She left soon after that (you've guessed it) to be married. She suggested that her sister, who was also a violinist, take her place, but that didn't work out too well. She was excellent with the violin but played poor piano accompaniments. Pugh had to send for Maxie to return and finish the season as per contract.

That accounts for "Fiddlers Three". Who was the Fourth?

Addymae Parsons from Chicago. Her mother, Fannie Church Parsons, ran The Parsons Normal Training

School in Handel Hall and Addymae was a well-trained musician in every way. She stayed with us until we disbanded at the end of the Chautauqua season – then married a Chicago music publisher named Hathaway. Aside from the fact that her mother's name was Church before she married Mr. Parsons, I remember that Mr. P had relatives named Sexton and Bell.

I think that quadruple coincidence of ecclesiastical nomenclature merits a mention here.

So do I—and I think LOVE & MARRIAGE must have been the number one tune on THE HIT PARADE even in the good old days.

No—in those days it was *(singing)* IN THE GOOD OLD SUMMER TIME.

You've mentioned CHAUTAUQUA a couple of times. How does it differ from LYCEUM?

To me the difference is mostly seasonal. CHAUTAUQUA means outdoor performance in the good old summer time—in pavilions, and in tents that weren't always satisfactory to sing in because of poor acoustics. For some peculiar reason a canvas top thoroughly soaked and heavy with rain, seemed to improve the sound.

And LYCEUM?

LYCEUM means performance during the Fall, Winter and Spring months in theatres, opera houses, churches and school auditoriums— even in gyms.

Were they just "one-night stands"?

In LYCEUM it was always a one-night stand unless we arrived early, then the music committee might ask us to perform at a school. On a Sunday we might take part in a church service. The home office expected us to comply with such requests as a gesture of good will. We never refused but we were not always keen about doing those gratis jobs.

Did you always give the entire program yourselves?

In LYCEUM, yes, but in CHAUTAUQUA we often shared the program with other entertainers, sometimes for a week at a stretch.

Do you remember any performances in particular?

There may have been some *Artistic Standouts*, but after half a century it seems like a routine in which we maintained a fairly high level of performance marked by unusual locations, like the desolate region in North Dakota where we didn't expect much of a turnout but had quite an audience— another was in the delightful Dells of Wisconsin where we stayed at a fine resort hotel enjoying the gorgeous scenery and boat trips between our two-a-day performances.

Did they always go smoothly?

Not always. We once had a terrifying experience in Ohio when a severe thunder-storm hit us. Everybody ran scared for home, getting out just before the tent collapsed in a violent wind, almost like a cyclone. It was put up again in time for the evening show.

I still squirm when I think of our performance one sultry night in a Mississippi church. We sat behind a three-fold screen in the brightly lighted chancel awaiting our turns. It was early in the season and the unscreened windows were left open for ventilation. June bugs, attracted by the light, swarmed about us, getting in our hair and down our low-necked gowns—front & back! We killed 22 of the pests before the windows were closed. Our audience—in the dark—was not so pestered. Such a situation is hardly conducive to smooth performance.

The location of the tent was sometimes unfortunate. Once ours was pitched too near the railroad and a long slow freight rumbled past as I was nicely started on an Irish lullaby. I used all the volume I had in order to be heard. The noisy competition was evidently enjoyed for I was generously applauded—for my *perseverance*, I am sure. I repeated the lullaby as an encore—properly *mezza voce* and with all the tender Hibernian *pianissimos.*

How were you "booked" by the Redpath Bureau?

I was recommended by James Stuart—a composer connected with The Cable Piano Co. Redpath's GHQ were in The Cable Bldg. Mr. Stuart sold my mother our first piano on time, and often when I went to make a payment, he would ask me to sing one of his songs. He made helpful suggestions and watched my progress for several years, then got me an audition with Redpath, and I was booked immediately for two years—believe it or not—to replace a soprano who had quit to get married!

While I appreciated his valuable help, I somehow could

not "cotton" to Mr. Stuart. He had a glass eye that I always tried not to notice for fear of being spellbound. Even so, Mr. S. had a tendency to make *passes*—perhaps because I didn't wear glasses, if Dorothy Parker can be taken seriously on that subject. Anyway I was always glad to have Mother along to see that all installments were properly paid. She was ever my guardian angel, best mediator and advocate.

God bless Mother! Speaking of "passes" what territory did your Circuit embrace?

With the exception of The New England States we zigzagged all over the Eastern half of the U.S.A. from the tip of Long Island, West and South to Austin and Houston, Texas, and North to Canada. That's a lot of ground to cover—even in two years.

Can you tell us something of the training that prepared you for the platform?

I began vocal and piano study when I was about eight years old.

BRYANT—does that name have any special meaning?

That was the name of the music school for children near our home in Austin, Illinois, now a part of Chicago. It was run by a Mr. and Mrs. Bryant and Mrs. Bryant's daughter by a former marriage. I had attended this school for about one year when Mrs. Bryant's second husband ran off with her first husband's daughter.

Was that ever a TRIANGLE!!! <Let's call it a "wrecked" angle.> Then what happened?

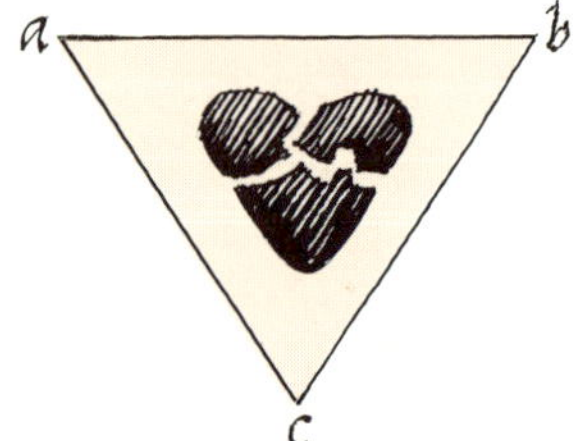

I was not studying geometry, but I remember that with those two teachers gone, Mrs. Bryant had to close the school. She told Mother that I had a voice worth cultivating, and recommended a Chicago teacher—Miss Mary M. Shedd—who was remarkably successful in training children to sing professionally.

The times were hard. Father had lost his position as a machine shop foreman and was making only a meager living at selling and repairing bicycles. I was the youngest of eight, the first four being stepchildren belonging to my father, who was a widower before he married my mother.

Little Irene
age 8

My brothers and sisters were still too young to help much financially, and then work of any kind was hard to get. The Welfare State was undreamed of, but kind neighbors proved a good substitute for it and somehow we managed to live. Household expenses piled up and some bills remained too long unpaid. I remember the firm but kindly man who came out from the City Water Department to shut off our meter—*but until the bill was paid*—how could we live without water? He *accidentally* fixed our kitchen sink faucet to leak *just a dribble*. We saved every drop.

Welfare—indeed!

IN that situation Mother wondered if the cost of vocal lessons would be justified. Father was not consulted. Nevertheless, Mother took me to Miss Shedd and, being German, was Fleissig in dickering as to the expense.

After testing my voice, Miss Shedd agreed to give me one lesson a week at 50¢—the same as we had been paying at the Bryant school. Soon, however, she was giving me three lessons a week *at no charge* because she was using me and several other talented youngsters to demonstrate her particular kind of voice culture which she called THE AMERICAN METHOD.

She must have been an exceptionally generous person.

She was indeed generous. From that time until I was sixteen and (unwisely perhaps) stopped studying with her, she made no charge for instructing me—paid all traveling expenses for both Mother & me and for all dress materials that Mother—who had been an expert dressmaker for Marshall Field & Company, made into a wardrobe suitable for my public appearances.

In addition to this, she provided instruction by tutors to make up in some measure for what I was necessarily missing in public school because of my concentration on voice & piano. The 3Rs rated second to Italian & French. Mother taught me German and some arithmetic.

Where was Mlle. Shedd's studio?

She had *three studios* in the Auditorium Building which housed the Auditorium Theater where the Grand Opera performances were held.

It sounds like big business.

It was and she was adept in getting publicity to promote it. Shown over leaf are some examples of it, effective then, dated now—ever precious to me.

These Singers are Available for CONCERTS, CLUBS And RECEPTIONS

7

Our health was carefully guarded, catching colds was a serious offense. Long heavy underwear—evident here, rolled down to ankle and up to shoulder {CHIC!}—was an important part of our apparel. Cotton stockings were the rule, silk was considered immodest! Nylon and such fabrics were still to come.

These Artists Can Be Secured for

Concerts, Receptions, Clubs, Etc.

By Addressing

MARY M. SHEDD,

Rooms 74, 77 and 79 Auditorium Bldg., Chicago.

IRENE GERTRUDE BRIGGS (AGE 13) SOPRANO (COMPASS 3 OCTAVES

MISS ANNA ROSENGREN MEZZO SOPRANO.

HAZEL LUCILE JAMESON AGE 14 MEZZO SOPRANO SCOTCH BALLAD SINGER.

HARRY CANEVIN (AGE 15) ELOCUTIONIST AND SINGER.

What an Entrepreneuse!

She surely was, and as I look back on it all, perhaps she was something of a hypnotist, too.

I remember a full-page feature article she wrote for one of the Chicago Sunday newspapers on the therapeutic value of deep breathing* which she was currently demonstrating in hospitals. This was entirely apart from any vocal instruction to the patients and dealt only with tone vibration and correct posture.

* There's an excellent article in last month's Reader's Digest {March 1959} on that very subject, that some of our audience may have missed or might care to review.

Did you youngsters get to hear any of the operas?

Occasionally Miss Shedd would give us tickets, but more often we'd sneak down from the 7th floor studio to the 5th, which gave entrance to the top gallery of the theatre, & stand there at the doors to listen. The ushers got to know us and if there were seats vacant at the rear, would let us go inside.

AMERICAN JUVENILE QUARTETTE

Extensive and
Varied Repertoire
Solos
Duetts
Trios and
Quartettes

We were young and critics were kind. Some said our tone quality was PURE, some called it ANGELIC.

HARDIE PITZELE
Contralto

ALONZO MOORE
Tenor (Boy Sop.)

IRENE BRIGGS
Soprano

HARRY CANEVIN
Baritone

What a break for you Kids!

Yes — That was in the so-called Golden Age of Song the heyday of such renowned artists as NELLIE MELBA, MARCELLA SEMBRICH, LILLIAN NORDICA, EMMA EAMES, and JOHANNA GADSKI. Except for LUCY POOD, sopranos were my greatest interest in life.

And who was Lucy Pood?

She was my pet poodle.

Oh . . . Were you fond of her?
I refer to Mlle. Shedd —

We were all devoted to her. We worked hard and breathed deep to please her. By my thirteenth birthday I was singing several quite difficult coloratura arias such as the SWISS ECHO SONG and THOU BRILLIANT BIRD from THE PEARL OF BRAZIL. I sang regularly three times a week at demonstration periods, usually held in the Auditorium Recital Hall on the 7th floor.

CY POOD — THE ONE AND ONLY PERFECT POOCH

With good results? I am sure our teacher thought so. We children made singing seem so easy that she gained many pupils, adult as well as juvenile.

Did she give demonstrations outside of the Chicago area? Yes—about that time Miss Shedd took Mother & me to New York and put us up for two weeks at the Waldorf Astoria where we demonstrated her AMERICAN METHOD and stirred up considerable interest.

Since your picture appeared on the cover of the MUSICAL COURIER at that time I shall show it here, if you please, along with the write-up of the previous week.

MARY M. SHEDD.

MISS MARY M. SHEDD, discoverer and teacher of the American Method of Singing, has been spending the past week at the Waldorf-Astoria. Miss Shedd brought with her from Chicago one of her pupils. Miss Irene Briggs, who, although only thirteen years old, possesses a marvelous voice. When she began to study with Miss Shedd her voice was weak, and her range covered but eight notes. Now the volume of her voice is so great that it will fill any hall, and she has a range of three octaves. It is a wondrously mellow voice, and every tone from the highest to the lowest is as nigh perfect as can be. Her repertory, which is astonishing for so young a child, is as follows:

Salve Regina.......Dana
Ave Maria.......Gounod
Air from Atilla (in Latin and Italian).......Verdi
Brilliant Bird, from Pearl of Brazil.......David
I Will Extol Thee.......Costa
Serenade à Juanita.......Jouberti
With Verdure Clad.......Haydn
Vesper Hour.......Bracket
Heaven Is Not Far Away.......Moler
Lord Be Merciful.......Bartlett
Nearer, My God, to Thee.......Wiegand
Shine On, O Star.......Sawyer
La Danza.......Marchesi
Woodland Lullaby.......Johns
Coming Through the Rye.......Parry
And other ballads.

Miss Briggs will shortly fill several concert engagements in New York and Cincinnati.

To a representative of THE MUSICAL COURIER who called upon her Miss Shedd said, in part:

"Through my method every voice accepted by me will develop into the same tones with which De Reszke, Patti and all great artists were born . . . . You have heard Miss Briggs sing and have no doubt perceived the difference in her singing and that of those singing by other methods. This difference is in reality a lack of noise, this noise being caused by use of the throat, or rather support of throat, and results in voices becoming harsh and unmusical, and eventually a failure. Everyone uses the throat for support, either in singing or speech. If they did not they would sing like Melbas and Pattis, and speak the same as Mansfield, Rehan and Bernhardt.

"Children's voices do not train more easily than adults, for the most easily trained pupils I ever taught were a lady, aged forty-nine, and a gentleman, thirty-two. Age has no effect on muscles that support pure tone. All is governed entirely by physical conditions."

"The American Method of Singing is the science of pure tone development, which was only recently discovered and at present is known to but few.

IRENE BRIGGS

I fear that so much repetition of this one example of my youthful affectation of easy grace may become boring to our long-suffering audience. If so, I am sorry.

I agree, however, that Miss Shedd was quite a promoter as well as a vocal instructor.

NB: Aside from the reviews on teacher & pupil there is so much of interest on page 30 of the October 2, 1901 issue that we are showing it on the opposite page to point out:—[1]that music critics at the turn of the century were no less outspoken than thos
cycle completed, ie: some type faces in vogue 70 years ago became outmoded and were seldom used ex
of these old types have been revived &/or redesigned with such names as VENUS and CRAW CLARENDON—

orchestration of the work often calls for high praise.

The performance was not of a high standard, although it gave a fair idea of the opera. The two chief singers were not vocally well disposed, and the chorus was often pitiably weak and ragged.

The program of the afternoon concert, September 27, included Mackenzie's "Coriolanus" Suite (first time in America), Liszt's "Concerto Pathétique" (arranged and played by Richard Burmeister), Beethoven's Second Symphony. Mr. Van Hoose sang with commendable breadth, tenderness and manly feeling "Adieu, donc," from Massenet's pornographic version of the sad story of John the Baptist.

The suite is dull. It is without color, without dignity, in a word without character of any kind. Ah, if Mackenzie had only followed his Scottish inclinations, and introduced a bagpipe, or thundered out with the full brass "Hoot mon!" before the dwelling of Tullus Aufidius!

Mr. Burmeister played delightfully. The concerto itself is one of Liszt's less important works, and it hardly repaid the loving labor of the pianist-arranger.

There is little to say about the evening and final program. It was a program of "Artists' Night," and such programs always remind me, when I see men and women armed with a popular aria, of Hamlet's speech: "Then came each actor on his ass." * †

Suzanne Adams appeared and showed her unblenching wifely devotion by singing a waltz by her husband. She also sang "Batti, batti," and sang it all too slowly. Mr. Bispham sang Iago's creed ("Otello"), and "Quand' ero paggio" ("Falstaff"); Miss Stein sang Bemberg's "Jean d'Arc"; and Mr. Williams sang inimitably the aria from "The Swan and the Skylark." The orchestral numbers were the overture to "The Merry Wives of Windsor"; variations from the Kaiser Quartet of Haydn; Bizet's "Jeux d'Enfants"; a march by Saint-Saëns. A motet by Mozart was the finale, a motet not worth the resurrection.

The weather was glorious throughout the festival; the prices of admission were higher than before; the attendance as a rule was smaller; the result, as I hear, was a pecuniary loss.

IRENE BRIGGS.

IRENE BRIGGS, whose picture appears on the front page this week, is the thirteen year old pupil of Miss Mary M. Shedd, discoverer and teacher of the American method. As was mentioned in last week's MUSICAL COURIER, Miss Briggs' voice and repertory are astonishing, and her teacher may well be proud of her.

Last April Miss Shedd read a paper on the American method before the Chicago University, from which the following extracts are taken:

"Ever since the world began new lines of thought have been condemned, and as this method is a new discovery, prejudice and doubt must be met with and swept away. It is for that reason that I am always anxious to have my pupils sing for those who doubt. They compare their singing with that of those singing by other methods and are convinced that the American method is the only true one.

"Think of the thousands who graduate in vocal music each year; many of them have studied from five to ten years. How many of them become great artists? Only those who are naturally born singers.

"To make singers sing naturally is just what the American method aims to do. There is no effort, no stress or wear and tear upon the throat. Children sing with full, round, mature tones, while men and women long past the prime of life have tones that are as fresh as those of men and women twenty years of age.

"Students of this method notice the rapid improvement in their voices and are encouraged thereby. They do not study for years and then find out that they never will be able to sing. In from one to two years earnest students are ready for professional work. Perfection is what is aimed for and perfect singing is the only kind worth one's time or money.

"A writer in a musical magazine says: 'In order to become a good vocal teacher one should have a pleasing personality and plenty of magnetism.' As long as teachers depend upon hypnotism, magnetism, pleasant smiles, finely equipped studios and social influence instead of actual knowledge, certain and systematic, just so long may we expect failures. Many men and women who began to study with us for the purpose of learning to sing are now preparing to leave their occupations and become teachers. This proves that the American method has a tangible, practical side which appeals to the business instincts of many.

"The testing of the voice reveals the condition of the muscles which determine the possibilities of the voice, and enables me to state with a reasonable degree of accuracy the length of time required to perfect the voice and place the singer in demand. Heretofore the element of uncertainty has been the most discouraging feature in the study of singing. Many of our so-called great instructors admit that success is by no means assured. To the earnest student who places himself in my care I can honestly state that success is certain. No matter how badly one sings or how small the voice, so long as the ear is acute as to pitch I can teach you to sing with as pure a tone as any artist ever sang."

In Chicago, where Miss Shedd is well known, her pupils are her best recommendations, and the following letter from the late George W. Lyon, of Lyon & Healy, also speaks for itself:

Mary M. Shedd will be the greatest voice teacher the world has ever known by reason of her appreciation of pure tone, which enables her to detect and remove faults in the beginning, in which she surpasses any one that I have ever known. On account of her wonderful knowledge of pure tone I do not hesitate to say that some day teachers of the piano, violin and other instruments will seek a recommendation from Miss Shedd, as any uncertainty in foundation work becomes apparent to her at once. Miss Shedd has often astonished me by her just criticism of cornet and flute players, her accurate ear detecting the incorrect breathing, made obvious by the uncertainty of execution and lack of even power.

BLAUVELT SAILS.—Madame Blauvelt sailed on the Kaiser Wilhelm der Grosse last Wednesday. She has already forty engagements in England and Germany, and will sing on successive Saturdays in London, leaving the Continent especially for the purpose. In London she sings at Albert Hall, at Queen's Hall and the Crystal Palace. She has had the rare distinction of six engagements with the Hallé Orchestra, under Dr. Richter. January 2, 1902, she returns to New York.

*In second Quarto, 1624

...ay:— [2] that there is evidence in the above ads of a typographic ...sively, perhaps to lend "Gay Nineties" flavor. Recently, however, several ...d as standard—even smart. As REMUS said to ROMULUS: TEMPORI

PARENDUM

‹ TR: WE MUST YIELD TO THE TIMES ›

[3] † That scathing line from HAMLET might mean each actor rode his favorite hobby. 'Tis said "Good Queen Bess," with tongue in cheek, relished such innuendoes. Victoria did not.

She sure was pushing your Stock.

And pulling strings, too. We even gave a demonstration at Carnegie Hall—on stage—but, of course, to an empty house, after which I was excused while my teacher talked earnestly with Mr. Somlyo, the manager, who had been so kind as to give us an audition. Miss Shedd was determined that some day I should sing there to a FULL house.

Alas for her ambitions and my childish dreams! It is not that easy to break into THE BIG TIME.

It reminds me of du Maurier's tale of TRILBY and SVENGALI.

That's why I suggest Mlle. Shedd was a hypnotist, although she vehemently denied that allegation.

But how thrilling for a Little Girl!

I was pretty young—but not so little.

You were PRETTY period.

Thank you, Ray.

What next?

We returned to Chicago and I continued with Miss Shedd until I was sixteen—seven years in all.

And then?

I bounced around from one teacher to another, becoming confused by different methods. Nevertheless I was awarded a scholarship by the Cosmopolitan School of Music which offered a complete musical education.

How Wonderful!

But the voice teacher I was assigned to was not S.M.K. Gandell, the one I had been counting on because my brother Robert, a baritone, was doing so well under his instruction. I also wanted to continue with my old piano teacher, Albert Fox, who accompanied me on the organ at Sunday services. The Cosmopolitan's directors decided it was a case of winner take ALL OR NOTHING AT ALL and since I was already earning some money, they withdrew the scholarship.

Shirley Mark Kerr Gandell

Albert Fox

How awful!

You mean how *willful* of me! I have often regretted it. I was pretty foolish, wasn't I?

You were PLAIN foolish, period.

Well—I did finally study with dear old "Gandy" anyway for 2 years, paid cash on the barrel head, too.

Did you have any other teachers?

Theodore Harrison of the American Conservatory and Mary Pierce Niemann—excellent as coaches and accompanists—were helpful. However, as often as I appeared professionally in recital & concert and occasionally in oratorio, it was somehow without the confidence and childlike naïvete of my earlier years with Miss Shedd. Then nothing ever fazed me. I simply did

Theodore Harrison

Mary Niemann

as I was told and didn't know what it meant to be nervous. I assure you that I do now.

Curtis Hall
Was he an old flame of yours?

Silly! That was the recital hall atop The Fine Arts Building in Chicago. I was soloist there for four years at the Sunday morning services of THE CHURCH OF THE NEW THOUGHT headed by Ursula N. Gestefelt, who broke with Mary Baker Eddy on some technicality and formed a church of her own but still quite similar to Christian Science. Sunday was my busy day. Sometimes I sang in churches of three different denominations.

I am told you were effective as an evangelistic singer.

I shall never forget singing in a series of revival meetings at Maywood Congregational. The evangelist was a real spellbinder and used the old SA technique to draw a crowd. Perhaps I should explain that in those days SA stood for Salvation Army. He hired a brass band to march through the streets playing gospel hymns, and the people followed into the church much as the children of Hamlin followed the Pied Piper.

Was he a good persuader?

Well—he was so persuasive that before the series was over he had persuaded me—

Halleleujah!!! . . . to the idea of becoming his wife—

What!! . . . I said—to the IDEE—YAH . . . .

Yah? . . . . as with many conversions, mine didn't last long; what with contrary persuasions by my friends in general and my father in particular, I realized I was not predestinated to become the wife of a preacher and with some reluctance, returned the engagement ring.

FATHER

That surely was another case of FATHER KNOWS BEST.

I can agree now— but it was difficult then.

And may I assume that your evangelistic boy-friend had this photograph taken for the cover of his revival program?

(over leaf)

That's right—the saintly expression with eyes turned heavenward was all his idea—so were the daisies in my hair. He said I would be a great help to him in his work.
I have no regrets—not now.

IRENE GERTRUDE BRIGGS

Who has been engaged as soloist at The First Congregational Church, is one of the most talented young artists in Chicago. She has won an enviable reputation through her elegant diction, dainty sentiment and flawless tone color.

Miss Briggs was formerly soprano soloist at The Warren Avenue Congregational Church, the largest of that denomination in Chicago. This brilliant young soprano begins her duties at The First Congregational Church Nov. 21, and will render two solos each Sabbath evening.

Circa 1910

Are you sure? Positive.

HELL & HIGH WATER—was that something you did in spite of?

Oh, that—that takes me back to when I first learned that *the show MUST go on*—my first Chautauqua.

How old were you then?

About twelve. Miss Shedd had booked our American Juvenile Quartette for two weeks of appearances in a large pavilion at the Methodist Camp Grounds on the bank of the Kankakee River.

Was there a flood on the Kankakee?

Nothing so tame as that. I had also been booked for a solo at a church in Kankakee, Illinois, about three miles downstream. We were to go by bus but the motor was balky & time was awastin'. Someone offer'd to take Mother and me in a canoe. We were deathly afraid of the water and had no faith in canoes—but that was the only way we could make it.

And you survived!

I'm here to tell you. Mother was even heavier then than I am now. We were so low in the water that we feared every minute we'd be swamped & drowned.

If we ever prayed, we did on that trip.

Carry Nation's Hatchet— Weren't you a bit too young to be bearing arms in defense of your country?

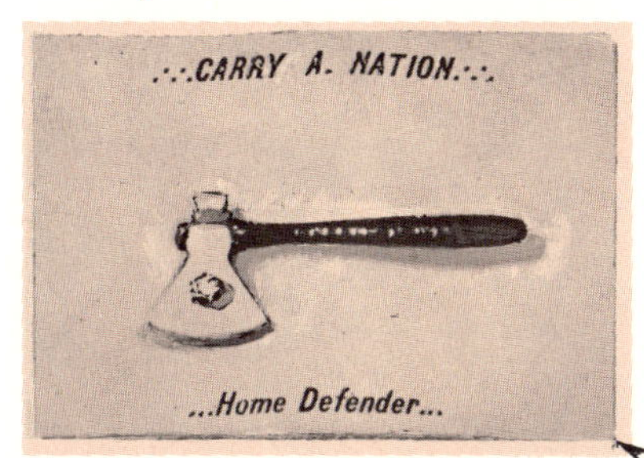

That's about the time I appeared with Carry Nation, the great *Saloon-Smasher*. An election was coming up and prohibition was the burning issue. She was a forceful speaker. My name is CARRY A. NATION, my initials are C. A. N. she shouted, and I CAN!!!

She liked my singing—warned me never to let lips that touch liquor touch mine and presented me with a little souvenir hatchet that I have always cherished. *And here it is!*

Indiana Saloon—Don't tell us you sang in taverns!

I certainly did NOT! After a recital in Hammond, we were invited to stay overnight at the home of a prominent citizen. Before retiring I was asked to sing for the family group. They liked it and made quite a fuss over me.

Someone said: I wish Pa could hear her—where is Pa?—Most likely over at the saloon. I'll go get 'im.

At that point Carry Nation's influence found expression as I impor*tuned in* with—*Don't go, I won't sing for anyone who has been drinking.*

You little prude! And so to bed—

Yes—I was rude. I should have said—or sung:

> *Father, dear Father, come home with me now,*
> *The clock in the steeple strikes one . . .*

It has been set to music, you know.

{Background music—Father, dear Father, come home}

Little Rock, Arkansas— This may be of local interest. Ah reckon you're a BIG girl now, back on the road with Redpath.

I recollect we arrived in Little Rock after dark. Shelley and I were directed to a basement storeroom to dress for our concert at the YMCA gymnasium. I was putting on my underthings when I heard a slight noise at a rather high horizontal window. Glancing up, I saw men's faces peering in at us. Shelley had even less clothing on than I did. She was practically stripped for gym.

Were these Young Christian Men Associated at the window?

I don't know about that. I ran to the window and, with rare presence of mind and a pleasant smile, lowered the shade—*gently*—so as not to tear it off the roller.

I'm glad you did that. In those days girls were really modest.

Today, girls perform in public with almost no clothing and think nothing of it (*sighing*). Ah me, how times have changed, no one would give me so much as a second look now.

Don't you care, Lady Godiva, I go along with THOMAS MORE who said

(singing) *It is not while beauty and youth are thine own*
And thy cheek unprofaned by a tear,
That the fervor and faith of a soul can be known
To which time will but make thee more dear.

Thank you—thank you, Ray—so much. You are so kind and consoling.

BAGGAGE CAR—hm-m—
Did you sing that old tear-jerker THE BAGGAGE CAR AHEAD—about the pathetic old man weeping in the day coach behind?

Never heard of it!

Well—it was popular then and every one enjoyed a good cry over it.

Here again it's a case of *the show must go on*. Our train was late and Pugh told us that unless we wanted to go on stage in our street clothes, we'd have to dress on the train. In fact, foresighted Pugh had already arranged for Maxie and me to go to the baggage car where we could dress from our wardrobe trunk.

Of course there were no curtains, so we asked the baggage men to look the other way. We don't know if they did or didn't. We didn't look at them.

(I wonder as eye wander)

SEVEN-YEAR-ITCH—
Do you remember that?

Can I ever forget it! At a hotel in Xenia, Ohio, where the name starts with X but sounds like Z (and also marks the "spots" for me) the wall between our rooms was so thin Shelley and I could talk to each other through it. The next morning I found myself broken out with an ugly red rash that itched. I was alarmed and called for Shelley to come and see it. She took one look and screamed—*Don't you dare to come near me!* When Pugh saw me so red faced at breakfast, he told me

I must see a doctor as soon as we arrived at our next stop but, *meanwhile don't sit near us on the train!*

The MD's diagnosis was SEVEN-YEAR-ITCH. He gave me iodine to paint on the red blotches. When I told him I had to appear on stage, he mixed up a whitish liquid to paint over the brown iodine spots. I could see myself playing the part of a pasty-faced clown!

You might have made a hit as CANIO in PAGLIACCI—all you lacked was a tenor voice. How did you manage?

We were making a long jump with no date for that night. I'd had a smattering of FAITH HEALING in THE CHURCH OF THE NEW THOUGHT, and decided to give myself a treatment, so, until I fell asleep that night in my Pullman berth, I "*held the thought*" that all would be well when I awoke.

Came the dawn—the rash was gone!

Marvelous!

That's what I thought until about one week later when Shelley came into my room to show me just the same kind of rash I'd had: it gave me a feeling of retribution, if not revenge.

This time, without the help of a doctor, we diagnosed the malady ourselves—correctly—Bed Bug Bites.

That must have shaken your faith in MATERIA MEDICA.

I realized that my doctor was a quack—and wished to have my money back.

IRISH COP! I hope you had no trouble with the police!

It was after midnight when we arrived in an Indiana town, and there was neither a horse-drawn nor a motor-driven conveyance to take us to a hotel. Pugh asked a policeman if he could recommend a decent place near by so we would not have to lug our baggage so far.

Suren' I'll be after helpin ye hoist yer suitcases right up this here hill to a foin roomin' house. I know th' lady that runs it. He was so genial & convincing that we, too weary to be wary, fell in line and followed him up the hill to the "foin roomin' house".

He rang the bell, and a sleepy, disheveled, barefooted "lady" appeared in a bathrobe. Hi, Maggie, these frinds o' mine ahr a lookin' fer a gud place tuh schlape. Give 'em the best ye've got Maggie darlin' like th' gud gurhl ye ahr, an' a gud night to ye ahl, said our friend as he returned to his beat.

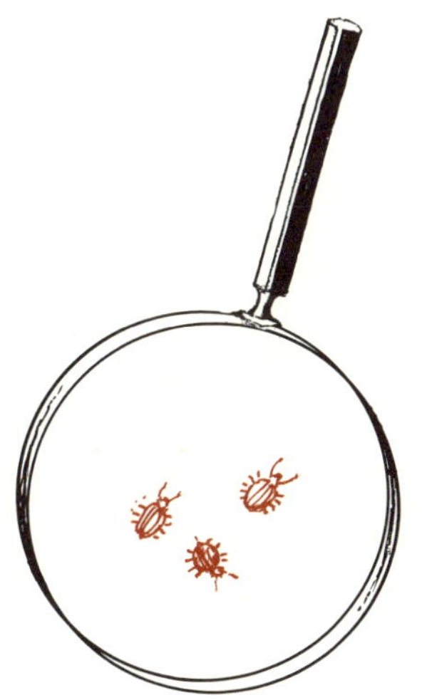

Did she accommodate you? She gave Pugh a room upstairs; Austin & me, twin beds in the parlor —

and they were Irish too? They were—and we were dead tired, too, & went right to "schlape." Next AM Austin's eyes were swelled almost shut. Bedbugs affected her like that. Every morning thereafter, her eyes were my bedbug bite barometer.

How did you come out? I had a few welts myself and was commiserating with Austin when Maggie knocked at our door and reported: Th' gintleman left airly an' ast me to give ye these here thrain tickuts. Pugh explained later that he just could not sleep there and, after pinning a *pointed* note to his pillow, got out and walked to the nearest hotel. We girls were furious because he hadn't taken us with him instead of leaving us to be eaten alive!

(Background music — Cockles and mussels alive-alive-o)

So much for bedbugs you have met; now our time is running out. . . . answer this question — Yes or No: How did you meet your first . . . and your last husband? *(Pausing)* That can't be answered Yes or No.

Were there any in between and if so, are you sorry? I take the 5th Amendment.

Was there ever any evidence of artistic talent among your numerous brothers & sisters?

SISTER PEARL had a natural soprano voice but never used it professionally. Brother WALTER, as a boy soprano won a gold medal in St. Martin's Episcopal choir, but ROBERT, our youngest brother, became the real *trouper* of the family. With his big resonant baritone, he was an outstanding raconteur as well as a singer—giving readings, pianologues and chalk-talks on the circuits and in University Extension work for 23 years.

At first BOB worked with a variety of entertainers—male & mixed quartettes, opera singers & lecturers, actors & ventriloquists, bell ringers, phrenologists & politicians. However, during the last 10 years of his career he was booked as an individual attraction. Well known on many of the circuits as *"Smiling Bob,"*

WALTER & ROBERT

Father was a widower with 4 children when he married Mother. Our stepbrothers & sisters: Will, Ed, Bess & May, developed no particular talent in any of the arts.

"The Platform's Most Versatile Artist"

he always received a hearty welcome on return engagements. One year, in making up its annual ratings, VARIETY, a leading theatrical magazine, placed him the highest in his particular field.

Wherever he went, Bob reveled in the various aspects of nature, sketching them in water colors & pastels. He often used these "quickies" as the basis for his more swiftly executed chalk-talk pictures, which looked easy to do but took a great deal of careful preliminary consideration. He had the knack of catching a mood of nature quickly, as in the sketch below, and his understanding of human nature was such that he could capture an audience in the first few minutes of a program and hold it straight through to the end.

See page 80

Among the last of the old circuit troupers whose services were still called for, Bob died in 1950.

Apparently you and Bob had much in common. Was he helpful to you in your work?

He did me many a good turn. The best was soon after I had finished with Redpath. We loved dumb animals as well as our fellowman, and took pleasure in training and domesticating both species — even the wild ones.

Please tell us all about it. ONE FINE DAY, Bob brought a shaggy, starving, stray scribe home to dinner. I didn't know it then but that

Ah-ha—ah-ha—ah-ha?— was a very important event in my life—I believe that young scribe liked me—right from the start. . .

I'm not surprised . . . .

I know he liked my cooking . . . . .

I'm sure he did . . . .

. . . because about two years & maybe a hundred dinners later, he asked me to marry him.

I reckon he wanted to be SURE— He *SURE* took a long time! In this age of Space & Speed I reckon a hamburger, a coke & a two weeks' acquaintance would be considered a long engagement.

How long was yours? November to April.

EASTER SUNDAY, 1917—
what's so special about that? Aside from marking the entry of America into World War I, it seems to me I had several heavy dates on that particular day.

Yeah? . . My DEAR DIARY—(*if I had only kept one!*) would surely have amplified the following items:—sang morning service—quartette—Austin First Presbyterian—shortly before afternoon vespers, married man of my choice—taking advantage of lovely Easter floral decorations at no extra charge. Dr. Beatty tied knot—good & tight—Jessie Adams at organ—Wagner to—Mendelssohn fro.

After singing vesper service, husband took me to dinner at not too expensive restaurant (he's ½ Scotch & very frugal). Thence, to Oak Park, Illinois, to sing Easter evening service at Cuyler Avenue Methodist.—
. . . What a day . . . .

I REMEMBER *that day, too, IRENE . . . .
One year later I wrote a jingle about it, and on each anniversary since then I've repeated it with a slight change in the third line to bring it up to date. On our Golden Wedding in 1967 it went like this:—*

What a day!

On April the eighth—nineteen-seventeen,
I stood at the altar beside my IRENE.
Fifty years have flown fast since that momentous day
And we still remain lovers, we're happy to say.

Thanks, Irene, thanks for everything—including and concluding this interview. You are most welcome, Ray. Goodnight.

Now let's all sing Good-night Irene— Just the refrain, please (The verse ain't fittin'):

CURTAIN

THE CHAUTAUQUAN.

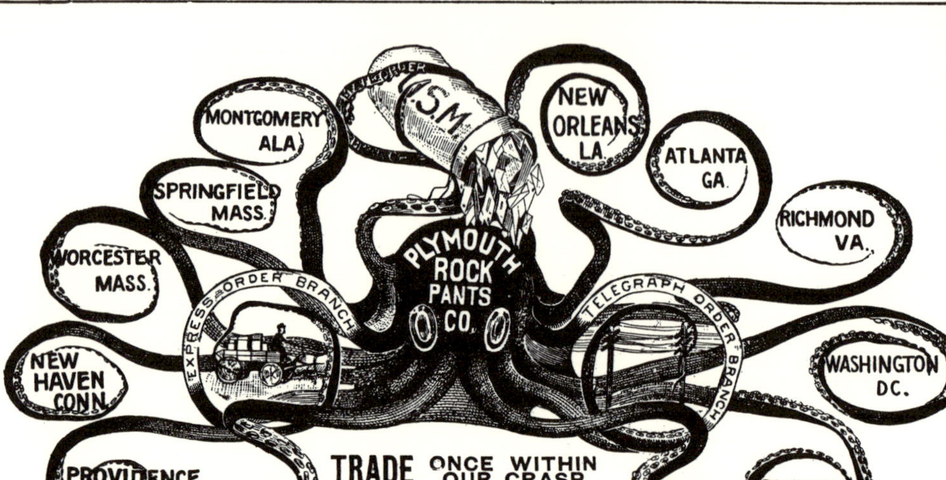

TRADE ONCE WITHIN OUR GRASP NEVER LEAVES US.
OUR ABILITY TO CATCH and hold that which we seek is like the
TENACIOUS & RAPACIOUS POWER OF THE GREAT
SEA MONSTER.

IF YOU CANNOT REACH ONE OF OUR BRANCH STORES, WRITE US AT ONCE AND WE WILL MAIL YOU, FREE, 52 SAMPLES TO SELECT FROM WITH SELF-MEASUREMENT BLANKS.
Address, PLYMOUTH ROCK PANTS CO., 11 to 17 Eliot Street, BOSTON, MASS.

BEAUTY

Wrinkles, Black-heads, Pimples, Freckles, Pittings, Moles and Superfluous Hair permanently removed. Flesh increased or reduced. Complexions beautified. The Form developed; Hair, Brows and Lashes colored and restored. Interesting Book (sent sealed), 4c. **Mme. Velaro, 414 W. 47th St., N. Y. City.**
Mention this Magazine.

COMFORT
For All
26,000 in use.
Send 6c. for Illustrated Catalogue, 24 pages. "Healthy Homes; How to Have them," 36 pages of valuable information, price 5c.
Heap's Patent Earth Closet Co.,
MUSKEGON, MICH.

YOUR CHILD MUST BE KEPT HEALTHY or she cannot be BEAUTIFUL. SENSIBLE MOTHERS BUY GOOD SENSE CORSET WAISTS
FERRIS' Patent Ring Buckle at Hip for Hose Supporters.
Tape-fastened Buttons—*won't pull off.*
Cord-edge Button Holes—*won't wear out.*
BEST Materials throughout.
BEST for Health, Comfort, Wear and Finish.
Thousands now in use in the United States and Canada. For Sale by Leading Retailers, or mailed FREE on receipt of price, by
Ferris Bros., Manfrs., 341 Broadway, N.Y.
Marshall Field & Co., Chicago, Wholesale Western Agts.

Satisfaction Guaranteed or Money Returned. TRY THEM.

FIT ALL AGES
Infants to Adults.
Childs .50 .70 .75
Miss. .70 .75 .80 .85
Young Ladies 1.00 1.10
Ladies 1.00 1.25 1.50 1.75 2.00

AVOID *Inferior Imitations.* *Be Sure* your Corset is stamped GOOD SENSE.
Send for Illustrated circular.

AGENTS Send for E. B. TREAT'S Catalogue of new books. Shots at Sundry Targets, by Talmage $2.50 Quick sales. Big pay. Also, Mother, Home & Heaven, 400 best authors Edited by T. L. Cuyler. $2.75. 192,000 sold 5000 Curiosities of Bible $2. By Mail. E.B.TREAT, N.Y.
CAT-A-LOG. TRADE MARK

DRY GOODS BY MAIL.
Prices Lowest.
Styles Best.
Samples and Catalogues Free.
Established 1840.
LE BOUTILLIER BROTHERS,
Broadway and 14th Street, N. Y.
K-Jan

THE CHAUTAUQUAN. 523

PACKING HOUSES AND FACTORIES,
Pittsburg, Pa.
La Porte, Ind.
Sharpsburg, Pa.
Walkerton, Ind.
Allegheny City, Pa.
Ross, Pa.

VEGETABLE FARMS.
Aspinwall, Pa.
La Porte, Ind.

RECEIVED 28 1st Medals and Awards. INCLUDING Paris Medal 1889.

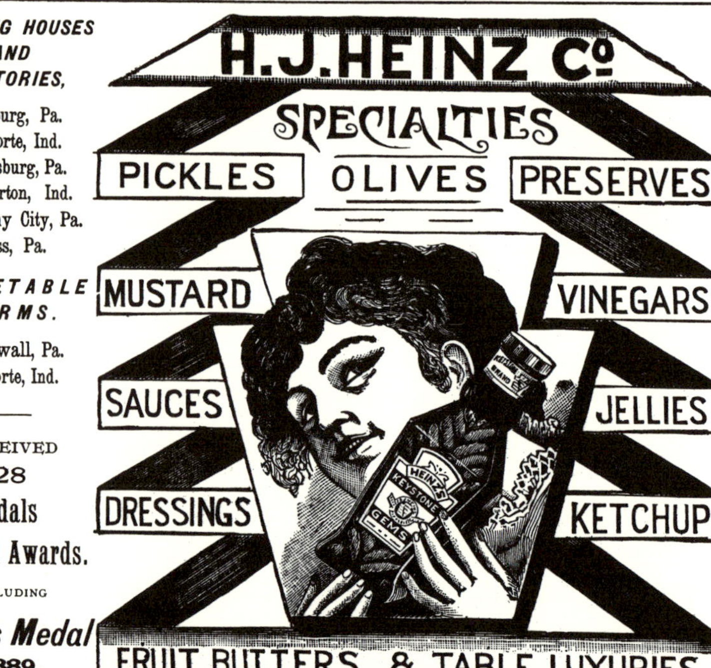

AGENCIES.
New York, N. Y.
Philadelphia, Pa.
San Francisco, Cal.
Cincinnati, O.
Baltimore, Md.
New Orleans, La.
Cleveland, O.
Louisville, Ky.
Kansas City, Mo.
Washington, D. C.
Indianapolis, Ind.
Dallas, Tex.
Wheeling, W. Va.
St. Paul, Minn.
Omaha, Neb.
Columbus, O.
St. Louis, Mo.
Denver, Col.
Wichita, Kan.
Boston, Mass.
Albany, N. Y.
Chicago, Ill.
Buffalo, N. Y.
Jacksonville. Fla.

GENERAL OFFICE—PITTSBURG, PA., U. S. A.

in THE CHAUTAUQUAN
—But there was no need THEN to persuade the user of any object, no matter how new or good, that it was already obsolete.

Part II
Historical

JOSIAH HOLBROOK
JOHN H. VINCENT
SCIENCE
RELIGION
THE FOUNDERS OF LYCEUM AND CHAUTAUQUA
IN THEIR RESPECTIVE ROLES AS TEACHERS
THE ONE OF SCIENCE + THE OTHER OF RELIGION
EVENTUALLY MERGED & DEVELOPED STUDIES
BY THE LIGHT OF GENERAL KNOWLEDGE FROM
THE LAMP OF LEARNING IN
THE ARTS & HUMANITIES + LITERATURE & PHILOSOPHY
AGRICULTURE & INDUSTRY + ECONOMICS & POLITICS

Notes—Historical & Personal on Lyceum & Chautauqua

with Calligraphic Overtones

Two men, each driven by a dream, had a tremendous, surging influence upon American culture during most of the century that began with the year 1826. The fanfare, the ballyhoo, the booster parades, the energetic pursuit of knowledge in tent communities, are no longer a part of the American scene. But libraries and museums across the land, ever-increasing university extension courses for adults, and a number of vigorous & independent chautauqua communities continue to testify to the sound principles of Josiah Holbrook, founder of the Lyceum movement, & John H. Vincent, founder of Chautauqua.

The latter was not yet born when Holbrook began to build the machinery that would only need a little oiling, when the proper time came, to railroad Chautauqua to wayside stations all over the country. Josiah's driving passion was to introduce the systematic study of the natural sciences into public schools. He envisioned in every city & town, and even every village, a healthy branch of what he called THE AMERICAN LYCEUM. Its public-spirited members would organize & keep up to date a library in which natural history specimens—mineralogical & geological—along with other cultural exhibits, would be catalogued and arranged in the best way to supplement the local public school class work.

Holbrook's plans called for delegates from a network of these groups to form state lyceums, which in turn were to send delegates to a national association. Fifteen neighboring villages quickly responded to his first call, following the organization of his original Lyceum Center at Millbury, Massachusetts, in 1826. Within a year 100 villages had joined, and during the next 12-month period, Lyceums were formed in nearly every state in the Union. New York was the first to organize a state Lyceum Association, followed closely by Massachusetts & several other New England States. Florida also climbed early on the bandwagon.

By 1834 some 3,000 lyceums were established in towns across the country. From coast to coast, men prominent in educational, literary, ecclesiastical and political life took an active part in the movement. They stimulated the broadening of school curricula & the purchase of new equipment for the schools. They took up the cause of more opportunity for girls and women to acquire better educations; and they helped to correlate the work of secondary schools with that of higher institutions of learning. They created a number of lyceum bureaus, which served as central offices, made arrangements for lectures or exhibits, and gathered information for lyceum centers within their orbit.

It was as though Josiah Holbrook had opened a dam above arid lands. Though his emphasis was upon natural sciences & chemistry & mathematics, once the gates were opened, every subject became part of the avalanche, from agriculture to philosophy & religion. People in the larger cities became as enthusiastically involved as those in the country towns and villages. Boston led the way in New England, with Daniel Webster presiding at its meetings for several years. At the beginning of the movement, weekly meetings were the rule in most communities, with home talent producing the programs gratis. Then some of the lyceums in the larger cities prospered to the point where they could offer a small fee for an outstanding lecturer to address them. Daniel Webster, who had already become known as a stimulating speaker before Holbrook touched off the lyceum flood, was paid $100 for a lecture at Concord. Fees soon began to climb for the speakers most in demand; and Emerson became one of these. William Lloyd Garrison, Wendell Phillips, Theodore Parker, Frederick Douglass, were familiar platform figures in the mid & late thirties, and slavery was the popular subject for debate.

This became the Inspiration for the Emancipation Proclamation, January 1, 1863.

It was at the Concord Lyceum in 1838 that Thoreau delivered his first public lecture. "Our least deed," he said on this occasion, "like the young of the land crab, wends its way to the sea of cause & effect as soon as born, and makes a drop there to eternity."

By 1839, a national convention of the American Lyceum Union received wide support from the press and had considerable influence upon public school systems. Bayard Taylor; John B. Gough, reformed alcoholic who plead eloquently for temperance; Edwin Whipple, Louis Agassiz, the famed naturalist; Henry Ward Beecher, often referred to in a jingle as "The Sunday School teacher" and rated as one of the greatest orators of all time; Oliver Wendell Holmes, urbane & witty physician — these were a few of the brilliant galaxy of American men of stature who spoke to lyceum audiences. They were soon joined by the intrepid Lucy Stone & Elizabeth Cady Stanton, who inaugurated the battle for women's rights. And shortly before the Civil War interrupted the progress of the lyceum movement; Anna Dickinson & Mary Livermore also fought for women's suffrage along with other reforms.

"Mr. Livingstone I presume?" . . . soon became the standard greeting to most men named Livingstone.

A NUCLEUS of active speakers remained available even during the war, and soon after it was over, lyceum bureaus became more active than ever. East competed with West for the best talent, and fees climbed so high that lecturing became a lucrative profession in itself. Early in his career as a speaker, Mark Twain received $300 per night. Beecher received $500; then to secure him & the best talent in the field, the Pond Bureau doubled that fee. Soon after returning from Africa, Henry M. Stanley signed up with Pond for a series of 100 appearances at $1,000 each, to tell the story of his adventurous search for Mr. Livingstone, the famous "lost & found" missionary. The Pond Bureau was justified in its fearless attitude toward high stakes, for the first Stanley lecture brought in nearly $18,000.

THE COMPETITION reached such fever pitch that the demand for good speakers soon exceeded the supply. It became necessary to press into service dramatic readers & musical entertainers. Lyceum opera & concert companies (such as the Jess Pugh group with which young Irene Briggs toured) became an accepted & popular part of all lyceum programs.

DOWN INTO the closing years of the 19th century, under the stimulus by then of both lyceum and Chautauqua, SELF-EDUCATION was the goal of an innocent era, through lectures, debates, classes for study & mutual help, essay contests, public readings & recitations, and lastly the so-called "Literary."

ALTHOUGH TOO YOUNG to participate (your scribe speaking) I remember that my parents & grandparents were active in meetings of their local "literary" during the winter months when farm duties were not too urgent. A "lumber sleigh"—ours or one belonging to a literary-minded neighbor—would be driven to our door, with seats improvised to accommodate others to be picked up on the way to the meeting. It was a jolly group in pursuit of good fellowship as well as self-improvement. How I begged to go with them—and cried because they wouldn't let me!

. . . and still would love to!

JAMES WHITCOMB RILEY's poems in Hoosier dialect were standard fare for all lyceum and chautauqua audiences; their appeal was universal. His verse was not all humorous; some of it was philosophic and fantastic—even epic in scope, for instance, "Flying Islands of the Night"—a monumental drama illustrated by Franklin Booth, a famous Hoosier artist and personal friend of Riley, who was himself a commercial artist & sign painter for a decade before his genius as a poet was recognized. (Good training for a "man of letters"!)

IT HAS been said that Riley was to American poetry what Mark Twain was to American prose. His "At the 'Literary'," in 6 rollicking stanzas, hit culture-seekers right where they lived—at home—and they loved it! Two stanzas should be sufficient to demonstrate just how such a goodly gathering of "Folksy" people was conducted—in this case by the poet himself:—

J. W. Riley
Poet & "Critic"

Very first night I was there
I was 'p'inted to be what
They call "Critic" — so's a fair
And square jedgement could be got
On the pieces 'at was read,
And on the debate,— "Which air
Most destructive element,
Fire er worter?" Then they hed
Compositions on "Content,"
"Death," and "Botany," and Tomps
He read one on "Dreenin' Swamps"
I p'nounced the boss, and said,
"So fer, 'at's the best thing read
At yer 'Literary'!"

Then they sung some — tel I called
Order, and got back ag'in
In the critic's cheer, and hauled
All o'the p'formers in:—
Mandy Brizendine read one
I fergit; and Doc's was "Thought";
And Serepty's, hern was "None
Air Denied 'at Knocks", and Daut—
Fayette Strawnse's little niece—
She got up and spoke a piece:
Then Izory she read hern —
"Best thing in the whole concern,"
I-says-ee; "now le's's adjourn
This-here 'Literary'!"

As an Afterthought—"le's's" add that 6th stanza to complete the picture which hints of Romance as a potent force that might well be included in that long list of Lyceum's "combination of basic subjects" which makes life more worth the living—and learning worth the effort.

—With due credit to Messrs Currier & Ives —
— and Memory —

They was some contendin'— yit
We broke up in harmony.
Road outside was white as grit,
And as slick as slick could be!—
I'd fetched 'Zory in my sleigh,—
And I had a heap to say,
Drivin' back — in fact, I driv
'Way around the old north way,
Where the Daubenspeckses live.
'Zory allus—'fore that night—
Never 'peared to feel jest right
In my company. — You see,
On'y thing on earth saved me
Was that "Literary"!

"Squire" Van Buskirk — Justice of the Peace & reporter for the Clyde Times — had quite a library of his own and was an important person at the literaries in our town. So was my grandfather, who had been a school master in "The Land O'Canaan", Litchfield County, Conn., before moving to the Finger Lakes Region in up-State New York. One of the programs required members to impersonate some famous piece of literature without saying anything about it—other members were to guess what it was, as in a charade. Grampa's "props" for his assignment were simply himself & 2 small letters, s-a, on the lapel of his coat. Classical as his name Homer and dignified—befitting his official title of Senior Warden of the Vestry of St. John's Episcopal Church—with stubby square beard & sideburns sans mustache—bald head topped with a black skull cap—6ft. tall & every inch a man, he looked more like a mandarin than an ecclesiastical potentate. Our readers, knowledgeable (as well as gentle) will, of course, have the correct answer already. Squire Van Buskirk pondered for half an hour before getting it—Pope's Essay on Man. None of the others came close but they had learned something and had exercized their powers of imagination. Once again, the lyceum concept had served its purpose.

Since "Gramp" & Holbrook were "down-east Yankees" as well as school teachers in the same area & era, they may have been (I like to think) Collaborators in the Common Cause of Culture.

THE STORY of Chautauqua in America is a classic example of the power of an idea in tune with its own time. A movement that was to mushroom into a source of inspiration for culture-hungry millions, the movement that Theodore Roosevelt was to call the "most American thing in America," began simply and earnestly in the mind of a young Methodist minister, John H. Vincent, who felt that Sunday School teachers should be trained for their calling as secular teachers were.

INSTEAD OF a revival camp-meeting style religion with its sapping emotional hysteria, he wanted to see an intelligent approach to the teaching of religious principles & history. No matter how idealistic the usual run of Sunday School teachers might be in the 1850's, they were ill-informed concerning biblical geography & history. Vincent also anticipated the modern "problem" of leisure, which few educators faced at that time. He believed that education was continuous throughout life and that it should be a joyous activity as well.

TAKING the advice with the support of Lewis Miller, a wealthy Sunday School Superintendent of Akron, Ohio, Vincent acquired a permanent site for the practice of a 3-year course of study he had been developing since 1855, and had formally proposed in the Sunday School Journal, which, as General Agent for the Methodist Episcopal Sunday School Union, he edited from 1865 to 1888. The fair grounds

Miller was also an Inventor & Manufacturer of farm machinery. He produced the first successful "Self-binding Reaper".

at Fair Point, on the shore of Lake Chautauqua, New York, had seen better days as the site of the tawdry sort of camp meeting that Vincent despised, but it presented his opportunity to achieve a marriage between dream and reality. And under his influence it soon lost all of the aspects of revivalism. ¶ THE FIRST two-week session of the National Sunday School Assembly, which began on August 4, 1874 (forty-eight years after the inauguration of Holbrook's first Lyceum Center), took place under primitive conditions at what is now "Miller Park." Backless benches faced a small covered speaker's stand surrounded by crude wooden cottages & tents. At night boxes of dirt held pine-knot torches for light. Normal classes, attended by 142 Sunday School teachers from 25 states, Canada, Nova Scotia, Ireland, Scotland & India, were held in 4 tents set up at the corners of the Park. At least 2,000 persons attended the opening outdoor meeting and the number increased rapidly. By August 12, some 10,000 to 15,000 were on the grounds to hear Rev. T. Dewitt Talmage. It showered often during the 17 days of the Assembly, but the outdoor audience merely put up their umbrellas and kept on listening... ¶ JOHN VINCENT was filling a very real need. His was an unorthodox approach to the growing divisive trend that was sapping the strength of the Protestant Churches, but it worked. By developing an international system of scripture lessons, he brought the denominations together. He taught the geography of Bible lands, the habits & customs of people of biblical times, by building a miniature Palestine on the camp grounds, an enlargement of the Palestine Class idea he had worked out with earlier groups such as his Galena pastorate. This practical means of studying & playing together soon took hold all over the country.

NB: The Map is upside-down

W — E

Actually, the Lake (Mayville • • Jamestown) extends diagonally from N.W. to S.E.

First Auditorium in Miller Park 1877

Seats face the Lake and boat landing.

AN EARLY PALESTINE CLASS

(Galena, Illinois - 1859 · 1861)

A Group destined to Grow

JOHN H. VINCENT at the age of twenty-seven with a back ground flock of students typical of the early groups that led to such scenes as those taken 60 years later in the Chautauqua New York Amphitheater, as shown on spread pages 86-87.

In Oval:—
No. 1 boy in front row back of J.H.V. is James W. Scott who later became the owner & publisher of the Chicago Times-Herald.

No. 2 boy in top row at right is George W. Swift who became the Mayor of Chicago.

A PALESTINE CLASS garbed in Oriental Costume to relive biblical times, is posed on a mount in the model of Jerusalem

IN 1875 community dining rooms and frame hotels were erected to make Fair Point the first permanent Chautauqua establishment. The Amphitheatre was added 3 years later. In 1883 courses in secular history and literature were developed by William Rainey Harper (later President of the University of Chicago), for Vincent wanted to broaden the intellectual background of the student body. Individuals who were successful in various walks of life were asked to address the Assembly each year.

EIGHTEEN SEVENTY-SIX saw evidence of the mushrooming hold that Vincent's ideas had taken upon the United States. Over 300 attended the New York Chautauqua summer session, and 3 independent assemblies were conducted successfully on the pattern of the original — the Bay View Assembly at the north end of Lake Michigan; one on Wellesley Island in the St. Lawrence River; and one at Clear Lake, Iowa. It was soon evident that the most successful chautauquas were those founded at some distance from large cities.

The Magic Lantern or Stereopticon was indispensable, then as now, to the Lecturer.

A Cleveland audience gathered for "The Chautauqua Idea in Picture & Story." The CLSC Round Table for December 1901 reports 6 sets of slides in use from New York to Colorado for this Lecture

THREE YEARS later, what might be called the first real correspondence course in the U.S. was officially instituted as part of the Chautauqua movement when Vincent hired a remarkable person, Miss Kate Kimball, then only eighteen, to be executive secretary of the Chautauqua Literary and Scientific Circle,* which he formed to provide a 4-year home-reading course designed to keep Chautauqua active the year round. From a course describing the chief civilizations on the globe, this was gradually enlarged to include reading courses in history, literature, sociology, & the sciences. A diploma or "Recognition" would be obtained by passing a comprehensive examination at the end of four years.

* Hereinafter referred to as C.L.&S.C.

Monogram of the 1890's

THE AGED concerned Vincent deeply. The Circle, he said in an address to the class of 1882, "without calling itself a university, is a university for the old, where the joys of youth may be put into the heart again. . . ." But part of his concern was also the family unit; Chautauqua was designed to bring age & youth together in many shared programs.

RECOGNITION DAY, for example, was soon established as one of the most moving special days of each year's assembly. (By 1900 some 50,000 men & women had earned their diplomas, and some 250,000 had actually enrolled for the course, even if they did not go through to the final exam.) And Children's Day was another high note each summer.

← At left . . . . "Golden Gate" Recognition Day at Chautauqua, N.Y. 1909 & Flower Girls on Recognition Day. Photo: Abbé Klein University of Paris. 1910

". . . When the 'Golden Gate' shall be opened and children with their baskets of flowers, conforming to custom from time immemorial, will strew with blossoms the pathway of pilgrims under the arches of the 'Hall on the Hill'." J. H. Vincent

The Hall of Philosophy was the imposing type of building Vincent early visualized & it became the model for many other Chautauqua Temples.

Sketches of early Chautauqua, draw

The Pattern was set and still goes on at Fair Point, renamed CHAUTAUQUA, N.Y.

The Landing Pier was a popular spot (1899).

What would these well-d
if a Bikini-clad M

...aham (1880) for HARPER'S WEEKLY.

Mama's Class in China Painting

A Sand-pile Party was fun for the Children.

...immers (1909) have said
...ppeared among them?

Papa & the older boys liked Baseball (1898).

Down by the Lake in the seventies

Model of Jerusalem – below Miller Park in 1874

" Combination frame and curtain Strong-box," called the Ark was used to house early Chautauqua Platform Celebrities.

TREE FOUNTAIN – a Pipe in the hollow Branch was the Source of watery Mystery to thousands of Children.

CHAUTAUQU

"A PLACE, an IDEA, a

Classes at Swedish Gymnasium around 1905

"Other Times, other Customs!" Do

Aula Christi Hall, Architect Paul J. Pelz, Washington, D.C.

This combined Cottage & Tent was the Summer Home of Bishop John H. Vincent.

C.L.S.C. GRADUATES in Procession on Recognition Day– the joyful Climax of four Year's Home Study & Reading

The Athletic Field — Hub of many Sports

NSTITUTION

" — John H. Vincent

beside the Amphitheater (1914)

Fun with a Motor Sled on the Lake (1914)

Everybody lived happily (1898).

A Diploma (Class of '84) "Which has lived up to its possibilities."
Each White Seal represents extra Effort in the Pursuit of Knowledge.

SUGGESTIONS FOR STUDYING THE "ÆNEID"

(Typical of Instructions to C.L.S.C. Students)

(1901)

Professor Miller in his account of the "Æneid" has called attention to the leading characteristics of the poem, and readers who keep these in mind as they read Vergil will find much enjoyment in applying the suggestions which he makes. Those who have time to refer to some of the books mentioned in the bibliography will gain a new idea of the dignity and importance of this great poem as it is looked upon by scholars. One member of the circle might be appointed to prepare a paper upon what the great critics have said of the "Æneid." But a still better plan, if possible, is to set all the members at work upon the poem. Here are some interesting points to be noted:

1. Religious significance of the poem. Idea of fate.
2. Its glorification of Augustus.
3. Characteristics of the gods; the part they play.
4. The character of Æneas; how he compares with the characters of Homer.
5. Other characters of the poem.
6. Views of the under-world in Vergil's time compared with Homer's. See the sixth book of the "Æneid" and the eleventh book of the "Odyssey."
7. Famous descriptions of scenery.
8. Fine passages of the poem.
9. Vergil's use of figures of speech.
10. The games of the "Æneid" compared with those of Patroclus in the "Iliad."

These ten points may be assigned to ten individual members, and each may study the poem either from the selections given in Mr. Miller's book or from the entire poem so far as they are able to secure it. Or each member may be assigned a single book of the "Æneid" and may report on all of the suggestions given above, in so far as they apply to his particular book. The best possible plan will be to secure some high school or college teacher of Vergil and ask him to guide the class (not to give a lecture), and devote two circle evenings to the study.

Still going Strong — A Sidewalk Session of the C.L.S.C. (1966)

THE CHATAUQUAN, "a monthly magazine devoted to the promotion of true culture, organ of the Chautauqua Literary and Scientific Circle," began its long career in October 1880, with Thomas Flood as editor. Fearlessly Volume One number 1 undertook a history of the world. Then, coming down to more immediate issues, it provided information on the C.L.&S.C.: its aims, methods, membership (beginners, whether they entered the 2nd, 3rd, or 4th year of a course, were counted as first-year students, and continued in a sort of gigantic Frère Jacques round from where they started).

George E. Vincent of Flood & Vincent Printers, was a Brother of Bishop Vincent.

THE STUDIES for 1880-81 were outlined in detail, and a list of required readings given (40 minutes daily), with an accent on recommended scriptural materials. Attendance at Chautauqua was not required, though it was desirable. However, quarterly reports of reading done, etc., were required. Individuals could take the course alone, but it was suggested that they form or join a circle of two or more who would agree to meet for regular discussion, as suggested on the opposite page.

LEAFING THROUGH The Chautauquan, which remained the faithful servant & counselor of the movement until it was taken over in 1914 by the Independent Weekly, is an education in itself in changing styles, emphases upon righteous causes, & advertising.

ILLUSTRATIONS reproduced by the photo-engraving process in halftone began to appear in 1891, in the era of fantastic Sapolio and Ivory Soap ads, with the evocation of household drudges on the one hand & delicate Burne-Jones-style drawing-room ladies on the other. It was an era also of small-print ads and solemn statements.

157

WHY ARE SOME PEOPLE ALWAYS LATE?—They never look ahead nor think. People have been known to wait till planting season, run to the grocery for their seeds, and then repent over it for 12 months, rather than stop and think what they will want for the garden. If it is Flower or Vegetable Seeds, Plants, Bulbs, or anything in this line, **MAKE NO MISTAKE** this year, but send 10 cents for VICK'S FLORAL GUIDE, deduct the 10 cents from first order, it costs nothing. This pioneer catalogue contains 3 colored plates. $200 in cash premiums to those sending club orders. $1000 cash prizes at one of the State Fairs. Grand offer, chance for all. Made in different shape from ever before: 100 pages 8½x10½ inches. **JAMES VICK, SEEDSMAN, Rochester, N. Y.**

GREAT OFFER!

PIANOS! ✣ $35. ✣ ORGANS!

Direct from the Factory at Manufacturer's Prices. No such offer ever made before. Every man his own agent. Examine in your home before paying. Write for particulars. Address

THE T. Swoger & Son Pianos & Organs
BEAVER FALLS, PENNSYLVANIA.

From Rev. James H. Potts, D. D., editor of *Michigan Christian Advocate*, Detroit, Michigan—"To say we are delighted with the Piano does not express the fact. We are jubilant. If all your instruments are as fine in appearance and as pleasing in tone as this one, your patrons will rise by the hundred." Mention THE CHAUTAUQUAN.

NOW READY

All He Knew. A Story By JOHN HABBERTON, author of "Helen's Babies," etc. 12mo., cloth, $1.00

Prices were unbelievably low.

348 THE CHAUTAUQUAN.

The Verse was done in a very Stylish Handwritten Script for that period for an Ad used in an outstanding Soap Campaign.

The "Soap Opera" was yet to come & the Efficacy of Cure-alls:— has always been exaggerated.

D. NEEDHAM'S SONS,
Inter-Ocean Building,
Cor. Madison and Dearborn Streets,
CHICAGO.
RED CLOVER BLOSSOMS.
And FLUID and SOLID EXTRACTS OF THE BLOSSOMS. The BEST BLOOD PURIFIER KNOWN. Cures Cancer, Catarrh, Salt Rheum, Eczema, Rheumatism, Dyspepsia, Sick Headache, Constipation, Piles, Whooping-Cough, and all BLOOD DISEASES. Send for circular. Mention this magazine.

"You see, young ladies, the value of a well-made soap depends very largely on the amount of *true soap* which it contains . . .

A WORD OF WARNING

There are many white soaps, each represented to be "just as good as the 'Ivory';" they ARE NOT, but like all counterfeits, lack the peculiar and remarkable qualities of the genuine "Ivory" Soap and insist upon getting it.-

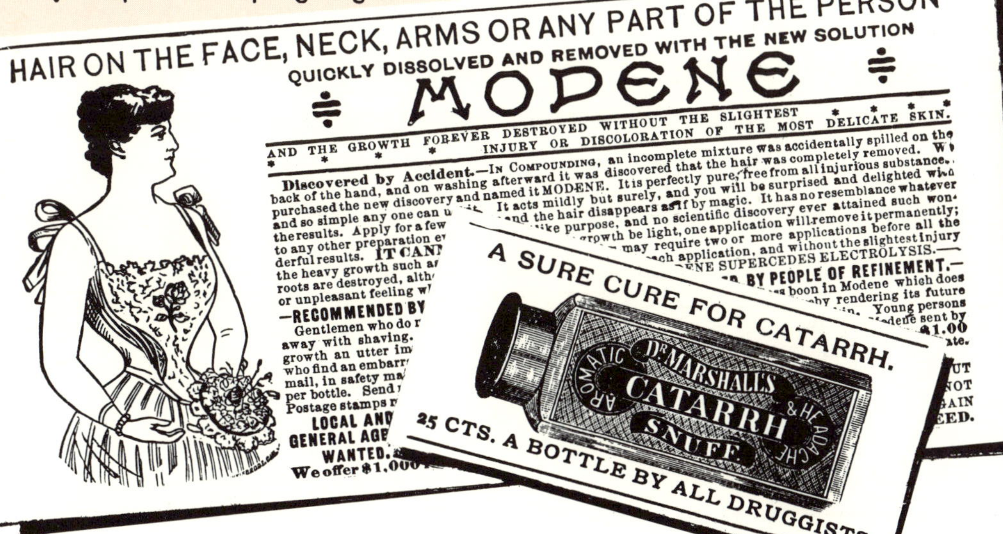

45

Ads for Soaps, Cosmetics & "Sure-Cures" were plentiful in THE CHAUTAUQUAN's front & back pages with lots of Good Solid Cultural meat sandwiched in between — So to Speak.

It Puts Off Old Age

by nourishing the entire system. Quaker Oats makes your blood tingle; nerves strong and steady; brain clear and active; muscles powerful. It makes flesh rather than fat, but enough fat for reserve force. It builds children up symmetrically into brainy and robust men and women. You can work on Quaker Oats. It stays by you. At all grocers

Perennial Youth & Good health were assured by Quaker Oats & Dowd's Exerciser with a bit of Vocal Culture thrown in for good measure.

D. L. DOWD'S HEALTH EXERCISER.

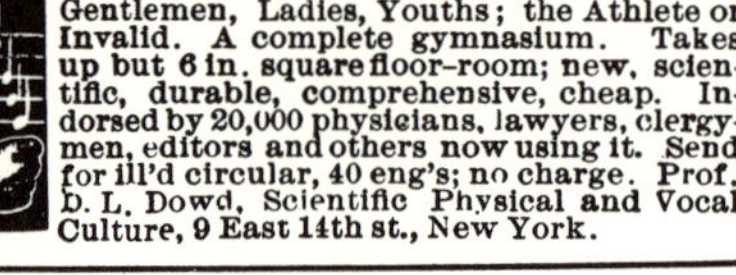

FOR BRAIN-WORKERS & SEDENTARY PEOPLE

Gentlemen, Ladies, Youths; the Athlete or Invalid. A complete gymnasium. Takes up but 6 in. square floor-room; new, scientific, durable, comprehensive, cheap. Indorsed by 20,000 physicians, lawyers, clergymen, editors and others now using it. Send for ill'd circular, 40 eng's; no charge. Prof. D. L. Dowd, Scientific Physical and Vocal Culture, 9 East 14th st., New York.

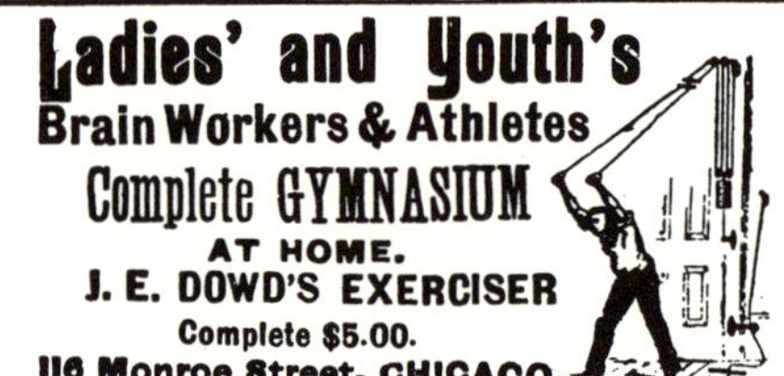

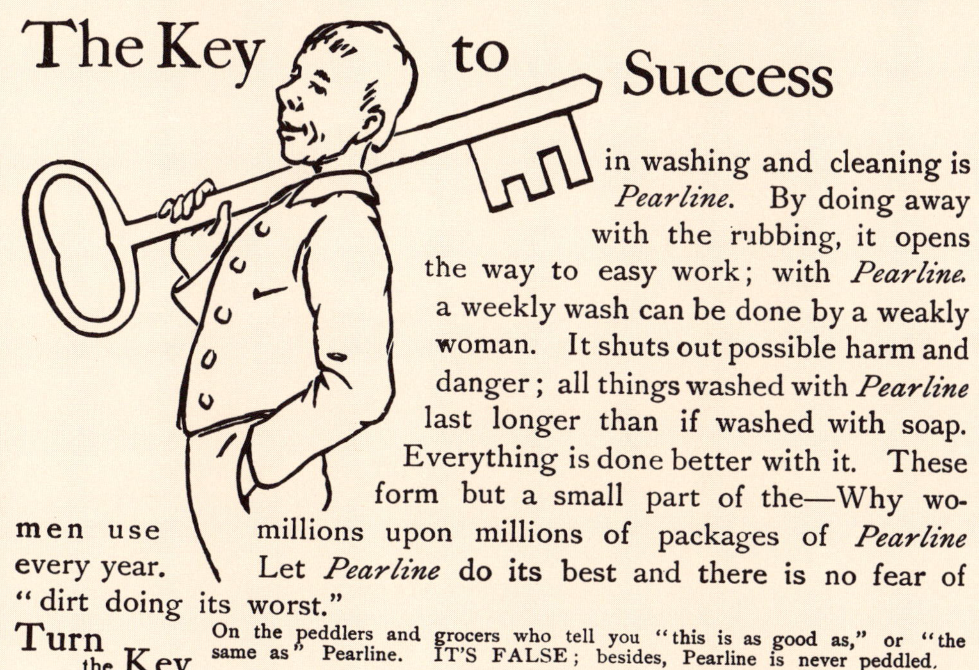

We are children who cheerfully join in the chorus,
When **PACKER'S TAR SOAP** is the subject before us;
Mamma tried all the rest,
So she knows it's the best,
And we laugh with delight when she lathers it o'er us.

Christine Terhune Herrick says: "No mother who has ever used Packer's Tar Soap for her babies would willingly do without it."

Egbert Guernsey, M. D., 526 Fifth Ave., New York, says: "I use Packer's Tar Soap in Children's eruptive troubles with the most marked benefit."

Packer's Tar Soap

Should be in every household. It is a luxury to use as well as perfectly pure, while many soaps are notoriously dangerous.

25 Cents. All Druggists. ☞ Mention THE CHAUTAUQUAN, and send 10 cents, stamps, for sample ½ cake, to

THE PACKER M'F'G CO., 100 Fulton St., N. Y.

ESTERBROOK STEEL PENS.

For Sale by all Stationers.

ESTERBROOK PEN CO., 26 John Street, N. Y.

James Pyle didn't see eye to eye with the claims of P&G's Ivory, nor with those of Sapolio and Packer's Tar Soap. As ever, Each was BEST.

And what a "Singing Commercial" that P.T.S. jingle would make today on Radio & T.V. if sung by Children's Voices.

Simple and Sedate — 1890. Good design, even by today's stiffer Standards

542 THE CHAUTAUQUAN.

Solid Vestibuled Trains

—AND—

THROUGH PULLMANS

TO

CHAUTAUQUA LAKE

FROM

CHICAGO,
St. LOUIS,
CINCINNATI,
CLEVELAND,
PITTSBURGH,
NEW YORK,
BOSTON,
BUFFALO.

People's Line of New Steel Steamers between Lakewood, Jamestown, & Chautauqua. Direct and sure connection with all Erie trains.

L. P. FARMER,	W C. RINEARSON,
Gen'l Pass'r Ag't, New York.	Ass't Gen'l Pass'r Ag't, Cleveland, O.

A Good Thing

in any line of business is always appreciated. I have a catalogue of 500 designs which I mail to any person on receipt of 25c. I furnish ideas to advertise any business and will be pleased to hear from any one who has an article they wish to bring before the public. No charge for ideas submitted unless accepted. Photo Engraving and Half Tone work in all its branches. Estimates cheerfully furnished.

H. C. BROWN,

35 and 37 Frankfort St., New York.

A CHOICE BITE.

Sedate and Simple — Among the first Ads to appear in Halftone, 1890

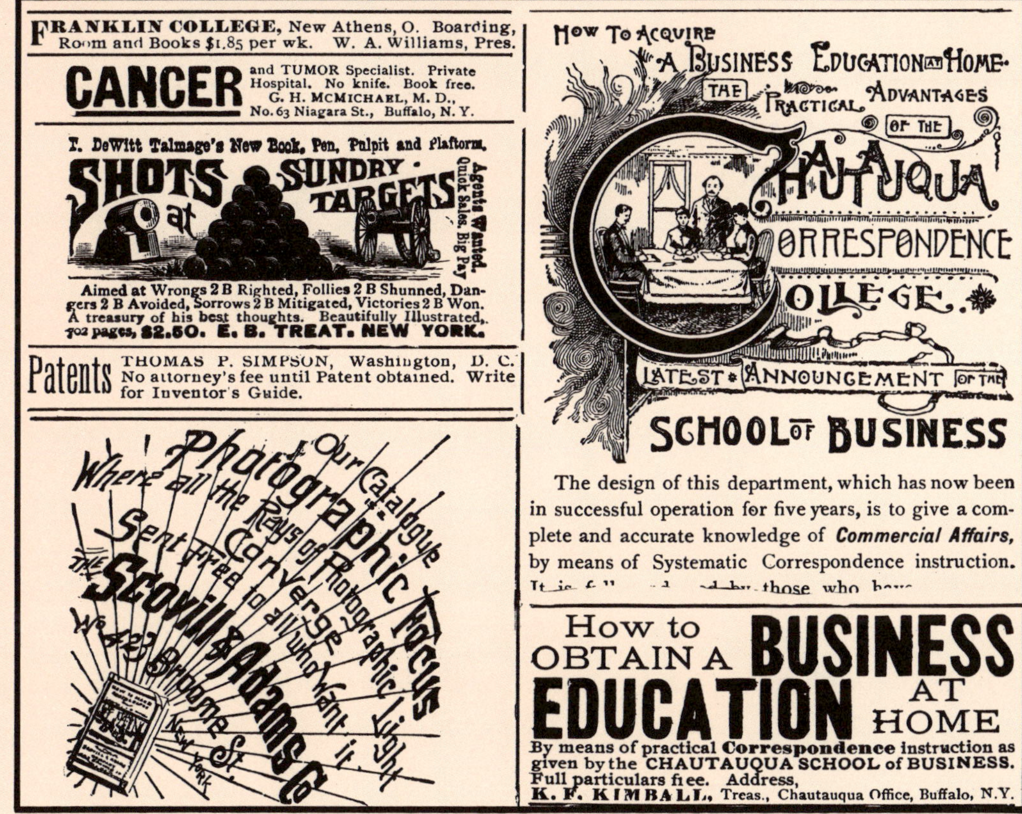

Top half of K.F. Kimball Ad, 1890, Initial 'C' serves for Chautauqua Correspondence College. It could serve for Clutter.

NB: Simpler arrangement of elements for K.F. Kimball Ad 1891

AN OFFSHOOT & direct result of the C.L.&S.C. courses were the public libraries established in small communities all over the country. Chautauqua's first public library was at Damariscotta, Maine, back in the 80's, the product of a club bearing the curious name of Skidompha. The Timrod Club—23 members—started with 200 books donated by towns-people of Summerville, S.C. Chautauqua members in a small town in Maryland took advantage of the state school law which required that for every $10 raised by the school, the county had to give $10. They rounded up all the talent they could find and put on enough entertainments to pay for an excellent library. In Havana, Illinois, Chautauquans unearthed a forgotten law that provided for town maintenance of a public library and made such a clamor over it that one was opened in the Town Hall to be followed eventually by a Carnegie Building, as many of the C.L.&S.C. initiated libraries were in time.

TYLER, TEXAS; Greenwood, S.C; Livingston, Montana; Guthrie, Oklahoma; Ridge Farm, Illinois; Blackwell, Oklahoma—we could go on & on with the list of small communities where a public library grew out of the hard work and enthusiasm of a small Chautauqua Literary & Scientific Circle.

The ladies of Aloha Circle, San Diego, California, believed in the virtues of big hats (1913).

WORD OF the success of the New York assemblies spread rapidly, and hundreds of independent chautauquas sprang up all over the U.S.A., consisting of a permanent year-round nucleus, which in summer was enlarged to overflowing by campers who flocked in to hear the lectures & join in the fun.

THE TYPICAL community was initiated when grounds for the chautauqua had been chosen by a committee and purchased through incorporation & sale of shares to interested citizens. In some cases a permanent community grew up around the chautauqua; homes were built by the original investors and tents were rented to visitors who came for the programs. Sometimes citizens of an existing community simply invested in a corporation to provide a camping ground adjacent to a railroad or seaside transportation, and through the sale of season chautauqua tickets were able to provide for the upkeep as well as for programs that would attract people from a wide area of outlying towns. It was also excellent business for the local merchants. Transportation to the chautauquas was usually at bargain rates, for both railroad & steamship companies found it profitable to cooperate with the organizers. A typical camping ground might look like this:

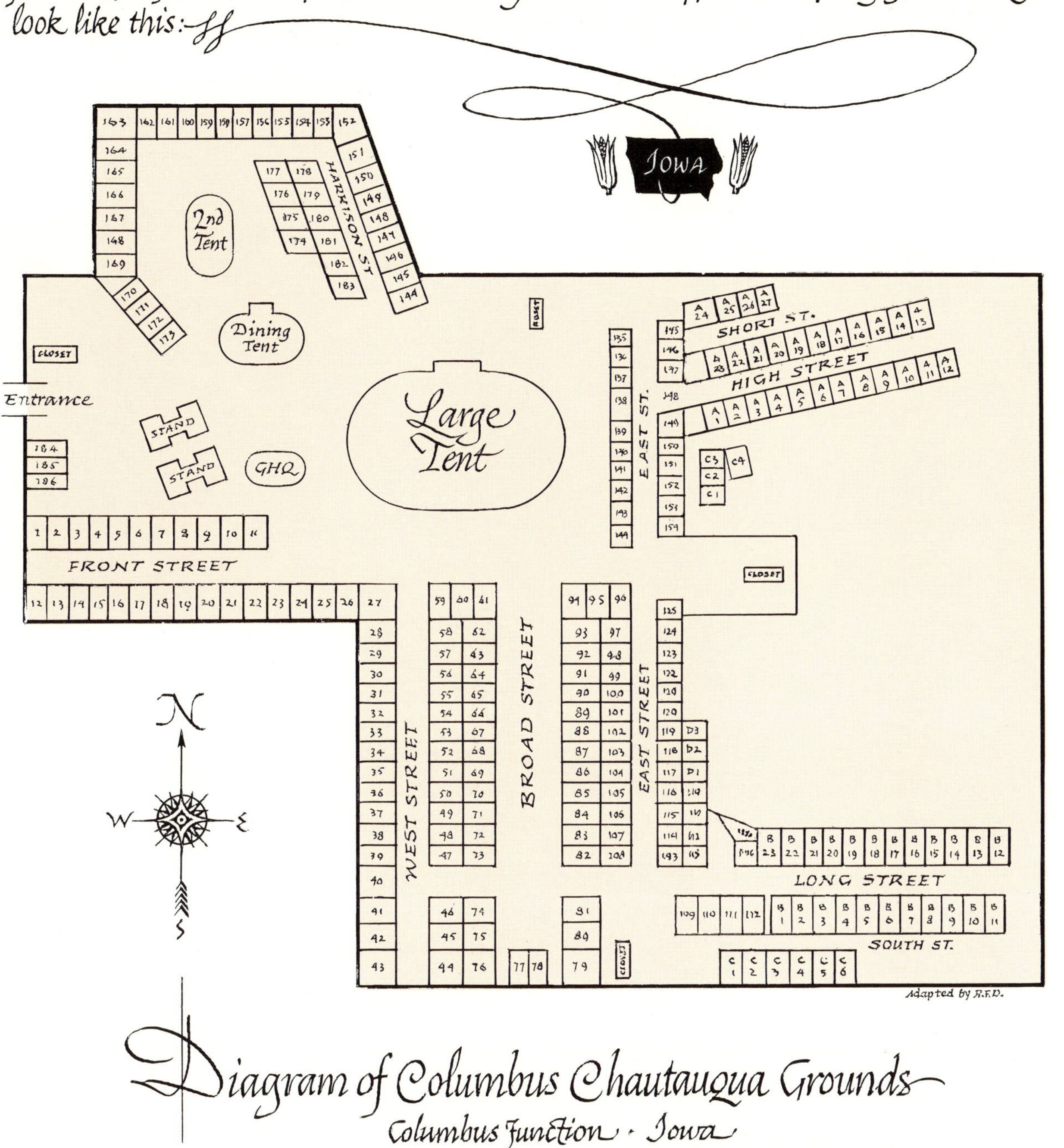

Diagram of Columbus Chautauqua Grounds
Columbus Junction · Iowa

·(FROM OFFICIAL SOUVENIR OF COLUMBUS CHAUTAUQUA · 1905)·

EARLY PICTURES and records of various Iowa chautauquas, now long defunct, are fairly typical of others established in the same era, including the still vigorous chautauquas at Ocean Park, Maine, and the New Piasa Assembly at Chautauqua, Illinois.

Nation-wide Chautauqua

Each dot represents an assembly — The flag indicates location of parent institution

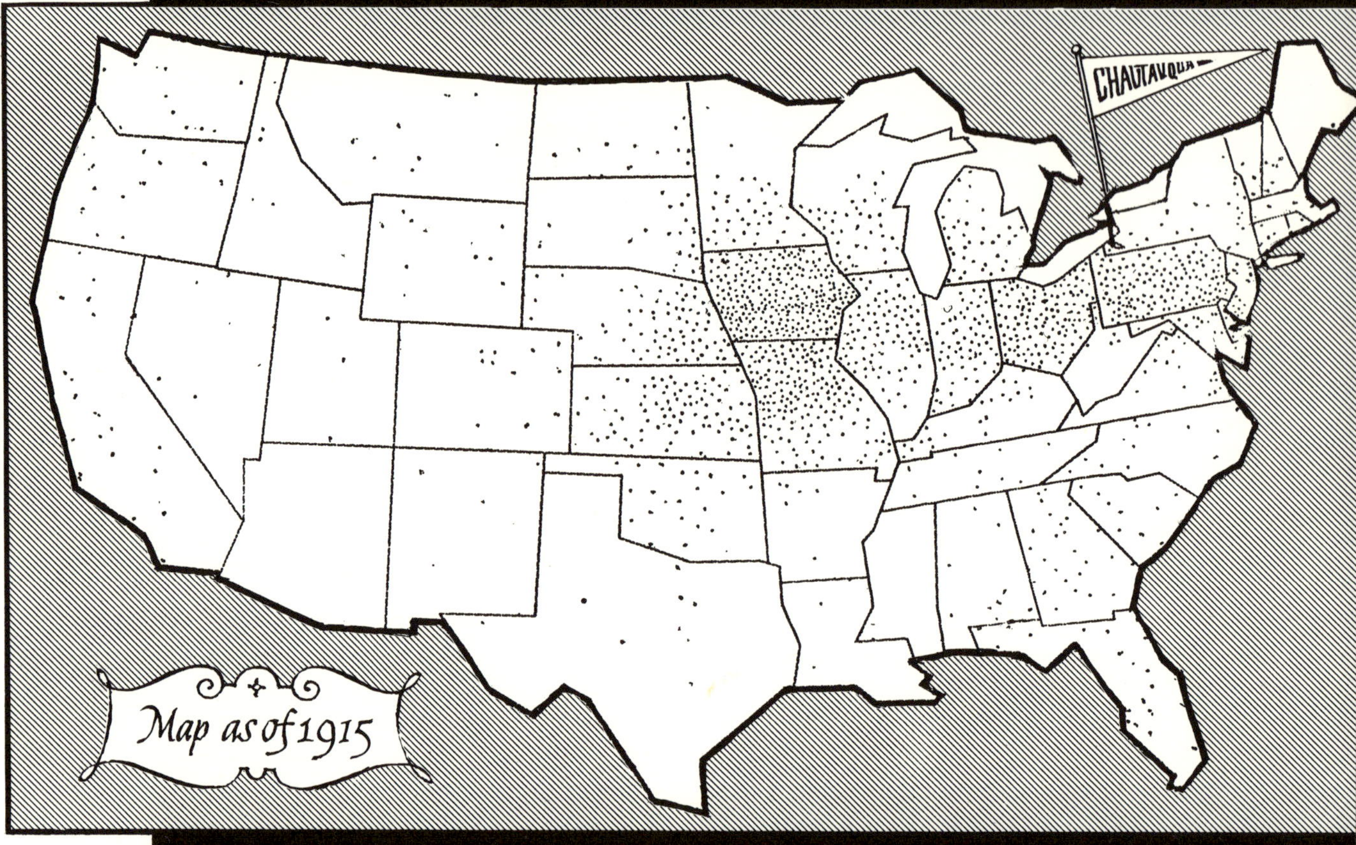

It will be noted that the largest concentration of assemblies is in the middle west. In view of this "is it any wonder that the renowned Walter Damrosch should have declared in 1928 after a careful analysis of the requests for music that had come to him over the years from all sections of the country that 'Iowa is the most musical state in the Union'".

Quoted from The Palimpsest

Chautauqua Parade in Fort Madison, Iowa (1914)
(The Shaeffer Pen Company had its beginnings in Jewelry Store at Sign of Big Clock)

William Jennings Bryan & ex Governor Buchtell of Colorado pictured with the Orchestra Ladies of Highland Park College.

Fiechtel's Yodlers

Lynton's Hussars

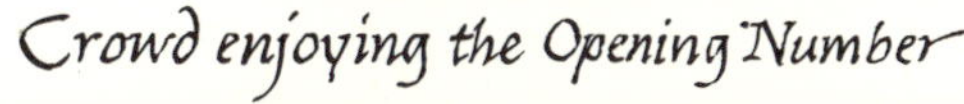

Crowd enjoying the Opening Number

Coming to Chautauqua (by Train)

A Saturday Crowd at Columbus Junction

George Yates Photo — Des Moines Register

The Chautauqua Salute at the Finale—waving handkerchiefs

Ocean Park Activities & Ceremonies have always been governed by the Sea.

OCEAN PARK grounds were included in 8 miles of solid beach, the principal factor in the choice of the area in 1881 by a Free Will Baptist group from New England.

VIEWS OF TH

THE STREETS and simple hou
have the sun-washed airy aspe
of a sea-side summer resort. Oce
Park early established building
zoning laws that successfully ke
out the highly commercialized a

THE OLD "DUMMY" LINE–
In the 1880's people could come by Rail or by the Portland Steamship Lines that docked at the steel pier at Old Orchard Beach, of which Ocean Park is a section.

TRAN

A

Be

Above: Early view of Temple Avenue around 1892

We are indebted for the Ocean Park material to Mr. Adelbert Jakeman, author of "The Story of Ocean Park", published on the occasion of its 75th Anniversary 1956

BEACH VESPERS sung at the foot of Temple Avenue at 5 O'clock every pleasant Sunday in July & August, have been a tradition for over 60 years.

ATER FRONT

osphere of shortlived chautauquas. odgings were 25¢-50¢ per night; ard was 75¢ to $1.00 per day. In e early days people could put their own tents and live free of arge in certain specified areas.

TION

s
e
ng

MILES S. YORK'S 6 WHEEL JITNEY took over when trains stopped running through Ocean Park in 1923. In the early days there were also board walks instead of cement walks; kerosene street lamps were lit every night by a boy who thus earned $15 for his Summer's work.

"Full House" at Billow

Ocean Park, as all Communities that follow t

CHILDREN'S DAY (EACH OF WHIC

Children's Day — Everything for Tots to Teenagers

Early Bathing Scene

A WATERFRONT HOUSE PARTY in the 1890's and a happy carefree Time for the whole Family.

attern set by the Mother Chautauqua, had its...

(...UP IN ITSELF) & RECOGNITION DAY

Recognition Day — Ocean Park Chautauqua style

"THE WAY OF TRUTH" An 1898 Group at the Gateway to the TEMPLE, Dedicated August 2, 1891.

B.C. JORDAN Memorial
Dedicated July 1915

The New Piasa Assembly, as

(organized in 1885) for Rest, Recreation & for Litera

Riverfront and Railroad approach to the Grounds

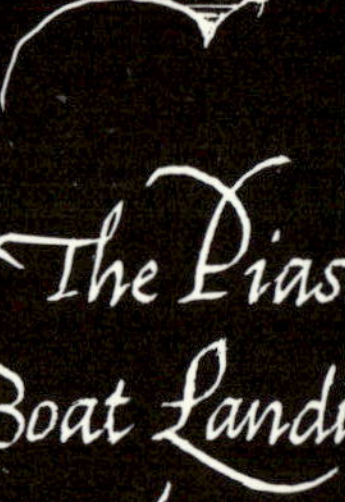

The Pias
Boat Landi
&
the Railro

now called. . ."an ideal summer resort
ientific and Religious Instruction & Culture"

Automobile Approach and Gate to the Grounds

tion were
convenient
to
Each other

We thank Messrs:— Ray H. Beyhmer, Joseph Voss & Robert Mahon for the Piasa material shown on this & the preceding spread.

MODERN BOAT HARBOR— To the right & back of this stood an Inn that burned down in 1919 (see large building in river-front photo, page 58). Admission to the grounds was 25¢ a day, or $1.00 a week or $2.00 for the season.

Concessions were rented to vendors of all kinds: lodgings and lunch stands paid $15.00; watermelon stands, butcher shops, coffee stands, photo galleries paid $5.00; news stands & barber shops $3.00 for the season.

WATER GIVES PIAS

BABB'S HOLLOW SPRING— "The theory is that the water, being so pure cures by filtering through the tissues of the body and washing out impurities which clog the canals." The main factor in choice of the site by a Methodist group was the spring which still feeds 410 acres of what was then forested land between two Mississippi River bluffs that overlook the valleys of the confluent Mississippi, Illinois, and Missouri—all-important in the river system serving America's great Hinterland.

Two-Story Cottage

TYPICAL LODGING, 1894— Visitors could also have tents at $4.00 (8x10 ft.) or $5.00 (10x12 ft.) and $5.00 extra for flooring. Or if they brought their own, they could set them up on the grounds for $1.00 per season.

A GROUP OF CHURCH LADIES— On the grounds ready for the week's activities.

THE PIASA SPRINGS HOTEL— near the center of the grounds and close to the famous spring in the 1890's. "Rates are $1.50 and $2.00 per day, reduced rates by the week. Special dinners will be served on Sunday for 50¢. The ladies {of the Civic improvement Association} solicit and will appreciate your patronage — they are entitled to it".

EALTH & PLEASURE

AMUSEMENT & RECREATION— Besides swimming, boating and fishing privileges, tackle & bait were available at reasonable rates. Piasa also boasted a lawn tennis court, two public croquet grounds, & one of the "finest baseball diamonds in the country." A director was on hand for the latter to organize teams & make all arrangements for games. A sketch class was organized each year, and many other activities — all of which still continue, but with highly modernized equipment.

CHILDREN'S DAY "FLOAT"— PIASA had its memorable days each Summer: Old Folks Day, Grand Army Day, Temperance Day, Book Concern Day, Music Day, and of course, Recognition Day.

THE SUNSET PAVILION— still a popular gathering place for enjoyment of the cool evening breezes from off the river.

N
W
E
S
TO GRAFTON
AUDITORIUM
Map of PIASA Grounds
Flint Park
MISSISSIPPI RIVER

Both the Piasa and the Ocean Park Chautauquas were interdenominational from the beginning though started by ministers, and perhaps this is the reason for their still thriving. There was never a lack of "something to do" for young or old. There must have been youths who rebelled & wanted to strike out on their own, but on the whole, family life, shared activities, happy occasions, were the order of a day that seems longer & longer gone in the face of the 1968 chasm between generations.

A signal of the truly American quality of John Vincent's movement lies in the choice of the Indian-derived name Chautauqua itself, as well as that of many of the individual assemblies. The name Piasa derives from an Illini tribal legend about a monstrous bird that lived in a large cave in the bluffs just below what is now the city of Grafton, Illinois. According to the legend, Utim, son of the beloved chieftain Owatoga, on a fishing trip early one morning, witnessed the capture of one of his young comrades on the riverbank by a huge flying beast.

The State was named ILLINOIS by French settlers after the Illini Indian tribe.

After that, every morning at dawn, returning later at times, the creature was seen to swoop down for another victim, be it brave or squaw or venturesome child. The Indians named it the Piasa, "the bird that devours men." It had all of the horrible aspects of a monster: it was said that its body was that of a horse, and its chin square, with great protruding lower teeth that clamped-in its upper lip, and a long, flowing, coarse beard. It shared some aspects of mythological dragons also, for flames roared from its nostrils, fire gleamed in its eyes; its body & its tail, which it could wind round itself 3 times lengthwise, were covered with a scale-like coat of mail that no arrow could penetrate. It had huge sharp talons, and pronged horns, and an enormous wing-spread. The motion of its wings made a sound like the moaning of a west wind.

The Illini tribe was stricken with terror, and Chief Owatoga was helpless. Finally, after three days of fasting & praying to the Great Spirit, he had a vision in which he was instructed what to do.

Six unmarried braves were chosen, the chief's son among them. All day long, old Tera-hi-on-a-wa-ka sharpened and poisoned arrows, and the tribe fasted & prayed. In the evening they left their village and made their way to the largest white bluff and watched while the 6 young men hid themselves and their brave chieftain stepped out in full view and stood humbly with folded arms, grateful that he could be the means of saving his people.

With the first streak of dawn came the scream of the Piasa as it swooped down over the treetops and saw the old man, then plunged toward him. Owatoga flung himself face-downward and clutched with all his strength the sturdy roots that grew there upon the bluff. The great bird sank its talons into the old man's flesh and lifted its wings to fly. At that moment 6 arrows were imbedded beneath its wings. Time and again the bird tried to lift itself, still clinging to the old man, but each time arrows sped to the mark, the only places where its body lacked a coat of mail. The poison was very strong. Soon the monster screamed in agony, loosened its hold on Owatoga, wrapped its wings about its body, and rolled down the bluff into the Mississippi River, never to be seen again.

THE OLD CHIEF was still alive though badly mangled. He survived to be feasted by the Illinis, and the arrow maker, the artisan of his tribe, painted an enormous bright-colored portrait of the vanquished bird on the side of the bluff where the chief had waited.

The Piasa legend has been adapted from material published in the New Piasa 75th Anniversary Booklet.

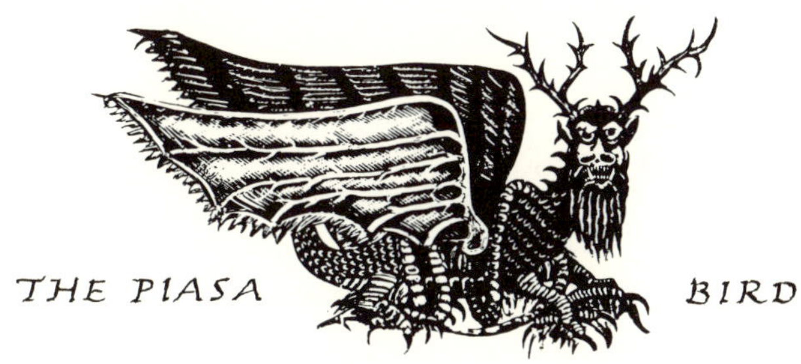

THE PIASA BIRD

THE PAINTING remained for many years after the Illinis had retreated from the area, for various mentions & descriptions of it exist, going back to one by Père Marquette—the Jesuit explorer. From these, Professor William McAdams, Illinois State Geologist, made a drawing which has been adopted by the New Piasa Chautauqua Assembly as its symbol. Hundreds upon hundreds of arrowheads were found on a flat ledge beneath the painting, and it is believed that an Indian never passed without shooting an arrow at its feet in token of Illini deliverance from the terrible bird. In 1836 John Russell, a professor from Shurtleff College, climbed up to the cave and found that it measured about 20x30 feet and was full of human bones. Thus it would seem that the legend was based upon some measure of truth.

By 1913 Mack Sennett & his Keystone Co. had the Formula for slapstick and the World laughed at its caricatured self.

Studio scene – said to be the best picture of one ever taken – at the Vitagraph Company's Studios, Brooklyn, New York, during the making of a spectacular picture & reproduced in THE CHAUTAUQUAN, June 7, 1913, which featured the rapid growth of the new industry. * * * * * * * * * * * * * * * *

THE OPENING THEMES of the 20th century tend to be forgotten. But no matter how extravagant some aspects of today's civilization are, clearly traceable motifs run through the whole period. For one, the necessity of dealing with the century's 2-edged gift of time – leisure time, which can be used creatively or become a burden. By 1914 it was the major concern of both Lyceum & Chautauqua, whose leaders were urging the use of public schools as centers of creative activities for the whole family. It seems odd that although the problem is now more urgent than ever, after a long period during which it was largely ignored— even in the face of evidence provided by park benches during the depression ☞ it is still not given the measure of concern that Vincent & his followers gave it then.

apples 5¢

some ex-executives had to get out & earn "WAMPUM" the hard way!

ANOTHER recurring motif is the search for roots, for identity. The use of Indian names in Chautauquas across the country had deeper significance than mere fascination with many hyphenated words. People wanted to know more about the countries whence their ancestors came, and about those whom their ancestors usurped. In the process of finding out, they contributed to history, or to the preservation of it, for many legends were carefully unearthed & recorded. ¶ Today folk festivals are crowded during the summer; town historical societies are on the increase; young people are as ever looking for roots.

THE AVERAGE small-town boy of the 90's had to struggle to attain more than a grammar school education, perhaps even for that much. There was little culture offered to the adult. He dug for it himself, or traveled long distances and put up with discomforts for weeks at a time to get it. There were few libraries and most of them were private. There was no radio or television, with its tremendous potential for bringing culture into the home. The movies were just beginning to grow.

WILLIAM J. PETERSEN noted in The Palimpsest (May 1962): "The American Vitagraph, by Frederick Held, of Brooklyn, New York, was featured the first 2 days of a chautauqua at Columbus Junction but probably no one even dreamed in 1904 that this would one day be a potent factor in causing the decline of Chautauqua."

OLD TIME

CHAUTAUQUA

DO YOU REMEMBER?

Well, then, dearie, you're much older than I!
The Parke County Sesquicentennial Committee
invites you to come to Parke County

AUGUST 19-21, 1966

and join the "way back when" crowd.

MUSIC, BANDS, HISTORICAL ENTERTAINMENT, LECTURES, PINK LEMONADE, etc.

All at BEECHWOOD PARK in Rockville

in the old Chautauqua Pavilion. Camp sites in the park available. Swimming pool nearby.

For tickets, room reservations, program, etc., write:

Parke County Historical Society
Rockville, Indiana

THE CHAUTAUQUAN
A·WEEKLY·NEWSMAGAZINE

The Marriage of Lyceum & Chautauqua

really began in 1904, when Lyceum agencies, which had been supplying lecturers & musicians to outlying towns during winter months, finally saw their opportunity in the summer chautauquas to consolidate their own efforts on a year-round basis. By this time Lyceum had already reached and descended from its first peak of activity; Chautauqua had widened its front to include every kind of healthy stimulus; the purely humorous subject vied with the solemnly intellectual; the common-sense practical aspects of home economics and national politics with the international scene. C.L.&S.C. groups were making their way around the world physically as well as mentally.

It was in 1904 that Keith Vawter, of Cedar Rapids, Iowa, began to offer communities along a planned geographical transportation route a "package deal" consisting not only of a definite program of cultural entertainment, but all the equipment necessary, plus management teams as well. The local community had only to guarantee the sale of a certain number of tickets, as had been done all along for Lyceum winter circuits.

"The Independent chautauqua introduced the idea of popularized education to select groups," wrote Hugh A. Orchard in Fifty Years of Chautauqua, but "it became evident, early in the 20th century, that the chautauqua could never be brought to serve the whole population so long as it consisted of isolated units of culture." Through circuit chautauqua, there was "a migration of thought and music from the old centers of population & culture to towns, villages, farms & fields."

By 1923 circuit chautauqua had reached its height, and it is estimated that some 40 million persons were attending chautauquas in 10,000 communities.

THOSE are very impressive figures! A chautauqua of a little over 100 acres with 200 permanent homes would have as many as 600 tents per assembly. The more successful chautauquas, many of which were initiated by individuals who had benefited from summers spent at Lake Chautauqua, would have as many as 11,000 visitors per day.

New York had provided them with an amazing precedent in 1875 when U.S. Grant visited Chautauqua: 10,000 present the 1st night & an estimated 25,000 the next.

UNLESS a chautauqua grew out of active cooperation of the citizens of a community, it did not live long: a graft from the outside dies on the vine unless fertilized by plenty of local zeal. However, many of the communities that had been started up by promoters who made a business of it were susceptible to the lyceum circuits when they began to take over.

BY HAVING sufficient talent to supply a 7 days' (later 9 days') program and selling it as a package to a sufficient number of independent chautauquas to make up a whole season's work on the part of each performer, Vawter was not only able to solve the problem of travel for individual performers, but to finance a better grade of entertainment at more reasonable prices. Instead of having to travel long distances because of widely scattered dates, often spending sleepless nights, back-tracking to keep an engagement in a town next to one where they had performed only a few days before, lecturers & entertainers on the Vawter Circuit could count on an orderly sequence of dates & locales.

IN ADDITION to the new concept of circuit programming Vawter also supplied better equipment, such as a complete stage, with curtains, arch, wings, dressing rooms, etc., in contrast to the previous awkward platform arrangements. His methods of publicity, such as hoisting overhead strings of gay pennants in city streets to announce

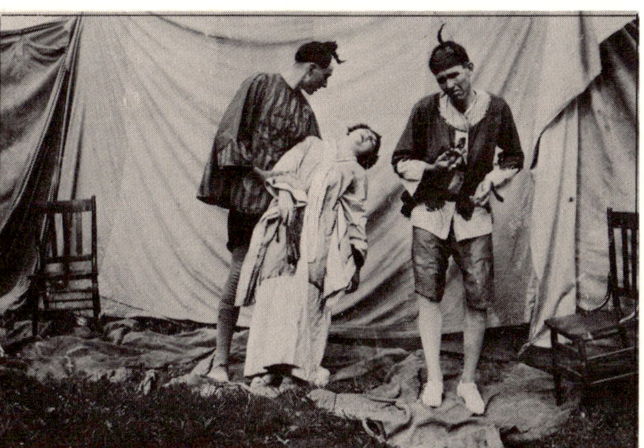

Sometimes the props were few and the stage was merely space between audience & actors

"Something goes ting-a-ling". . . ". . . Played her guitar."

"There are smiles . . . . that make you happy . . ."

"Show's over." Northboro, Ia. (1921)

Any old vehicle would do as long as it made the next date.

approaching program dates, were soon copied by other managers, eventually causing a revolution in chautauqua management.

SEVERAL circuit chautauquas developed quickly along the line of Vawter's, with individual touches of their own. The Midland established a program for children of chautauqua families. Chas. F. Horner (who bought out Vawter in 1912) was particularly successful in stimulating local community interest, with lasting results. In the wake of the Redpath-Horner circuits would appear such permanent establishments as a playground, or a golf course, or a community center. Harry P. Harrison was the first to introduce grand opera to the circuits, hiring Alice Neilson in 1915. He also employed the famous Ben Greet Shakespearean company.

Another lasting result of them all was the education of thousands of students who earned their way through college working on the circuits.

ADVANCE preparations for a circuit took a long time & were an extremely complicated business. Contracts between community leaders & the circuit had to be made so long in advance that the date clerk's job was the most ticklish of all. He might start out with an ideal series of programs leading up from Florida in April to arrive in the Great Lakes Region in the hottest weather and down again as far as St. Louis by the cooler autumn weather; but upon submission of the schedule to the towns along the way, he might find it necessary to consult innumerable time-tables and draw up 5 or 6 plans before he could satisfy a majority.

THOUSANDS of young men & women college students earned their tuition during the summer months, working in the chautauquas as advance agents, tent boys, ticket sellers, managers, superintendents, and even as "talent."

EVERY single worker had to be accounted for every single day of the season. For a 7-day circuit this meant around 50 employees, including 8 superintendents, one of whom would arrive in the first town one day ahead of the opening date. He would be assigned every 9th town on the circuit (allowing him an extra day to get to his next assignment at the end of a week's program). The cashier & tent boys would arrive the evening before opening date and be assigned every 8th town. The girl who tended to the children's programs would have to be in a town 3 days before the opening, & would be assigned every 10th town. Thus at the end of any one assembly, the employees would all go their separate ways.

EIGHT large well-constructed tents would be required (by 1923 the average size was 125 by 175 feet). Arrangements had to be made not only for their being set up in time, but repaired quickly in case of storm damage. Fencing had to be provided & set up. Seating arrangements, suitable lighting, the printing of rolls upon rolls of tickets (of which a careful record had to be kept), the proper timing of display ads—all had to be completed well in advance for each town on the circuit. Ten days before the opening in the first town on the circuit, a complete setup rolled out of the warehouse, and for every day thereafter until the first 12 towns were covered. The amount of scheduling involved—and that before the days of computer help—is staggering to contemplate!

Booster Parade:
the Fanfare Begins.

A portable chautauqua showing entrance, enclosure, and tents, as used by the chautauqua circuits run by Professor Paul Pearson, Swarthmore.

Typical scene at a Western Tent Chautauqua, circa 1914

WE HAVE a first-hand report from an advance agent for the White & Myers Bureau, which managed 322 midwest assemblies by 1922.

Anna Marie Averill speaking:

THEY HIRED WOMEN because of the shortage of men during World War I. I sold Lyceums as well as Chautauquas. The contract was a typical form with space for sponsors' signatures, and the "talent" was written in because it varied with the cost. There were briefing sessions by the Lyceum & Chautauqua Companies before we were sent out to cover our allotted territories as agents. First there was a thorough run-down on the commodity we had to sell—TALENT. Here I met up for the first time with the "package deal". Only one celebrity to a customer! If I sold William Howard Taft, Ida M. Tarbell, or Irvin S. Cobb, Tobey's seals or Knachenbrecht's Alpine yodelers went with them.

EACH AGENT was provided with a list of the people who ran things in the community. It might be the bank president, the school principal, or Hank Munger, dealer in real estate. The agent stopped at the main hotel and casually dropped the name of his prey as he registered.

"HANK MUNGER? Sure I know him! Comes in here every day right after lunch. That's his chair, the green one."

BEFORE I'D BEEN on the road a week I was adept at finding "Hank's chair." I'd pop up, when he came in, with a toothy "keeping your chair warm, Sir." I could feel him thinking, "Nicely raised child." It was a good firm step toward getting his signature on a contract.

THE AGENT WAS TAUGHT never to disparage his competitors. Rather, one spoke of them with compassion, "poor things, they can't be blamed for not having the fine talent we have to offer."

MY TERRITORY was Wisconsin, North & South Dakota, Iowa, and Minnesota. I covered a lot of South Dakota by caboose. Though I was often the only passenger, I was never lonely. As soon as we were under way, a freight handler would come in with an apologetic: "Say, would you mind? This is SO and SO's rooster Billy who is going to Grandma's. Goes twice a year while his folks go to St. Louis. He doesn't like box cars . . . wants company! Just talk to him & stroke him now & then."

THE PASSENGERS who shared that long bench with me besides Billy were a lamb, dogs, and cats, a raccoon, and once a burro named August. They were charming traveling acquaintances, and luckily for me, often their destination was the home of a person whose signature I needed!

ONE NIGHT in Richland Center, Wisconsin, as I alit from the train, I saw a lighted sign ". . . . . . HOUSE." Naturally I took it to be a hotel. As I registered, the proprietor looked me over carefully, asking several times: "Are you alone?" He called two boys (bell boys, I thought) who, one on each side, took me to my room. They gave me a gentle shove, closed the door, and locked it from the outside. "How sweet" I thought, "to be so considerate!"

SOON there was a knock on the door. I was told dinner was being served. The same two boys escorted me to the dining room and seated me politely, and one asked, "Are you allowed silverware?"

I LAUGHED HEARTILY, trying to show a cozy readiness to enter their game. I ate a lovely dinner, then went to the lobby to call the head man on my list, who was the school principal.

"OH YES!" He was most cordial. "The board is meeting tonight, I'll pick you up. Where are you stopping?"

I HEARD HIM GASP when I gave him the name of the establishment. He asked me to call the proprietor to the phone. After a few words with him the proprietor hung up, shook his head and waggled a finger at me with "Pack your bag, Kid, you've been sprung!" I have always felt that that principal and his board gave me a signed contract that night to mitigate any after-shock I might experience from having unwittingly committed myself to an insane asylum.

THIS ITEM from The Unterrified Democrat (Linn, Missouri, 1/21/1926) gives us, as The New Yorker would say "in its entirety" the account of one Chautauqua agent who was NOT successful:—The lady who spent several days in Linn this week in an effort to arrange for a Chautauqua here next summer made a failure.

SANS SIG

AMUSING INCIDENTS such as these could be multiplied by all circuit workers from tent boys to platform personnel & superintendents who by some feat of legerdemain made the whole operation tick & come off according to schedule.

IN A TYPICAL season after the initial contract with a community was signed, an advertising crew drove into town in autos decked with colorful pennants. They put up posters, distributed leaflets, and stirred up anticipation generally. Six weeks later, the platform superintendent and an energetic group of college boys who served as roustabouts, drove in to be with the chautauqua all week. On the morning of the first day, the boys rolled up their sleeves and got to work without wasting time, setting up the equipment, which according to the schedule would have rolled in the night before. When they had driven their stakes, hoisted canvas ropes, clapped their stage, arch, wings and dressing rooms into place, shifted the chairs or benches into serried lines, and hurried through the thousand and one tasks of that first day, the servicing of the week's engagement had just begun. There was always something about the tents or grounds to repair or shift about for the performances. One can imagine also that in every town in every season there'd be a percentage of emotional problems to handle—bewildered children in search of lost parents who were equally distraught in search of their progeny; or tussles between teenagers that had to be broken up before they'd erupt into anything more than an exhibition of animal spirits at play.

THE CIRCUIT was a lot like a musical chairs game: A big Florentine Band, which played at Town A the first day, might have met the chalk-talk artist taking his first look at Town A as they boarded the train for Town B. The artist might have met performers for the 3rd day's program in turn as he left for Town B, and perhaps would run again into the Florentine Band about to train for Town C.

SO IT WENT, covering 102 towns in 7 states through the summer. Wherever the band played, it was the first day of Chautauqua in that place; wherever the artist appeared it was the 2nd day... and so on... and on... The game was carried out *with such clockwork precision that it is said not one man was reported late for his appointed hour in the 1,271 events of the entire season.**

*Adapted from an article in THE CHAUTAUQUAN 8/28/13, on The Chautauqua Association of Pennsylvania, run by Dr. Paul M. Pearson, backed by 250 business men in & around Philadelphia. Each assembly week provided 31 events for the sum of $2.00

The Redpath Special is met by a crowd at Hopkinsville, Ky. (1914)

Model T Fords Were the Favorite Car in the Corn Belt, and the tents were often patched as a result of severe wind storms.

Intermission time and chance for personal assessments of the audience while performers rested.

FRANK LUTHER MOTT* wrote of his chautauqua memories in The Palimpsest (May 1962). He served as platform manager on Circuit A of the Midland Chautauqua System of Des Moines during 4 summers, starting in 1917. He not only had to introduce the talent, but try to arrange contracts with each community for the following summer's chautauqua — and in some cases he had the grueling task of collecting the guarantee when a sale had fallen short of the contracted quota.

* Dean Emeritus of the School of Journalism at the University of Missouri.

"TALENT'S LOT was not easy," he wrote. "Nightly 'jumps,' often on the inferior trains that served the small towns; daily appearances, excused only in cases of serious illness; all kinds of weather, including torrid midsummer heat, rain beating on tent tops, wind that often made the quarter-poles dance a frightening jig; country hotels & irregular meals — such were the hardships of the Chautauqua troupers. . . .

"I REMEMBER one blistering hot afternoon in a small Iowa town when Bryan was on the platform, and the orator was drenched with perspiration. He stopped in the middle of one of his rounded periods, dipped both hands into a large pitcher of water that stood on the table before him, and came up with 2 handfuls of crushed ice, which he held to his throbbing bald head for a minute or two. There was no laughter; I think we all felt sympathetic with the old man. . . .

" . . . . With some of the talent I made lasting friendships & looked forward to my weekly reunions with them in happy anticipation of little parties & much good talk.

"STRICKLAND GILLILAN, the humorist-author of the famous 'Off ag'in, on ag'in, gone ag'in, Finnegin' verses — was a rare soul & a good companion . . Governor Leslie Shaw, of Iowa, was an inexhaustible treasury of homely wisdom . . . One evening, when I saw him pacing back and forth behind the curtain, I bantered him a little about his obvious tenseness . . . . 'My boy,' said the Governor, ' . . . No man ever yet made a speech that was worth hearing without getting wound up tight enough to bust a spring before he went on.'"

Wm J. Bryan in Action

Chautauqua Culture was manifestly absorbing.

THE PROGRAMS themselves varied—but the better ones usually included one full Drama performance and evenings of Band Music, Choir singing, Orchestral performances, String and Voice Quartets, Violinists, Pianists, Bell Ringers or Debates on public Issues, & various Lectures.

Jess Pugh and other Entertainers supplied Music & "Wonders" of every sort

Photo courtesy Mrs. Floyd E. Cooley

Ewing's Ladies Band—made up mostly of Iowans -(1915–19)-

Bergen & Charlie McCarthy—& Who?

The Booster Parade always brought out a Crowd—

Just inside the Entrance

For most of the photos on this spread we thank Redpath's Gay MacLaren Backman & Messrs Frank L. Mott & Howard Preston

Dunbar's Bell Ringers took up plenty of the Stage.

More Women than Men showed up.

Enjoying the Chautauqua

The Mozart Quartette – also known as "The Little Coquettes"

Children & Grown-ups mingle.

Charles E. Booth was among the First to make the Circuit in a Car, sometimes arriving just in time to go on.

Irene Remembers vividly the first time she appeared on a program in which Richmond Pearson Hobson debated some question of current national interest. Each debater was successful in creating the impression that he actually hated the other's viewpoint. Each shouted & raved & tore the other's arguments to shreds. What the subject was or which side won is immaterial now, but Irene remembers her shocked surprise when after the performance the 2 men got into the same car & drove away, laughing and talking good-naturedly, with Hobson's arm thrown affectionately around his erstwhile opponent's shoulders. No doubt they were concocting more vituperative arguments with which to confound each other in their next encounter. No matter how unreal they were, such debates did have the merit of presenting both sides of truly important issues.

Hobson was a popular hero of the Spanish-American War, so popular for having sunk the Merrimac that a common jingle of the day claimed all the "girls" wanted to give him a merry "smack"; & the term—"Hobsonizing" became briefly a synonym for flirting.

People came by horse & buggy in the more remote communities. Farm families, the men especially, seemed to favor Saturday night performances. The women and children usually came in first and took the front seats, while the man of the house found a place for his "rig" and, coming later, had to take a rear seat, which he seemed to prefer anyway. As a precaution against the loss of his precious lantern, he might bring it with him, smelling of kerosene, and set it on the floor beside his chair. There were other odors, distracting then but recalled now with nostalgia—the well-scrubbed children with their aroma of naptha soap, pop corn and peanuts; the cracking noise of the shells was always an irritant to the performers. There were pranksters, too, in those days but few real delinquents. A lantern or a lap robe might be reported "missing" after a performance—a buggy whip "gone" or "borrowed"—never a whole rig.

Concert Tour 1911-1912

JESS PUGH CONCERT COMPANY

MR. JESS PUGH Characterist in Readings and Songs
MISS IRENE GERTRUDE BRIGGS . . Soprano
MISS SUSAN SHELLEY . . . . . Violiniste

Under the Direction of Redpath Lyceum Bureau

PROGRAMME

A Madrigal in May .... Ernest Newton
MISS BRIGGS, MR. PUGH

Airs Russes .... Wieniawski
MISS SHELLEY

A New Love Song .... Lynnes
MISS BRIGGS

Reading Selected ....
MR. PUGH

Berceuse—Lullaby and Prayer .... Ovide Musin
MISS SHELLEY

(a.)—An Old Love Song ....
(b.)—Selected ....
(c.)—If I Knew You and You Knew Me.... MacDermid
MISS BRIGGS

Readings, Solos, Characterizations, Duet ....
MR. PUGH, MISS BRIGGS, MISS SHELLEY

(a.)—Optimism ....
(b.)—Woodman Spare That Tree ....
(c.)—My Ain Folk .... Lemon
(d.)—Love and War .... Cooke

Typical program with some "selected" numbers chosen by Irene DaBoll (then Miss Briggs) and co-performers as they sized up the taste & temper of their audience.

THE BIG BUSINESS of stealing cars came later with our more advanced society. A farmer might find himself going up hill all the way home before he could get a wrench with which to reverse the substitution of the large front wheels for the small rear wheels of his vehicle—a teen-age trick carried over from Halloween mischief. That was better than if it were stolen or finding it wrecked in a roadside ditch, or wrapped around a tree—so often the fate of cars "borrowed for kicks" by unrestrained teen-age vandals of our present mechanized era.

THE MODERN MOTEL with comforts & luxuries that would put to shame the simple accommodations that even the best hotels in rural areas of the middle West could offer, were still far in the future. Jess Pugh used to take special delight in telling (off-stage) about an accident that befell Irene one night at a small-town hotel and led to an embarrassing scene the next morning.

IRENE was subject to spells of earache and had found that she could get relief by heating water over Sterno* and applying hot compresses to the painful area with a washcloth. Awakened by such an attack, she got up & groped her way to the washstand, which held a pitcher of water, a washbowl and a shaving mug. It also served as a cabinet for 2 other pieces of crockery—a squat-looking vessel called a "thunder-mug", & a larger one called a "catcher"§. Irene lit the sterno, poured water into the shaving mug & heated it over the flame. Sleepy & fumbling with the compress, she dropped the mug. It recocheted from the padded "midnight-husher" wrapped around the lid of the thunder-mug (to prevent the clang of earthenware ☞ "things that go bump in the night") and hit the catcher with a resounding crash, shattering it!

* A waxy fuel with an alcoholic base called "canned heat"

§ More aesthetic than "slop jar"

IRENE spent the rest of the night worrying as to whether she should report the damage or keep still about it. Finally, she decided on silence, so after paying her hotel bill in the morning, she hurried off to the railroad station without having said a word to any one about the accident. The train was late & a good-sized crowd was waiting for it. Shortly before it arrived, a bellboy from the hotel came hell-bent on the scene paging: "Miss Briggs—is Miss Irene Briggs here?" He presented her with a bill for the damage, and she had to pay it, much to her chagrin & the amusement of the crowd—particularly that of her 2 traveling companions. Jess Pugh's talent as a raconteur eventually built the incident up to what he called "The Story of the Pitcher & the Catcher."

CHALK-TALKING was one of the features of circuit entertainment that always seemed to go over well. Proficiency in it required arduous practice. The artist had to know exactly where to make each stroke; the harmonious relationship of the colors had to be carefully worked out in advance—every movement coordinated to go with his line of "patter". The right-handed chalk-talker worked from the left side of his easel with a long free-swinging arm so his audience could have an unobstructed view as he developed the picture, usually on a thick pad about 30 x 36". If he found it necessary to work from the right side, he would cross over & assume the stance of a tennis player making a back-hand stroke.

IRENE'S BROTHER BOB often talked & sketched with the help of a musical background. He also recited poems and rendered songs appropriate to his subject—the trick being to end the song or the reading with the last stroke of the chalk. Another trick that added an element of dramatic suspense was to draw the picture upside-down or sideways so to puzzle the audience until he turned it rightside up with a final flourish & a deep bow. The whole performance was almost as rhythmic as a dance.

The art lent itself particularly well to the rapid, eye-catching chautauqua programs.

THERE is nothing more ephemeral than a chalk-talk sketch—a thing of the moment, showy & startling up to 10 paces, but not for close inspection. The chalks were soft and easily smudged, so Bob made a practice of tearing each sketch off the pad, dropping it on the floor, and walking on it as he proceeded with the next one. This was intended to spoil it as a souvenir, but in spite of that, people would often stand in line after the performance to pick one up, and ask him to sign it, and take it home, believing they had retrieved something of real value.

BOB WAS always having to explain that he was not Clair Briggs, one of the Chicago Tribune's widely syndicated cartoonists, famous for his "When a Feller needs a Friend" and "The Days of Real Sport"; Clair was not a chalk-talker. Alton Packard was one of the best on the circuits—"tops" as a humorist & satirist.

SOME of the younger circuit talent went on to make names for themselves in radio & television. One day, at the office of a booking agency that had just signed him up for a series of chautauquas, Bob met a young ventriloquist who gave him a sampling of his art. Bob was sure it would go over big. Whether the agency booked the young man that day, or at all, Bob didn't know, but he certainly did go over BIG eventually on radio & TV—his name was Bergen and his dummy was Charlie McCarthy!

"Come Where the Campfire is Gleaming", rendered vocally as well as pictorially, was always a popular number....

While Bob was away from home much of
he time, he dwelt on Home & Mother as
favorite subjects. Most people are incurably
sentimental; that's why Edgar A. Guest's
"It takes a heap o' livin' in a house to make
t Home" was always well received....
And when Bob had finished "I'll take you
home again, Kathleen" <a sure-fire tear-
jerker> there were few dry eyes in the tent..........

Such a lot of paraphernalia to tote around!

Transportation was often impromptu...

those who had lived through the terrors of a tornado-(many had)-watched his deve lopment of "The Coming Storm" with mixed emotions.

The pictures shown here and on the following spread are from the albums of Robert Briggs, a versatile performer in Lyceum, Chautauqua & University Extension from 1917 to 1940.

Bob didn't always label his pictures with names & dates . . .

1 – might have been taken in the FAR WEST.

2 – could be a Scene in the SOUTH WEST, or maybe in the "BAD LANDS" of South Dakota; the Plaque suggests that the site may have Historical Significance

The Circuits Covered AMERICA

By Railroad & by Highway, the Performe

Always Waiting Coming or Going on trains or on buses, no matter how wearisome, TALENT TRAVELED.

Irene, waiting + A Group of TALENT – just arrived + Ready to depar

3—is labeled simply "Mary, Mary, Quite Contrary"!

4—is "Homer Abegglen—'Tent Man'": typical of the clean-cut College boys who served as "Roustabouts".

5—Evidently some folks in the "Deep South" will hear a fine Orchestra & see a "legit" drama: "Cappy Ricks".

HE BEAUTIFUL
ne to know their Country

Sometimes the Seamy Side of It

"Temple Of Fine Arts". No matter what the weather, or how crude the "Oprey House" they headed for, Performance was the Aim.

ighway open for a Winter's Date in northern Mich. + At St. Benedict, Pa.

CHAUTAUQUA'S romance & color now & then strayed beyond the tent walls into the homes of the receptive townspeople. Before he was billed as an individual attraction, Irene's brother traveled with a number of musicians. One Italian couple whom we shall identify only as Rudolfo and Mimi (not the names given them by their sponsors in baptism), if they were not pleased with local hotel accommodations, would walk through the best residential section of the town until they came to a house that look simpatico. Here they would explain to the owner, in broken English, that they were grand-opera singers, just arrived for the chautauqua, and had found the hotel accommodations insufferable; but if these kind people would take them into their lovely home as paying guests, they would be oh so grateful!

AS FAR as Bob could determine, they were invariably invited to stay as distinguished guests by the flustered & flattered homeowners. Since the hosts were apt to be music-lovers—and even sponsors of the Chautauqua—the walls would soon reverberate with song—and a most enjoyable time was had by all. . . And Rudolfo and Mimi paid not one cent!

SMALL TOWNS still have their 4th of July parades, their occasional invasions by Shriners, or potentates of some other order, but by & large radio & TV have taken over and created a totally different, isolated, media of culture & amusement. The fanfare and ballyhoo and particular variety of "togetherness" created by Chautauqua are over, but the powerful stream of enthusiasm created by Chautauqua has left its mark everywhere.

SMALL TOWN · U · S · A · circa 1921

PERHAPS one of the major accomplishments of the circuits was the fact that they were traveling forums for great political ideas & lasting causes. Robert M. La Follette, Booker T. Washington, George W. Norris, Maude Booth, Champ Clark, S. Parks Cadman, not to mention Carry Nation and Billy Sunday, were among the most popular speakers who came to the audiences with important ideas.

WILLIAM ALLEN WHITE is said to have remarked: "The Progressive Party was formed from a dozen chautauqua speeches in Iowa and Kansas." (Perhaps a bit exaggerated.) The Baltimore Sun in 1914 stated that no prominent public official could afford to neglect the Chautauqua program.

CIRCUIT CHAUTAUQUA could also be said to have carried "The Little Theater Movement" across the country, and to have forwarded the Civic Theater idea proposed by the dramatist Percy MacKaye, who was keenly concerned with the problem of leisure. MacKaye believed in participation by the people in the arts, not merely spectatorship, and in the dedication of the arts to the service of the whole community "involving a new expression of democracy" used to "fill time, not to kill it." His bird mask "Sanctuary" was offered as an attraction of more than passing interest by Redpath Chautauquas.

SANCTUARY
A BIRD MASQUE

"In 'Sanctuary'" commented the local press, "Mr. MacKaye has found a subject of congenial interest—the care and protection of our wild birds. At the close of the performances a goodly number of the local children get upon the platform, dressed to represent the different bird spirits, and the masque comes to an appropriate close."

ONE of the widest traveled & best known orators at chautauquas was William Jennings Bryan. While Secretary of State he wrote an article on "The Nation-wide Chautauqua" for the Independent Weekly (7.6.14). "The Chautauqua," he wrote, "is a forum where all have agreed to meet and discuss on common ground.... The 'loaded,' or 'stuffed,' or 'doctored,' or in any way adulterated Chautauqua with special personal or group interests to serve has little chance of success.... The privilege and the opportunity of addressing from one to seven thousand of his fellow Americans, in the Chautauqua frame of mind, in the mood which almost as clearly asserts itself under the tent or amphitheater as does reverence under 'dim religious light'— this privilege and this opportunity is one of the greatest that any patriotic American could ask.... (They) carry with them a peculiar responsibility of which no patriotic American with a conscience can remain insensible....."

A Great Throng of America-minded Folk at a Symphony Concert in the Amphitheater at Chautauqua Lake, N.Y.

July 4, 1929

Though slightly blurred, the faces of THOMAS EDISON, HENRY FORD, ADOLPH OCHS, *& Chautauqua President* DR. ARTHUR BESTOR *can be seen in the podium group*

As seen standing a

um (from Stage point of view) [1]Dr. Arthur Bestor, [2]Henry Ford, [3]Adolph Ochs, [4]Thomas Edison

Broad-pen musical notation of the 18th Century & that of today

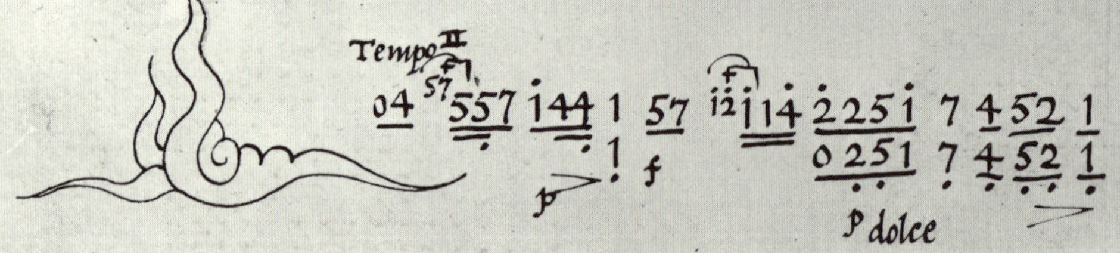

Tune thusly: 1/1 to 3/2, 1/1 to 4/3, 3/2 to 9/8, 3/2 to 9/5, 1/1 to 2/1
Do to Sol, Do to Fa, Sol to Re, Sol to ♭Ti, Do to Do↑

In Mvt. one, Grave, the gracenote to an octave does not make a repeated tone – only a "broken" octave.

May–June, 1964
Author's Private Edition by Lou Harrison

Calligraphic excerpts from the Contemporary WESAK SONATA

Old or new, music & calligraphy live on together

Part III
Calligraphical

Poem No. 1—From Songs of Innocence & Experience by William Blake—Poet-Painter-Engraver-Mystic

Notes on Calligraphy & Scribal Writing

(With a few musical overtones)

FIRST—why bother to write a book about the old Lyceum & Chautauqua Circuits—and SECOND—why write it out by hand? Those methods of communication were by-passed and buried years ago—even centuries ago with regard to calligraphy, which is, by definition, beautiful writing. The passing of the circuits was fore-shadowed late in the 19th century, before their heyday at about the quarter turn of the 20th, by Edison's inventions in the field of reproduced sound and moving & talking pictures. The bell began to toll with ever increasing volume when Marconi invented wireless telegraphy in 1900. By 1935 the circuits had run their course and for all practical purposes were dead. Calligraphy also suffered a lingering fate following the invention of printing from movable types, uncertainly attributed to Gutenberg about 1440. During the next fifty years the craft of printing, nursed by calligraphers, established the principles of typography and produced some of the finest examples of printed design & press work ever known. This was called the incunable period from the Latin word incunabula meaning swaddling clothes, cradle, or anything pertaining to infancy. Meanwhile, the infant has grown to gianthood and that was the greatest thing that could have happened for the development and spread of civilization—the multiplication in minutes of what the scribe could not have accomplished in a lifetime spent in copying texts that only the rich could afford.

BUT IN GROWING UP the giant has had spells of delinquency in which he has lost, to a considerable extent, the human touch that the scribe (or calligrapher) gave him in his youth. This is perhaps to be expected in times of intense mechanization such as the Industrial Revolution.* The worst period of delinquency (degeneracy may be the better word for it) developed in the last half of the 19th century and caused William Morris to lead

*And today in the age of electronics.

a crusade for good taste in printing by advocating a return to incunabular styles based on the manuscript hands of the earlier masters of writing & illumination. His approach to the problems of reform was not entirely successful but he is credited with lighting the torch that has been carried on by Edward Johnston in England, Rudolf von Larish in Austria, and Rudolf Koch in Germany. Many of their pupils have become teachers dedicated to making handwriting legible — and even beautiful — as the student becomes, shall we say engrossed with the subject. It is encouraging to find that many of them do, and that it tends to make them more careful of what they write as well as how they write it.

Mimi & Mimi — both somewhat Bohemian

ANSWERING QUESTION 1: Why write a book about the old lyceum & Chautauqua Circuits? We write about them because they played an important role in the cultural development of the American people. They are a part of American history and there is more that may well be added to the excellent but limited information now available. Irene Briggs, as one of the old troupers, tells of her experiences & of how she was prepared for circuit work, feeling that some of today's young singers who may never have heard of such things as lyceums and Chautauquas, might have a passing interest in how she got that way. And avid collectors of Americana will find it here — pure and unadulterated.

ANSWERING QUESTION 2: Why write it all out by hand? One very good reason is that I am part of a movement to promote Italic handwriting in the public schools, and wherever people can be persuaded to improve their own handwriting. I have also been frequently importuned to set down my thoughts on matters calligraphic. Another reason is my belief that the entire text of this book, written calligraphically, will attract sufficient attention to prove my point that Italic is easy to read and not difficult to write, given the proper instructions and tools to work with. Further, the "hand-done" job invariably conveys a more heart felt indication of personal regard & warmth where sentimental attachments are concerned, for instance, love letters & poems, marriage certificates, wedding & birth announcements — to mention such documents in their logical sequence.

AND FINALLY, a book such as this may also contain incidental tributes to various persons (including the Lady who has shared the scribe's joys & sorrows for over fifty years) who have been sources of inspiration & help. Surely they deserve nothing less than a personal acknowledgment — albeit reproduced — that may be reckoned as a labor of Love in which the cost in time and effort is never counted.

THERE IS NO QUESTION but that the quality of American handwriting has fallen to a deplorably low level. This is evident with very few exceptions today, for everywhere we see handwriting that is notable mainly for its lack of legibility & penmanship. The reason for this is first, the lack of proper models & instruction in the public schools, and later, the disregard of any standard of legibility—let alone beauty—due most likely, to the prevalent urge to speed things up all along the production line.

HANDWRITING can be done with varying degrees of speed, still be legible, and retain commensurate beauty. I shall demonstrate those points with excerpts from the letters of several calligraphy-conscious people with whom I have corresponded since my own interest in better handwriting began some 30 years ago. I have kept many such letters on file and regret there is not space for more of them to show that each has its own individuality although based on a common model—the 16th century chancery cursive letter (littera cancellarescha corsiva). Originating in Italy, this slightly slanted letter, written with a chisel-edged nib without pressure, is well related to the Humanistic Roman letter made with the same kind of nib. Eventually this cursive style became the common handwriting of all countries that had adopted the Latin alphabet for book work. The first Italic types of ALDUS were based on it. ALL Italic types ARE based on it. 'Time has tested and proved their fundamental rightness.

Substantially Quoted from Printing Types by D. B. Updike, Harvard U.P. Vol. 1, page 55. —the capitals are mine.

IT WILL BE NOTED that the following examples show diversity within a certain area of conformity. This has to be, for, after all, what is there about an A that makes an A an A? To parody a well-known line§, an A is an A is an A whether it's written or read or sounded. What does a symphony orchestra do before it begins its program? It sounds its A! to know that all of the instruments are on pitch. Then follows a great variety of sounds and we hear music.

§ About a rose by Gertrude Stein.

SOMETHING LIKE THIS PROCEDURE can be applied to the use of our Latin alphabet from A to Z and the result will be handwriting that is at least legible, and at best, beautiful; for it can be developed from a craft to an art—from what is merely practical to what is truly beautiful, as practice brings facility and appreciation to the use of line, mass, tone and color on the page. As such, handwriting (calligraphy) has been found worthy of attention by art critics.

THE MOST COMPREHENSIVE EXHIBITION of Handwriting ever held in America—TWO THOUSAND YEARS OF CALLIGRAPHY—required 2 years to assemble and the space accommodations of 3 Baltimore Museums in June & July, 1965. Reviewing the Baltimore exhibition in THE LITERARY SUPPLEMENT to THE LONDON TIMES, Beatrice Warde summed up the case for Calligraphy with Churchillian force & brevity as follows:

THE ART OF CALLIGRAPHY has twice been killed stone dead by a mechanical invention, and has twice found a new life and a new set of justifications. As an industry for the copying of texts, it was destroyed by the printing press. As an essential tool of commerce and finance, and as an evidence of gentility, it flourished for three centuries and was strangled by the typewriter. In its third and present life it stands with the fine arts, safe from any further technological threat.

—Be that as it may—

Early Printing Press from Moxon's Mechanick Exercises, 1683, & an early typewriter, c 1890 called, of all things "Calligraph"!

NONE of the examples shown on the following spreads were included in the Baltimore exhibition. They are simply letters &/or excerpts from the ordinary correspondence of people who have adopted the Italic Style and are shown here as evidence that individuality in handwriting is neither lost nor limited but rather enhanced, whether based on 16th century chancery cursive or on modern Italic adaptations of it. Since the size of one's writing gives an indication of personality, these examples are reproduced in actual size except for a few that were too large to fit the page. Most of them were selected to show that speed in writing is compatible with legibility & good style—often with beauty.

UNTIL some 35 years ago calligraphy was to me an unknown technique—even the word itself was unfamiliar. I find it still is to most people. It is encouraging, however, to learn that, along with the ever increasing emphasis on more & better education, our institutions of higher learning are beginning to recognize the need for legibility as well as speed in this age of instantaneous electronic communication. Several universities, colleges and art schools now have highly competent teachers of calligraphy in their design departments, occasionally augmented by professional calligraphers to help in conducting seminars. Rochester Institute of Technology with its School of Printing & College of Graphic Arts, and Carnegie Tech, Pittsburgh, lead the way in training students who will carry on with the most advanced techniques & machines in America's third largest industry—Printing & Publishing. Realizing the importance of the 26 letters of our alphabet and how they were developed by scribes until the printing press took over, these schools instill an appreciation of calligraphy, showing how it has influenced type design, and teach the use of the broad pen & the flat-edged pencil in making type layouts.

ABCDEFGHIJ
KLMNOPQR
STUVWXYZ
1234567890
ABDEFGHJKL
MNPRSTUWZ

abcdefghijklmnopqrstuvwxyz
abcdefghijklmnopqrstuvwxyz
abcdefghijklmnopqrstuvwxyz
abcdefghijklmnopqrstuvwxyz

The capital letters, digits, and top line of lower case are Zapf's Palatino type. This second line is a free tracing of the top line illustrating the strokes used in Italic writing. Third line shows the addition of loops. Bottom line is full of common writing errors which make reading difficult. This demonstration was prepared by Fridolf Johnson.

THE most outstanding designer of type faces today, Hermann Zapf, is also acclaimed by typophiles & calligraphiles as the most accomplished calligrapher. No doubt his types are outstanding because of what he knows about calligraphy, which established the principles on which typography is based. The names of three of his types—Palatino, Sistina and Solemnis—bespeak their scribal & sacred influence. Even the subtle swellings of his Optima (smart and strictly modern) hint at rudimentary serifs with a charming grace that offers welcome relief from the severely mechanical Futuras and other sans serif types of the Bauhaus era, particularly for book composition.

SO MUCH for the support of calligraphy from the top technical schools and the finest type designer-calligrapher. But what do the public schools offer the youngsters who are just beginning to write? Not much. In most cases they are off to a poor start that indicates the lack of guidance & discipline that is necessary to orderly coherent writing which can also be helpful in reading and to clear enunciation.

IT IS ARGUED that children must be allowed plenty of freedom to express themselves as individuals. Of late years they have done so in various ways, some not so good. Just where to draw the line in handwriting is perhaps not as debatable as in other areas of conduct. However, as the youngster writes his way—his progressive-permissive right-of-way—up through the grades into high school, it often becomes apparent that "Johnnie can't read" therefore he must attend a class in "remedial reading", otherwise his chances of entering college will be considerably lessened. That is a point constantly stressed by press, radio, TV and by colleges that are already over-crowded with students that CAN read & write.

So, at that stage, or preferably long before, it would seem that Johnnie's studies should include remedial writing as well as remedial reading; and, better still, that such classes should be regular—right from the beginning.

BRITISH *schools have done more for the improvement of handwriting than we have, and their children write much better than ours. Italic is popular with the British. The following lines from a bulletin of the Society for Italic Handwriting of England suggests the importance of an early start:*

> Teachers have found italic handwriting to be simple enough for infants of five & six years of age & to be a substitute for print-script. Writing can begin, therefore, as it is intended it should go on.

Beryl Lindop
Age 7 years.

Nig and Nog were two gnomes. They wore green and brown suits, pointed caps, and very long beards. They were mean and miserly, and never gave a penny to anyone, and now you know just as much about them as their neighbours in Apple Tree Village did

Beryl, at Cholmondley School, handles a pen with commendable dexterity.

The following is from one of the S.I.H. annual reports:

> . . . . . More & more schools, both at home & overseas, are using Italic as the basis for teaching handwriting with very satisfactory results. In some schools the children are so enthusiastic that they run their own Italic Handwriting clubs in their free time. The results of the Society's activities are considerably greater than can be formally assessed.

This Information is substantially from "Can You Read Your Own Writing?" by Fridolf Johnson in American Artist, September, 1963.

SURELY, *for economic reasons alone, the improvement of handwriting is imperative. Illegibility is an expensive nuisance! Several years ag it was estimated that ambiguous handwriting was costing this nation over 70 millic dollars annually. We may safely assume that figure is now considerably higher, b what ever it is, it's too high a price to pay for our careless "chicken-scratches"! And eve today while billions are being tossed around with seemingly reckless insouciance — and a mere 70 million dollars is counted as "chicken feed" it might be well to paraphras Poor Richard on the advisability of saving a penny, for* a million saved is a millio earned. *Indeed, several hundred manufacturers, educators, calligraphers & busines men recognized that truth in 1954 when they marshaled their forces behind a Handwriting Foundation in Washington, D.C. The U.S. Post Office & other government agenc recognized that truth also: upwards of a million pieces of mail go to the Dead Letter Offic every year because the specialists who spare no effort to unscramble the inscrutable names addresses on envelopes cannot find any other place to send them.*

DEPARTMENT STORES *report losses in the thousands of dollars because of illegible sales slip that make merchandise undeliverable. While the invention of the Charga-Plate has help* this bind by identifying the buyer at the point of sale, the cure is far from complete; the real rhubarb begins when the sales slip goes to the warehouse to be filled for delivery, and the mother who ordered "sleepers with creepers" for her baby gets "slippers with zippers."

UNITED PARCEL SERVICE, which delivers packages for stores in thirteen cities, reports that every day in one city alone, about 1,200 packages are undeliverable, mostly because of illegible writing. To combat this costly nuisance, United Parcel has carried on an educational campaign since 1940 by sending to its clients every month 6,500 comic posters. Each month the poster carries a different message—one may show a crystal-ball gazer trying to fathom a C.O.D. sales slip. Another, a museum curator deciding that the hieroglyphics on a parcel may be Sanskrit. While such striking posters have been facing store employees for 28 years, United Parcel sadly admits that they have had "no discernible effect" on handwriting. A black chiffon nightgown, size 12, still mysteriously finds its way to a 70-year-old-200-pound lady who ordered a hot-water bottle.

One of the United Parcel Posters.

THE DRUG INDUSTRY, finding it increasingly difficult to read orders for pharmaceuticals taken by salesmen, commissioned a handwriting publisher to prepare a 152 page text for drug salesmen that takes them through a 27 week "refresher-course."

THE RAILWAY EXPRESS AGENCY whose claim prevention department wages a continual crusade against undecipherable labels, has made this comment: "Our general observation is that the quality of penmanship has declined, perhaps due to the fact that it is not sufficiently stressed in the grade & other schools." To cut down losses and time-killing annoyances, Illinois Bell Telephone Company brought in handwriting expert Ruth Kettle to organize classes for the 4,100 Chicago toll & long lines operators. In spite of the automatic dial, operators still manually record millions of completed calls every day. Most teachers colleges today believe that in the grade schools, handwriting should not be taught separately, but as part of the language arts which include reading, spelling, composition and grammar.

DR. LUELLA COLE, a former professor of educational psychology at the University of Indiana, issued this stinging indictment of handwriting teaching: "Everywhere I have gone I have found handwriting the worst taught, most neglected & least understood subject in the elementary school. The progressive movement has done wonders in teaching children to manage themselves... but it has not been able to cope successfully with handwriting."

The Facts stated on this page are from "Why Our Kids Can't Write" by Theodore Erwin in the Saturday Evening Post of September 10, 1955.

{A check of the telephone situation in September 1968 finds it essentially unchanged.}

This is an example of the highly individual Italic hand of the late George Salter (1897-1967), one of the outstanding teachers of Calligraphy and Graphic Design. A specialist in book design, he taught at Columbia University, Cooper Union Art School, New York University and was Art Director of Mercury Publications. The message herewith (by György Kepes, a professor at M.I.T.) is taken from one of the Harvard Lectures on Graphic Forms (1949). This excerpt is from the Typophile Keepsake: Calligraphics—Hands & Forms, 1955.

MAN AND MASS PRODUCTION

by György Kepes

Within an ever-increasing wealth of products, man himself became worn out, incapable of benefiting from his labors. Limited to a conveyor belt, he rarely feels the joy of creation. Unable to encompass the metamorphosis of things which take shape under the work of his hands, he forfeits the sense of accomplishment, the unity and thus the harmony in the doing, which might give him true satisfaction. Limited to the mechanical details of one or another singular movement, within the complicated cogwheels of the production machinery, he gradually loses those sensibilities which are the guarantors of his perceiving the richness of life. Drained of the nourishment which is essential to his growing to full human stature, he loses the measure and meaning of his deepest aspirations. Through mass production, which could only be achieved through mechanization, man's sensibility, emotional unity, has been killed or at least dulled and deformed.

Dear Mr DaBoll //

I am 20 years old and have become very interested in calligraphy in the last year. Any information regarding the art of calligraphy would be gratefully accepted. . .

Thankyou // Rick Cusick

12 XI 1967

9 West Mayfair Ave. Stockton, Calif. 95207

These lines were excerpted from a letter that Mr. Cusick wrote just before leaving his home for military training. We wish him luck and a long life as the calligrapher he says he wants to be. He surely can, because his talent & flair for it, plus his unusual verve are quite evident.

46, Firsby Avenue,
Shirley, SURREY.

Dear Raymond da Boll,

. . . further to my recent letter on the subject of "Peter Heywood", I am having the loss of the other copy investigated by a cousin in a senior position in our LONDON Post Office.

Sincerely,
S. John Browne

S. John Browne is a most accommodating British dealer in books that are out of print & hard to find. His handwriting is appropriately bookish and very legible.

up together and carry over from one year to the next, as long as we live.

"The verb TRADO never meant 'I worship the past.' A tradition, to be worthy of the name, must be a carrying-forward of some freight... that is too valuable to be jettisoned." — I quote that from this year's Cambridge U.P. Christmas Book, which Brooke Crutchley has sent to the 500 Friends of the Press. It's a book of Walter Nurnberg's camera-studies of the hands of craftsmen at work in the Press. I had the honour of being invited to write the Commentary. Here is one passage, where I tried to explain the feeling we all have, that a great printing-office is mysteriously unlike any other manufacturing plant in the world:—

"For this is a very special kind of factory: a kind which is built only on the very shore-edge between the world of physical things and the River Ocean that girdles it: between the Dry Land of whatever is tangible, measurable, describable, and ... the liquid intangible immensity beyond. On that mysterious beach the printers set their Houses, waiting for what shall come into their hands, when the impalpable brine of Thought has first crystallised into Words — measurable, describable sound-patterns — and then, in due course, into visible word-shapes on paper...

"[And meanwhile] ... everything that the printer calls his 'material' waits ready to encapsulate something radically other than itself: something that no Materialist has ever quite been able to account for..."

Good Luck to you in 1965, from

Yours sincerely

Beatrice

 · From a New Years greeting by Beatrice Warde, "First Lady of Typography" and Publicity Manager Emeritus, Monotype Corp., Ltd., London

Dear Mr. DaBoll: I thank you most heartily for the panoramic Christmas Card heightened with a specimen of your fine chancery cursive. The card arrived when I was helping myself to American weather — first in Miami and last in Chicago — though not for long. Newark looks handsome and I don't wonder that you like it. . . . Let us join together once more before long. Meanwhile Oremus pro invicem.

Yours,
Stanley Morison

9 March 1954

 · *Stanley Morison Top Authority on Typography and Calligraphy Designer of the widely used Times New Roman type face*

25. Newcastle Road, Stone, Staffs, England
2 · June · 55.

Dear Mr. Da Boll,
I would be most obliged if you could let me have a specimen of yr. cursive handwriting, for my small collection — I already have letters from Alfred Fairbank & Wilfrid Blunt & a letter from you wd. be most proudly welcomed by me. I do hope you will forgive the liberty I have taken, & reply at yr. convenience,

Every best wish for yr. continued success in calligraphy.

Yours faithfully. John West

John West Free lance design Consultant specialising in Graphics, Interior and Furniture design He is a self-taught Calligrapher.

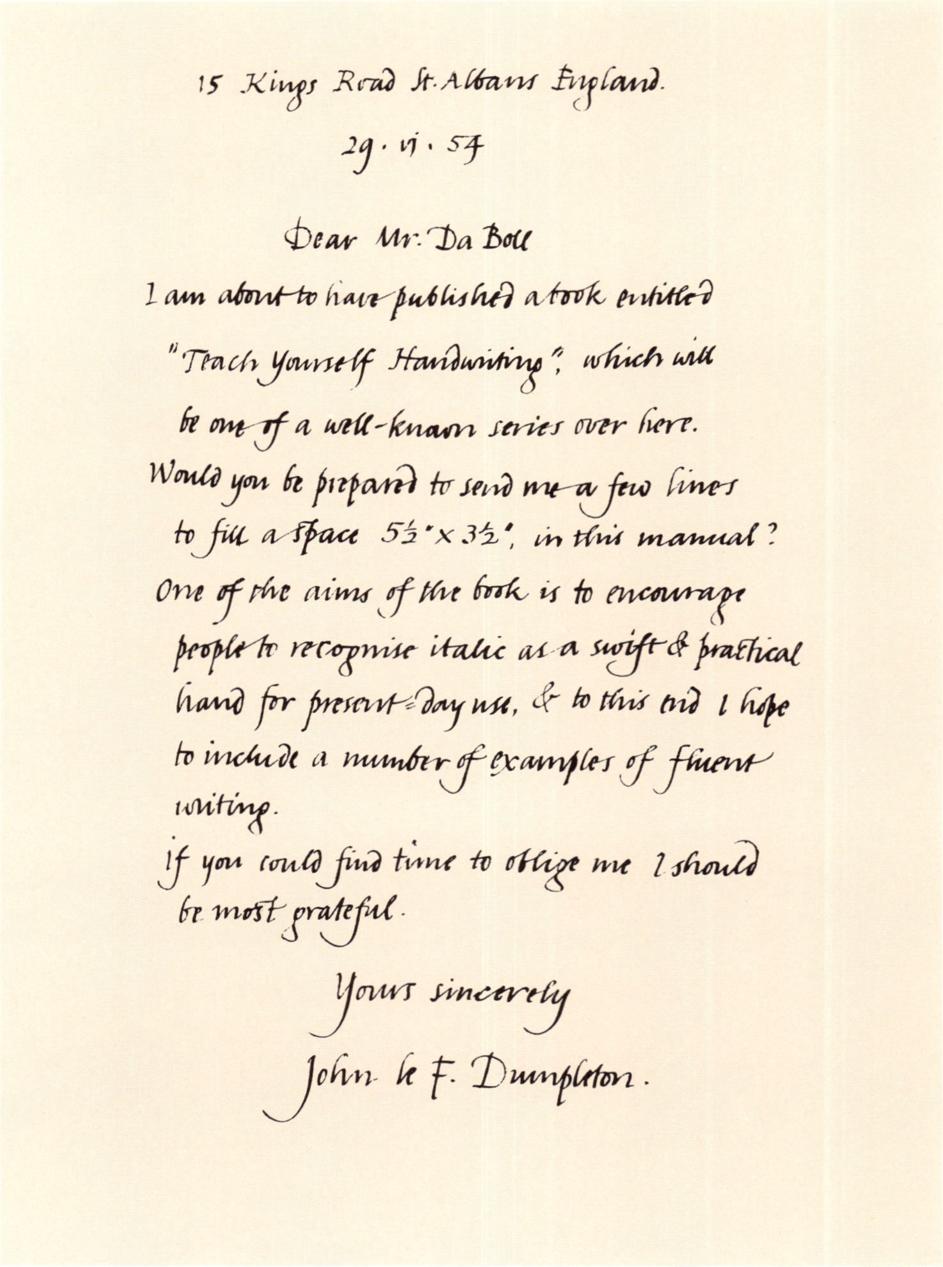

15 Kings Road St. Albans England.

29 · vi · 54

Dear Mr. Da Boll

I am about to have published a book entitled "Teach Yourself Handwriting", which will be one of a well-known series over here.

Would you be prepared to send me a few lines to fill a space 5½" x 3½", in this manual?

One of the aims of the book is to encourage people to recognise italic as a swift & practical hand for present-day use, & to this end I hope to include a number of examples of fluent writing.

If you could find time to oblige me I should be most grateful.

Yours sincerely

John le F. Dumpleton.

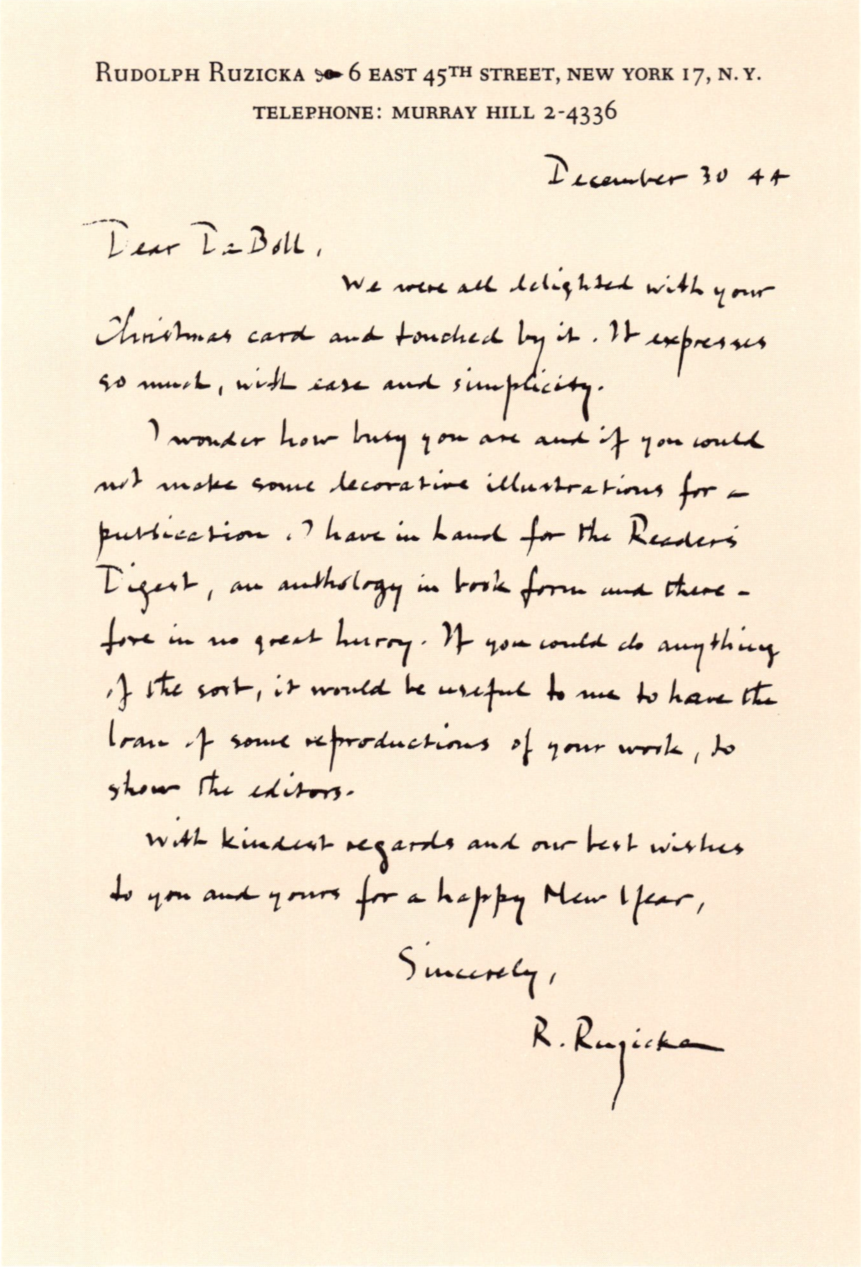

RUDOLPH RUZICKA · 6 EAST 45TH STREET, NEW YORK 17, N.Y.
TELEPHONE: MURRAY HILL 2-4336

December 30 44

Dear DaBoll,

We were all delighted with your Christmas card and touched by it. It expresses so much, with ease and simplicity.

I wonder how busy you are and if you could not make some decorative illustrations for a publication. I have in hand for the Readers Digest, an anthology in book form and therefore in no great hurry. If you could do anything of the sort, it would be useful to me to have the loan of some reproductions of your work, to show the editors.

With kindest regards and our best wishes to you and yours for a happy New Year,

Sincerely,

R. Ruzicka

 Aside from his manual, Mr. Dumpleton gives lectures on handwriting supplemented by a film strip on calligraphic techniques, showing actual movements of hands & fingers.

Rudolph Ruzicka—a wood engraver & book designer, also designs type. His FAIRFIELD face is shown above. His fluent handwriting has fine scribal character.

SUMMER SUN

The crowing cock espies the creeping glow
And wakened man remembers yesterday —
The undone hopes of endless weeks ago.
Now take good auspices, with fond delay,
And rise again to woo auspicious day.

This flaming orb is adamant and just, —
Resume travail or fortunate caprice.
What talents we may use to serve, we must,
And grasp rewards of beauty, love and peace.
The afternoon will throng with zeal and cease.

Beholden in his rest at eventide,
Man now recalls each sensuous delight
And rues the pithless feats of skill he tried
But smiles at ebbing sun in golden might
And sleeps beneath the star-lit, silent night.

— by Walter H. McKay

Walter McKay of New York designed the Columbia Family of Type Faces. A versatile artist with poetic talent, he loved to write his poems calligraphically.

63 Victoria Road, Leeds 6. Yorkshire. England.
13th October 54.

Dear Mr Da Boll,

Thank you for yr. most generous letter. I am very grateful. You were, by strange coincidence, mentioned in a letter I received only last week from our friend Bob Andrews, to yr. credit, I must quickly add. . . .

Thank you also for information abt. Arrighi's "Operina". Mr Andrews told me abt. it & I have asked him if he can get me a copy; there is nothing yet abt. from English publishers. I have the 1926 edition done for Warde. I shd. like this new very much.

Very sincerely yours,
Frank Taylor.

Lustre has now been added to what is becoming a very fine collection of contemporary hands.

Frank Taylor's extra-condensed Italic, like that of Mr. McKay is definitely based on the 16th century chancery cursive. A meticulous layout is characteristic of both of these designers.

3 Douglas St, Kirkcaldy, Scotland. Jan 10, 1968.

Dear Ray,

. . . My recent book, A Guide to Better Handwriting has got off to a reasonably good start and I am hoping that it will help to improve handwriting generally. There is a very real need for an all-round improvement in handwriting throughout the world and this could be done if there was more concerted action amongst calligraphers and knowledgeable teachers — unfortunately too much say is in the hands of the calligraphically ignorant. . . .

Best Wishes from Yours Sincerely
Tom Gourdie

Excerpts from the letters of 2 Scotsmen widely separated geographically, but closely joined calligraphically in the common cause of legibility · Tom Gourdie of Kirkaldy literally joins "hands" across—

· RALPH DOUGLASS · · 3308 LOMA VISTA PLACE · N E · ALBUQUERQUE · NEW MEXICO ·

April 13, 1966

Dear Ray:

. . . . sorry to learn that you have to face a cataract operation. On the other hand, I can testify if it is any comfort to you that I had a cataract removed from my right eye four years ago (at the age of 66) with completely satisfactory results. My vision had deteriorated to 20-80 and it is simply great to have 20-20 sight again!

Good luck on the book. That is a real chore! Thank you for your help on mine.

With all good wishes!

Sincerely,

Ralph.

the gutter with Ralph Douglass of Albuquerque · As teachers of Italic handwriting they wield chisel-edged broad pens instead of the traditional 2-edged broad swords to promote quick comprehension and friendship by means of written words —

Harold Adler 3530 Coy drive, Sherman Oaks, California

the 4th of July, 1960

Dear Ray;

In writing this letter I can truthfully say I am starting the 4th off with a bang in answering your letter received a long time ago. Last month at my wife's repeated . . .

Mr. Adler is a versatile designer of letter forms for Hollywood motion pictures & advertising design. NB: the Oriental character of his monogram.

In Braune in the Bourgogne our hotel menu was so beautiful & simple an object of printing as to make me call on its printer, Jean Dupin, 75 yards from our Hotel de la Poste. He had no English (what Frenchman has? or needs it?), but in five minutes my French was enough to make us both conclude we were born with typographic tastes just alike!

Yrs. sincerely
Paul

Paul Standard's fluent informal hand is shown here in contrast to his formal chancery style as used for the inscription below. As an author, teacher and lecturer, he has pioneered in the promotion of Italic handwriting in the U.S.A.

Dedicated to all who remembered the needs of future generations through bequests to
Federation of Jewish Philanthropies of New York

Rubbing from inscription in brass plaque (6" x 36") mounted upon enclosed electronic lectern containing paper scroll (6" x 120 feet) listing Federation Charities benefactors. Plaque's original, & scroll itself, written out by Paul Standard, New York 1962

Miss Daisy Alcock . ATD . ARCA . FRSA . NRD
Studio 27, Kensington High St. W8
London . England .

THE CORONATION OF QUEEN ELIZABETH II 2ND JUNE 1953 · ELIZABETH II BY THE GRACE OF GOD OF THE UNITED KINGDOM OF GREAT BRITAIN & NORTHERN IRELAND AND OF HER OTHER REALMS AND TERRITORIES QUEEN, HEAD OF THE COMMONWEALTH, DEFENDER OF THE FAITH.

July 28th '57

Reproduced Lettering with Compliments .

Miss Alcock has executed many distinguished calligraphic commissions for England, such as the above medallion, and the RAF Roster of The Heroes of The Battle of Britain in Westminster Abbey.

Our first speaker was Sir Francis Meynell who gave a most delightful talk which he called "According to Cocker". Alfred Fairbank spoke about the revival of italic handwriting from William Morris up to the present, & Miss Hooper spoke on italic handwriting from the point of view of teaching it to the junior school child. . .

Yours sincerely,

Anna Hornby

Three widely different Italic hands—two by famous British scribes & teachers—the third by an equally well-known American typographer & type designer.

Daisy Alcock writes speedily & large—sometimes on pages 16" x 21"—& with lines close together.

Anna Hornby, Sec'y., Emeritus of S.I.H., writes a small hand—lines far apart.

Oswald Cooper, best known for his widely used **Cooper Black** *(which he didn't like), here offers advice on what to avoid when designing a type face. In so doing, he reveals his own delicate but firm Italic hand.*

Oswald Cooper

(His signature)

MAN
dwells
GLORY
impends

From the highly condensed instructions presented on 3 pages of Cooper's insert for the 27 Chicago Designers Annual of 1939 using Roman type of 1468 as models.

a Traditionally, a type designer is able to devise a complete font to match a single letter you furnish for style. Tradition does not say where you get the single letter, but the designer, who has been waiting for it, takes it and makes all the other letters resemble it and a new type is born. If you want a type of enduring charm made from your single letter, you will caution the designer to tincture the new font only lightly with the theme, to make the resemblances mild. Types get along better if they are not too clever, if their excellences are not too evident. You will find no diagrams for good types on these pages. This is a tour amongst the hindrances to good design! For instance, this variation on the Sweynheym & Pannartz a tries too sedulously to look like the model. You will not want your new type to be so preoccupied.

Speaking of students of Calligraphy — The enclosed poem was written out (& composed) by a monk from our mother house, St. Joseph's Abbey, Mass. He has been with us for several months & will be going back to his own monastery next month or so. He is a poet and was interested in some kind of a script to write his poetry in. To make a long story short — I converted him to the Italic hand & gave him a few lessons. He has taken to it with great zeal & as you can see has become a real penman. I gave no more than a word of direction on letter forms here & there & left him to develop along his own which he did to my great delight. He composed this poem in honor of Oregon "the land of the Plum trees". He wrote it out quite rapidly — he sends it to you as his gift. His name is Fr. Martin Wheeler.

I must thank you once again for all your kindness toward me as it means a great deal when you take time out from such a busy schedule to send me such a full & enjoyable letter & so many samples of yr. work.

Yr. friend in Christ,
Robert

An excellent example of "disciplined freedom" swift & direct — in tune with our times

The fine Italic hand of Fr. Robert Palladino, OCSO. Like his brother monks, he wastes neither time nor words in idle chatter — but serves as the cantor & director of the choir of Our Lady of Guadalupe monastery at Lafayette, Oregon

In The Land of The Plum Trees

From the halberds of the tributary mountains
dark winds purple the roses
fragrant as a shaft of thunder.

A dream lingers in the naked garden
in cadenza of honey and cymbals
but the night prevails like wine.

From his cinnamon cave
an owl drums the night-watch.

Fr. Martin's poem, originally written as a spread, 22" x 8½", is shown here as two separate pages to gain size. Another facet of his genius becomes apparent when, on temporary leave from his monastery, he argues despondent people out of the urge to commit suicide. His office is aptly named Rescue, Inc.

Spume of the far sea shrieks in my mind
like doves
fluttering against the ruby windows
hanging in the blackness.
(Light and the green sky are swords
within me)--

longing for the moon
in the land of the plum trees.

Martin.

His pupil demonstrates unusual aptitude.

Chicago 2/20/67

A brief report, Ray,
on things here: We've changed our name from One-Leaf-Books, Inc. to Educational Graphics, Inc. —more fittin, eh? / Of the 2 o-l-bs you "calligraphed," The California Missions still has no sponsor but is selling at retail thruout Cal. — The Printed Word & Books, sponsored by Bruning [div. of Addressograph Multigraph Corp.] is at last getting into free distribution to schools, etc.

Now in work: The Story of Flight — for a top Airline (to give to schools) — and two on Sports: The 48 Years of The Green Bay Packers — sponsor, Kimberly-Clark! and Baseball, Its Records since 1905 (will tell you sponsor later).

So — there's our story!... What's yours, besides your labors on the Chautauqua book?

Very best to you and Irene / Bill

E. Willis (Bill) Jones — formerly a free-lance art director, now a publisher of educational material for public relations — disapproves of endearing salutations in letters except to loved ones — makes his salutations a first sentence the main point. He started a crusade to "drop Dear in Business Letters" in the early 30's and actually got twenty-two companies to adopt the practice at that time. At least a dozen still hold to it. He is also an advocate of simplified spelling.

. . . I cannot exaggerate the importance to me of my files, for whatever typographic, etc, knowledge or skill I enjoy has been learned on my own, through observation and study, interest and experiment, through the years. Saving all manner of clippings and filing them under "type-design, typography, book-design, lettering, calligraphy, advertising, photography, architecture, interior decorating, etc etc etc" has given me an art education I did not get in my liberal arts (philosophy major) schooling. . . .

Howard E. Paine

Mr. Paine, Editorial Layout (Chief)-National Geographic Society, supplements his liberal arts education by intensive observation.

For the past year I have had a lot of fun adressing various service clubs, giving a half-hour talk on the origins of the alphabet, with a quick lead-up to Italic handwriting. Similarly I have talked to PTAs and one Board of Education meeting (with 40 teachers present), on the same subject supplemented by a showing of John Dumpleton's film-strip on the "Development of Italic." Great interest, enthusiasm, determination to "do something" but very little action so far.

Mr. King died on November 11, 1967

Richard King

For some 12 years Mr. King enthusiastically and unselfishly devoted himself to the development and expansion of his traveling exhibit, A BETTER HANDWRITING, circulated among museums and libraries, schools and universities all over the U.S.A. The only expense to the sponsoring organization is the cost of transportation from the last exhibitor. This activity continues in memory of Mr. King. Requests for the exhibition may be made to Dr. Edwin Gould of Johns Hopkins University.

Richard Nall—an art director—admired the handwriting of one of his college professors, and copied it, there by gaining a legible Italic hand of his own. So, imitation induced by admiration & flavored with originality is an excellent way to learn. Mr. Nall's name as used on his stationery is evidence of my own frank admiration of the inimitable style of the outstanding English wood engraver & designer, Reynolds Stone.

A Comforting Quote: "He is most original who adapts from the greatest number of sources". Robert Herz: Ideafile.

3814 CHAMBERLAYNE AVENUE, RICHMOND, VIRGINIA • TELEPHONE 6-0743

November 1, 1954

Dear Ray & Irene:

We enter the pre-Xmas season with the co-incident peak free-lance-load. In our church calendar it will soon be "Advent", and I wonder if the Anglican cleric who so named it was not also a church scribe who laboured with a flock of pre-Xmas design jobs. I much prefer getting off a note to you than tackling the job on the board above: another Xmas Card. Man wants to show (his) white building in a snow scene and wants it in 4-color process. Counting green pines, two colors remain begging for attention. Looks like red-yellow holly berries are going to have to be dubbed in somewhere! But better too many than too few...

"Tell the Archbishop he's darned well got to OKay this one—its got to go to Alcuin for production and he's all the way over at York!"

Best to you both: drop several lines when you can!

Helen & Dick

Street, New York City 2

8th September 1949

George Trenholm both
ite you and say hello &
what Stanley Morison
envy you the Newberry
collection. Mr Morison feels it superior
to the Metropolitan Museum's collection
of writing books, that I am familiar with.
One of my dearest hopes would be to combine
a lecture series in & around Chicago with
some weeks of study at the Newberry,
with you perhaps. But this would have
to be at the end of the school season,
in June for example, and that is difficult
to arrange. . .

Cordially,

Arnold Bank

Mr. Bank, born in New York City, 1908, is a calligrapher, letterer, typographer, teacher and lecturer; coming up via sign painting, typographic shops, advertising agencies & publishing offices to be art director of TIME's advertising department, 1941-47. He has taught since 1935, & lectured since 1947, at leading art schools and universities from coast to coast. He is professor of Graphic Design at Carnegie Tech since 1960. He has designed several architectural inscriptions for public buildings, most notably for the "CREDO" monument in Rockefeller Center, New York City.

123 North Dubuque Street, Iowa City, Iowa. 31 March 1955

Dear Mr. Da Boll,

As a graduate student at the University of Iowa I am working on a thesis concerned with the historical models which have been of greatest interest to a number of American calligraphers in the formation of their informal, personal styles. I should very much like a specimen of your hand. Would you care to make a brief statement about the influences upon your writing? This would be of great help to me and I will appreciate it greatly. Should you find a few moments to favor me, I of course will not publish your writing without your consent.

Respectfully,

Roy E. Langley

before / after

Since writing you I have had the pleasure of meeting some fine calligraphers, Paul Standard and Hollis Holland in New York and Edward Karr in Boston, to mention a few . . .

This fall I'll teach a writing class in Pittsfield, Mass., a stones throw from Cranes Paper Mill. Wish me luck.

3 IX 59

Sincerely,

Roy E. Langley

It is important for the student to study the classical calligraphic forms — to know their "bones" — before attempting to develop a style of his own — as Mr. Langley has done during a period of four years.

. . . recently received your publication of Calligraphy's Flowering, Decay & Restauration by Paul Standard. Frankly, I found a distinct thrill on every page of it.

Here's to a more universal apprecia=tion of calligraphy!

Sincerely yours,
Dale Ballantyne

P.S. I teach lettering at the University of Iowa and will appreciate any news from time to time on new publications

[Dashed off but legible!]

Dear Ray,

I wish you would let me know any time you plan to visit N.Y. Meanwhile have a happy=writing new year.

Jerry Craw

(Freeman Jerry Craw is a graphist-letterer, calligrapher, lecturer, and designer of type faces.

Calligramme

Dear Ray: It is nice to have your informative letter and to learn from it of the change of your residence. I salute you for this act of wisdom and do not admit to even a mite of choking envy, the news is that good. Christopher Morley, with his particular gift of insight, once wrote that the holiest ghost we will ever know is the creative imagination. Seeking that personal divinity, all of us, by some token, seek a privacy that is key ingredient to any creative performance. Today is the best of times to melt our candles away before they ever can burn at both ends. I admire the iron of your resolution to diminish the accumulating pressures of the times.

Be certain to provide us with your new address. It is doubtful if there is any advantage to being on our mailing list save for the occasional stuffing of your RFD box.

Doing another designers article for DA - does any of your recent work suggest itself as illustration?

Best regards to Irene and to you.

31 March 1951

Yours, Herb

P.S.: Another day and another pen. Your package came in the day's mail with its gift load. Thanks for the stimulation. The Michael bookplate sweet indeed and an exemplary model. I would indeed like to see the Hamlet. Send at your convenience — no hurry!

Newest of our fancies herewith. Maybe you might consider a collaboration next year in observing Will's birthday. I've always thought Hamlet's instructions to the players a fine text for you to provide a calligraphic obligato.

Simpson learned to write calligraphically entirely by himself through observation. Admiring the work of W.A.Dwiggins, he corresponded with him and they occasionally collaborated. The Feather Vender mark is by WAD, based on a figure by the famous 17th century French etcher, Jacque Callot.

Herbert W. Simpson of Evansville, Indiana, is a "Jack of Black Arts" in all phases of Graphic Design production

Thought of Irene, the other evening, when I was reading Dard Hunter's autobiography, "My Life with Paper" & his experiences w. his brother Phil Hunter as fellow magicians on the Chautauqua circuit.
Also some wonderful reminiscences of this past great American cultural enterprise in a new biography of 'Ould' Bob Ingersoll (by Ovid Larsen). How I can now recall the waving of a sea of handkerchiefs - as a standard gesture at every finale - remember?
It was almost 50 years ago, when my Dad took me to a recital at the County Fair, in Natick, Mass... so long, long, ago.
Keep well - Irene + Ray, and all of the best to you & yrs.

As ever,
'Don' EGDON H. MARGO

20/XI/61

TENET INSANIBILE MULTOS SCRIBENDI CACŒTHES
& The incurable itch of writing possesses many.

at the sign of the Pen & Press 4961 Murietta Avenue, Sherman Oaks, California 91403

Egdon H. Margo—teacher of Graphic Design, Calligraphy and Typographic Design at U.C.L.A., Lectures in Calligraphy & Palaeography at many schools, colleges and museums • "Typochondriac" and Proprietor of a 90-year-old Private Printing Press. He became a Calligraphile while serving in the U.S. Army Air Force as a Liaison Officer and Interpreter with the French Air Forces.

Rochester Institute of Technology

SCHOOL OF PRINTING · COLLEGE OF GRAPHIC ARTS & PHOTOGRAPHY

65 Plymouth Avenue South, Rochester 8, New York

January 5, 1965

Ray,

Fred will no doubt be after me soon to get to work on another edition of the Newsletter. This one is well under way—

Al. (Alfred Horton)
Teacher of Printing

---

such complete command of your pen that you can use it in different ways appropriate to a variety of purposes, and in each of them there is the sense of viability even in the most formal examples. But in your personal letters, such as the one I have before me — free, cursive and informal, but with the essential unity of style, there is the quality of audible speech.

There is so much more – and better – that I want to say; but I regret having to tell you that your kindly and thoughtful surmise that writing is somewhat of a strain mentally and physically for I can only do three or for lines before my eyes lie down and refuse to focus, and the results are all too evident. But I hope the effort, awkward as it is, makes it also evident that I am still –

Sincerely

TM Cleland

January 7, 1962

Elder Statesman of the Graphic Arts wrote the above lines at 82 following a paralytic Stroke.

Italimuse, Inc. CALEDONIA · NEW YORK 716-KE8-6257

zip 14423

June 5, 1964

Ray, ... it might save time if you Air-Mailed a photostat of these changes, roughed in: one to me & one to Miss Amy Vanderbilt, 438 East 87th street New York, 28, N.Y., if it is easy for you to get photostats locally (quickly). If not, air mail sketch to me and I'll employ Mr. Xerox again.

Bestest, Fred

The Tinhorn Press has aquired a fout of Mr. Cooper's alleged favorite, original old Style Italic, and we are presently enjoying the luxury of it.

Yours sincerely,
Chuck Robertson

Above — Fred Eager President of Italimuse, Inc. Teacher of Music & Italic Handwriting.

Charles Robertson — Printer-Craftsman-Calligraphile-Proprietor of The Tinhorn Press, Atlanta

Three excerpts from a letter by Gil Hanna

Swift & fluent, his Italic hand is larger than most people write. It is shown here somewhat smaller than actual size. For further comment, see pages 139 through 147.

We suspect that Mr. Hanna's pupils actually learn a lot from his teaching. We like his attitude as a teacher.

Your move to Arkansas brought a flood of memories - having been stationed at Fort Smith for two years during the war - plus a year at Joplin, Mo. It is wonderful country, and the people are genuine -

I have two classes in lettering at The Society of Arts & Crafts which have been a wonderful experience I feel I have learned so much more than I possibly could have taught [I mean the youngsters have taught me more than I have taught them!]

We are agreed with regard to Mr. Benson's excellent translation & rendering of the Operina by ARRIGHI (Vicentino) and make further comment on page 162.

Mr. Benson's new translation of Vicentino is very interesting, and beautifully done - but I wish he had arranged the pages so that his page was opposite the 'original' -

Mr. Hanna's fine Italic hand in a more formal mood.

the babe wrapped in swaddling clothes, lying in a manger." And suddenly there was with the angel a multitude of the heavenly host praising God and saying, "Glory to God in the highest, and on earth peace, goodwill toward men."

May all the Joys of Christmas be yours — Gil & Pauline Hanna

TEL. 254·4050 · A.C. 704 **Spuehler's Little Alp Studio** 41 Rash Road · Asheville · North Carolina

Dear Ray,

We sold the farm in Elgin as of March 1st. '64. It got to be a chore to look after it long distance, and it's a great relief not to have to make those trips North.

Just made reservations for us to go to Switzerland for a month in July. Will take my water colors I think, but have to brush up on it as have used only oil for about 3 years.

Also have started a series of prints (silk screen) of North Carolina, finished two, and have more in the works.

If it gets too warm in your country – come and see us; we will be back End of July, and I can show you this gorgeous country and you'll get the benefit of our cool mountain air. We've got a nice guest room, and I sure would like to see you again.

With our best wishes to you and Mrs DaBoll.

Yours,

Ruth & Ernst

Ernst Spuehler — Versatile designer, painter, calligrapher, ex farmer, good friend

Mr. Milliken is a graphic designer & calligrapher blest with a sense of humor.

RD1 Annandale
New Jersey
December 18
1961

Poor Penman Writes Off Loss

United Press International.

LOS ANGELES, Nov. 8.— An unidentified man handed a note to teller Alice Davidson, 30, yesterday in a local branch of the Bank of America.

She could not read it and turned to consult another bank employee.

As the two puzzled over the note, the man fled from the bank.

They decided later the note read:

"I have a gun. Give me $1000 and be quiet or I'll kill you."

Dear Mr Da Boll — I have been carrying the attached clipping around in my pocket for a month, intending to send it on to you. What a pitiful fiasco! If the unidentified gunman had been a student of Arrighi, Palatino & the rest, & had developed, through constant practice, a legible, attractive hand, he might today be enjoying his $1000.

What better argument for those educators who say handwriting is "not important"?

& Greetings of the Season from

Steve Milliken

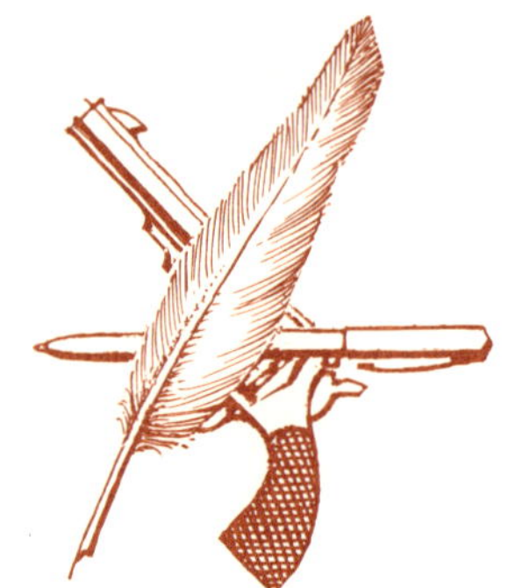

Steve is right; from the gunman's standpoint, his failure to collect proves the pen is mightier than the gun.

PEANUTS® By Charles M. Schulz

+ + + Maybe you saw this strip, but I can't resist sending it, because it reminds me so much of you & your preachments . . . . . Anent margins— to me. Now I am pounding the same thing into my students. (120 this year).

Ruth Justus is a teacher in the Minneapolis School of Art.

Also I am sending a story which appeared in our Sunday paper. The response to it was amazing. People from as far away as Montana wrote to ask "where can I get that book?" & "where can I buy one of those pens?"

Kindest regards,
Ruth Justus

Ruth is right— and will prevail! Margins often present tough problems—difficult to solve.

Lloyd Reynolds,
Professor of Fine Art,
Reed College, Portland, Ore.,
a leading exponent
of Italic handwriting,
Vice President of
the British S.I.H.

I'm delighted with your remarks on Chinese white. But there's much to be said for your originally calling it a betrayer. It tempts one to go on touching up—until the life goes out of the work (Lettering is another category!) In direct work, Chinese white, if used just too much, produces smug prissiness rather than improved form. May I show your comments (the first page of your letter) with the broadside?
Thank you again.

Yours,
Lloyd

Dear Mr. Dal

Here, happily, is my contribution to the Typophile Chap Book, probably the last one to be submitted, for which I apologize. Too much work, the problem of getting settled in a new city, and possibly old age, delayed me. The original I handed to Paul Bennett at lunch and he seemed to like it. . .

Fridolf Johnson,
Executive Editor of
American Artist,
Proprietor of
The Mermaid Press

April 21, 1960

Dear Ray:

A little book on German Modern Calligraphy is being sent to you which I think has quite a nice feeling. I'm sure you will agree that the handling of the initials is very well done (Always a problem). The book is printed in Gold, Red and Black and the printing is superb.

. . . Trust, you've benefited by the recent rains!

Regards to all

Rodney

Rodney Chirpe's family name was originally Zscherpe from the German meaning scabbard. For American use, his father's generation found it expedient to drop the first two letters and change the following 'e' to 'i', creating a name that suggested the symbol of a bird chirping on a twig.

Calligraphers often devise trade marks appropriate to their work & names.

Theodor Jung · 1101 Chautauqua, Norman, Okla

7 January 1957

Dear Ray,

. . . And by the way, if it's not revealing trade secrets, was the lettering on your card drawn, or calligraphed or mixed?

Regards,

Theo

Theodor Jung

Teacher – Calligrapher – Typographer

Am off early in the morning for a visit with Hermann Zapf in Cleveland. He is here in this country for a few weeks and is going on to California tomorrow night. Sent me a packet of nice things from Germany a few days ago. — Boy! am I on Cloud Nine!

Cheerio /

Harvey

G. Harvey Petty, Calligrapher – Typographer and Writer on those subjects for trade publications –

Dear Folks –

We have gone and done it! Attached is a copy of the listing sheet of the house we have contracted to buy in Glen Ellyn. As you will see, it has a few years on it, but lots of possibilities and is in a beautiful area . . .
. . . Believe we have the financial situation well in hand, but just in case we don't sell our present house before we have to meet the closing on September 16, it would be comforting to know that, if necessary, we might draw on your bank account (with repayment as soon as the house is sold). Would this be possible?

Love to all –

Briggs, Mair + the kids

Briggs developed his own Italic style by studying the SIH Quarterly Bulletins & CIH Newsletters. As a salesman of electrical cranes and hoisting machines, he has found that Italic handwriting is a real business asset, attracting interest & favorable attention.

OK son! Glad to help.

\+ + + ¶ The show is very popular, beyond all my expectations, and the campus has blossomed into an Italic rash. ¶ Could you drop me a note soon recommending amounts of insurance on your work?

Thank you again for all you've done —

Bob van Kluyve

Professor Robert Van Kluyve, English Department — Duke University, Durham, N. C.

A THOROUGH study of letter forms can get one involved in epigraphy, palæography, calligraphy, typography, lettering & even photography. Frederic W. Goudy, America's most renowned type designer and teacher of lettering said, "... I found that the study of the alphabet was no mere

... Ever since my delightful visit last October I've been wanting to send you one of my Peruvian photographs. So here is a favorite of mine...

Sincerely yours

Frank Kofron

Mrs. Theodore Cyrus Childs

has the honor of announcing

the marriage of her daughter

Doris Elaine

to

Mr. Leo A. Hodroff

on Wednesday, the twenty-ninth of December

one thousand nine hundred and sixty-five

Minneapolis, Minnesota

Frank Kofron's hand can be formal or fluent as the occasion requires. He developed his scribal talent by observation & study at home. Primarily an art director and typologer, he gained his knowledge of type under the tutelage of F. W. Goudy, and is a traveling photographer as well as a calligrapher-typographer.

dry study of ancient forms, but rather that the history of writing is, in a way, the history of the human race, since in it are bound up the development of thought, of expression, of art, of communication and of mechanical invention".

Mr. Goudy wrote several scholarly books & articles about typography, and he also frequently lectured on that subject. It has been told that after he had finished one of his erudite lectures, a somewhat dilettante lady gushed: "Please, Mr. Goudy, tell me just how you go about the designing of type." He replied, "I just think of a letter and draw a line around it."

It's really not quite as simple as that.

March 6, 1951

Dear Ray:

John Benson asked me to forward the enclosed letter to you. I have been out of town and I'm afraid the letter has been around for a few days.

What's this about you pulling out for the frontier? I shall have to learn more about this later.

Sincerely,

Bob

Dr. R. Hunter Middleton, type designer, proprietor of the Cherryburn Private Press, a founder, Fellow & Honorary Member of The Society of Typographic Arts, served as its secretary for 15 years and is a past president. As art director of Ludlow Typograph Co., has designed some 84 typefaces for Ludlow matrices—more than any other American type designer except the dean of them all, Frederic W. Goudy. In retirement he expects to give full time to printing as an art form at Cherryburn and to the practice of the nostalgic art of punchcutting.

7 August, 1957

Dear Mr. DaBoll,

Arrived home last Sunday, had a safe trip. I wish to thank you again for the very interesting vi...
am very prou...
to marvel at y...
plus your supe...

Had an inte...
"Shepherd of the...
many other st...

Once again, ...

We begin with the rectangle of paper. It is empty, a no-thing awaiting the pen. It is not even empty space, but only potential. But the moment the pen touches the paper and leaves even a slight trace of ink, the paper does become space and the boundaries of the area are awakened and respond across the space to the tache of ink. It is as if a magnetic field were suddenly activated. Each subsequent stroke pours added energy into the field, modifying all previous relationships within its edges. Since this activation of tensions and directional movements occurs if only spatters of ink are dropped on the page, how vital is the effect if the pen sweeps across the paper in an ordered sequence of rythmical, dance-like movements, whether free or forming

Entry by Mr Bob Bowman, Minnesota (reduced slightly).

Bob Bowman is a commercial artist with a flair for calligraphy. His informal Italic hand is shown above in contrast with his more carefully written page that won him third prize in the 1966 International Handwriting Contest sponsored by the Society for Italic Handwriting. Examples on the opposite page show his versatility as a lettering man.

BULLETIN OF THE
MINNESOTA
LIBRARY
ASSOCIATION

Polaris Arabian

DIGRESSING from handwriting to the closely related process of hand-lettering, we see why the commercial scribe has to consider the practical value of drawn as well as written letter forms. While the professional lettering man is sometimes an accomplished calligrapher, he is naturally much more interested in the variations of lettering styles than one who takes up the practice of Italic merely to improve the quality of his personal handwriting. It would be a dull & tiresome world indeed, if the letterer were limited to classical forms. The competitive field of advertising, which is his chief source of livelihood, calls for a wide variety of lettering styles. Here each new assignment seems to require some measure of inventiveness. What are the essential differences between hand-writing, hand-lettering, and printing with type— and what are the advantages of each from the standpoint of reproduced copy wherever the printed word is used?

HAND-LETTERING is used mostly for display in headings because it is more flexible and expressive than the cold rigidity of type allows, whether set by hand or by machine, but the manifold advantages of printing from movable types, either cold or hot off the press, are indispensable because of the necessity for speed & quantity. Handwriting is not only warmer & more personal than the other two processes, but has a more intimate quality. Except for the spoken or sung word, it is the most direct way to convey the subtleties of intellectual and emotional meaning between human beings.

ACCORDING to information from Printing Industries of America Inc. "it is impossible to imagine how our country could exist without printing. Indeed, on a broader scale, it is so essential to so many facets of life throughout the world that we could not exist without it. Of the twenty major manufacturing groups in the U.S. covered by the U.S. Government's 1963 census of Manufactures, Printing & Publishing ranks third in number of Establishments. Its size is exceeded only by food and kindred products, and by apparel and related products". So—if there seems to be some kind of a Renaissance of scribal handwriting afoot of late, we trust that Government Reports will reassure our Gentle Readers & Friendly Printers that scribes pose no serious threat to the printing business.*

* It is quite reassuring to find that the PIA Stationery calligraphically designed by your scribe 17 years ago, is still being used today.

WHAT THIS SCRIBAL ACTIVITY aims at is legibility with some attention to design & beauty in American handwriting and the possibility that, through emphasis on humanistic quality, we may even suggest the desirability of softening up some of the harder "too mechanistic" features common to typography today. After all, we still are human beings—not Robots! Why should machines usurp our human rights—including the one to write legibly?

TO BE CONQUERED and driven underground by machines that MAN—"Homogrub—has created in His own Image" with almost human attributes, was a fearful possibility that first occurred to William A. Dwiggins in 1911. Thirty-four years later he wrote a play$ about it. WAD—independent thinker and essayist—founded the mythical Society of Calligraphers and, as America's foremost scribal designer, created an alter ego-Herman Püterschein-with whom he argued matters of importance to the Graphic Arts-with particular attention to the designing of books-which he found to be hopelessly mired in a deep traditional rut (circa 1924). He fooled the unwary as to his identity, not as to his sincerity. Then, single-handed, he proceeded to pull book design out of the rut with strikingly original graphic forms & ideas which have brought him the highest honors the Graphic Arts can bestow.

$ Millennium I

ASIDE FROM MILLENNIUM I, WAD wrote several other plays, all enacted by marionettes of his own design and make. All were animated in his Little Theater at Hingham, Mass. Yet he found time to design type faces that we all read every day.

DWIGGINS had something in common with ARRIGHI, DAVINCI, and DÜRER—three great masters of the Renaissance. We add WAD to round out a famous foursome which has made important contributions to the design of our alphabet as we know it and use it today. Subtle changes are often introduced to keep it in step with the times. Some of the changes have not been so subtle, nor even necessary, as time has a way of proving. It seems to be a matter of evolution in which many styles are called for but few are chosen.

From the age of six I had a mania for drawing the forms of things. By the time I was fifty I had published an infinity of designs; but all I have produced before the age of seventy is not worth taking into account. At seventy-three I have learned a little about the real structure of nature, of animals, plants, trees, birds, fishes and insects. In consequence, when I am eighty, I shall have made still more progress; at ninety I shall penetrate the mystery of things; at a hundred I shall certainly have reached a marvelous stage, and when I am a hundred and ten everything I do, be it but a dot or a line, will be alive.

I beg those who live as long as I do to see if I do not keep my word.

Written at the age of seventy-five by me, once Hokusai, today Gwakio Rojin, the old man mad about drawing.

From Hokusai's preface to his Hundred Views of Fuji

WAD 1949

WAD
1954

Here Dwiggins has taken the text of the famous Japanese artist, Hokusai, & rendered it in his own peculiar combination of caps & lower case letters.

A line of WAD's characteristic Italic is shown at the left.

1

Leonard E. Read

Sincerely,

4

Harry J. Owens

Yours sincerely,

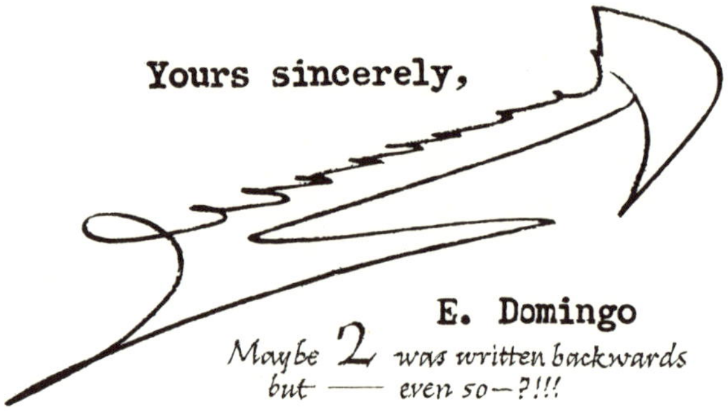

E. Domingo

Maybe 2 was written backwards but — even so—?!!!

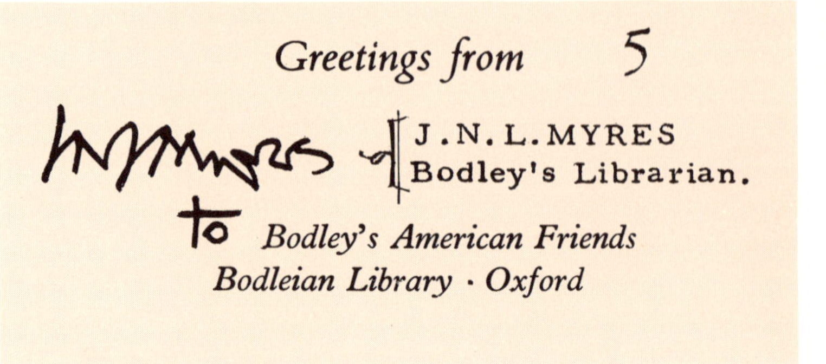

Greetings from 5

J. N. L. MYRES
Bodley's Librarian.

to *Bodley's American Friends*
Bodleian Library · Oxford

Paul McCain

3

6

"THE MOVING FINGER WRITES AND, HAVING WRIT, MOVES ON" — On to what? As to the above signatures, it moves on to the typewriter to make identification possible. There is no question as to the literacy of these signers—each is a VIP who has made his mark, literally & figuratively, in his chosen field, be it Education, Diplomacy, Big Business, Charity, Government... All seem to be pretty much a matter of Government now. Permission to reproduce the first 5 signatures was good-naturedly granted. The 6th was taken for granted. No.1 names the president of The Foundation for Economic Education, Inc., No.2—one of his foreign correspondents. Before asking permission of No.3, I showed his signature to a cashier of the bank that handles his finances, and asked if she knew whose it was. "Yes", she replied, "that's the way Paul McCain signs his name. He's president of Arkansas College. All of our tellers recognize & honor his signature but I doubt there are many outsiders who would without typewritten support". No.4 names the retired advertising manager of America's largest printing plant—R.R. Donnelley & Sons Company. No.5 identifies the librarian of the renowned Bodleian Library, Oxford, repository of ancient manuscripts—Greek—Latin—German—Hebrew—Arabic—Oriental... The name of No.6 has so many distinguished titles attached and is

Time was (in some cases still is) when an illiterate person's signature was required on a document, he made his "mark", a cross, + to be followed by the signature of a literate person who would vouch for him.
+
Here it seems that the typewriter has usurped the place of the cosigner.

so well known that it has become the common property of all people who love freedom. Thomas Jefferson is, perhaps, best known for the Declaration of Independence that he wrote with his own hand & for which he is chiefly credited as the author. His signature is shown here as he wrote it on that famous document. It is not as legible as his regular writing of the main text but is easily decipherable.

Jefferson is said to have written over 25,000 letters. Surely he could have made good use of a typewriter & carbon copies, for it is also a matter of record that he experimented with a pantograph that made several copies as he wrote the original letter.

AFTER ALL, one's signature is a rather personal thing and one has the right to do with it as he pleases—to make something special out of it—perhaps a sort of hieroglyphical symbol or even an abstract design. That's the way many outstanding trade marks are developed but, as such, a personal signature usually requires the help of a few typewritten characters as a common denominator to gain general comprehension.

THE BASIC FORMS of the Roman capitals have come down to us through the centuries, not only glorified but often persecuted. They have been expanded & condensed beyond reason, afflicted with the third dimension and lately threatened with the fourth! They have been illuminated by mediæval monks & modern electricians—"blown up" by engravers—lined up & "shot" by photographers. We have seen them juggled and TORTURED—even EXPLODED on movie and TV screens with the fragments reassembled to make sense again before our startled eyes! We have seen them stretched for miles across the heavens in what is called sky writing only to be dissipated by the vagrant winds. In short, these basic forms have served as the FRAME upon which every lettering man who has had a new idea could hang his more-or-less successful innovations. As these experiments have been developed and proved to be of value, they have created the modern trends of their own times. The evidence is in the typographic sample books weighing up to 20lbs., and in the spate of specimen sheets for ever flowing free from foundries & "foto lettering" studios.

. . . even drawn and quartered!

LOWER CASE ROMAN & ITALIC were the most important developments of all and created the greatest permanent value but during a period of 2,000 years no abstract form has had sufficient merit to displace a single one of the original capitals. Cap I once stood for J, and V for U, so 2 Vs made a double U, or VV. Z stood 6th in the Greek alphabet. The Romans omitted it but later restored it to good standing as No. 26, at the end of the line.

SO FAR, THE COMPUTER has sacrificed beautiful figures on the altar of speed, substituting in their stead, crude, cryptic cyphers in lines of astronomical length extending into outer space. Rectangular Zeros seem out of step with the alarming round numbered totals of today! However, the Space Age has just begun! We look hopefully forward to the beautification of computerized digits that will be more in tune with the music of the spheres {And a balanced Budget.

Now we find that our personalized checks must bear this latest indignity!

CORPORATE PU

The Mead Library of Ideas invites you to the 2nd Corporate Publications Exhibit. Included will be recruitment, facility, product, employee and research brochures as well as house magazines, newsletters, tec

A HIGH DEGREE OF IMAGINATION is evident in the choice & arrangement of the various corporate insignia that spell out the name of an Exhibition held at the Mead Library of Ideas. The above announcement of it, designed by Norman Litman, demonstrates that abstract symbols which stand for capital letters must be based upon their original counterparts in the Latin Alphabet in order to be understood. The more abstract they are, the more they will lack significance as letters if taken out of context and used singly. However, when the familiar "Seeing Eye" trade mark of CBS Television, for example, is used along with the other more-or-less abstract insignia of corporate identity, the "Seeing Eye" becomes an O and we have a line of capital letters that functions perfectly for the purpose of the announcement—making it a most effective eye catcher! And much more exciting than if set in classic capitals.

Identifying the above abstract symbols of corporate significance can be a test of one's memory. Each is an important trade mark in the field of advertising.

THERE ARE VARYING DEGREES of abstraction in letter forms, ranging from the simplest single stroke Futura cap I to the most complex illuminated initial that can be found in the famous 9th century Book of Kells, said to be the World's most complicated, beautiful & valuable volume.

TRACING THE GENEALOGY of our letters from the Phoenician Alphabet, c 1300 B.C., through the Greek, c 700 B.C., the Roman, 50 B.C., the evolution of the small letters, 300 to 800 A.D., the Gothic, 1500 A.D., and the Italic, 1600 A.D., we find that all letters are but abstract symbols for sounds to be spoken & heard as well as signs to be read. In prehistoric times, Man conveyed ideas by means of crude pictures of objects, animals & events—pictographs—(some were not crude but

ECATIONS EXHIBIT

nark manuals and direct mail. From September 6th to September 23rd at the Mead Library of Ideas • 200 Park Avenue, 43rd Floor, New York 10017 & 20 North Wacker Drive, 41st Floor, Chicago, Illinois 60606

truly artistic) — sketched on the walls of his habitat — usually a cave. Such graphic symbols date back thousands of years. This leads us to speculate on the possibility that the beautiful letter S, for example, may have been derived from the sinuous curves of the devilish old snake that tempted the first Woman — Even the sibilant sound of the serpent's hiss may be read into that most curvaceous & alluring of all letters!

WHEN WE CONSIDER that there are nearly 400,000 words in the English language & that they are all made from combinations of the 26 letters of our alphabet, and that by following their alphabetical sequence we can find the definition of each one of those 400,000 words in "jig time", we begin to realize what a simply marvelous and marvelously simple thing our alphabet is. Considering further that each letter is a purely abstract symbol, it is indeed a miracle that those 26 letters, when made into words, can convey the most subtle nuances of meaning — capable of swaying the human emotions and behavior toward acts of love or hate, war or peace.

This also brings to mind the infinite number of musical combinations in the 12 note scale & its development on the 88 keys of the pianoforte.

NO DOUBT there will soon be electronic dictionaries that will give us definitions instantaneously. Your scribe will be content to find them, himself, in "jig time". And if asked for his opinion on abstract design in general, would say he is for it as applied to the reasonable disposition of decorative motifs in their relationship to calligraphy & typography for the purpose of adding a lively note or to give a meaningful lift to text matter — as in the highly distinctive style of W. A. DWIGGINS. When the extremes of abstraction invade the field of Fine Arts, your scribe knows only that he lacks what it takes to comprehend the utter confusion of a Jackson Pollock painting on the one hand & the magnificent nothingness of a huge framed rectangle of almost solid blacks by Ad Reinhardt on the other. The raves & the awards they win must come from the "distillers of quintessences"— to borrow a term used in a "confession"— later proved false— but not before it spelled TROUBLE! Let me tell the tale as a scribal exercise:

We prefer the works of such modern artists as Andy Wyeth & Georgia O'Keeffe.

NOTA BENE:— IN OCTOBER 1967, your scribe, unaware that the document alluded to was spurious, used it in an Italic variant of the Chancery Script, for the text of the original version of this page, which he also embellished with two small Picasso reproductions in the margin. Anyone in the graphic arts field who has sweated out the composition of an illustrated calligraphic page & arrived at a pleasing balance, can imagine our frustration when it recently became necessary on two counts to do it over— (1) The Museum of Modern Art, fiercely protective of Picasso, was annoyed at our rather facetious approach, & refused us permission to use such small reproductions— adding a brief lecture on VALUES in modern art, for the benefit of your obviously "square" scribe. (2) LIFE itself— at least we were caught in good company— also used the document in a double-issue (12/27/68) devoted entirely to Picasso's art & life.

THE IMMEDIATE RESPONSE was a flood of letters, some condemning, some praising Picasso— and some pointing out that his so-called "Confession" had been contrived years before (in 1951) by one Giovanni Papini, who so despised modern art that he had tried to discredit it through one of its foremost exponents. LIFE apologized handsomely in its issue of January 17. We mutely do the same by sweating this page out once more.

Scribe before mirror reflecting frustration

. . . with "Confession" proved spurious, Scribe becomes furious . . .

Expresses Vexation— & then, to emote— with a reed pen, he cuts his own throat! "Out— damned spot! . . . Who'd a tho't the old man had so much blood in 'im?" For inspiration, all credits must go to Pablo Picasso & Edgar A. Poe.

ANY FORM OF HERESY is the life of argument, & in our democracy it would seem that everyone can express his own opinion and everyone, apparently, has his own reasons & angles— some SHARP— some OBTUSE! Who will be the final judge? Dispensing justice in these matters is as ticklish as wearing a hair shirt!

THE GENERATION GAP no doubt enters into it. Among the most ardent admirers of Picasso whose grateful letters appeared in the January 24 issue of LIFE was Governor Nelson Rockefeller, but with regard to at least one of Picasso's works, the Governor's mother, Abby Rockefeller, was adamant. Although tolerant as well as generous toward artistic innovation, especially in her support of the Museum of Modern Art, she was precise in her likes & dislikes . . . She refused to choose personally the formidable MINOTAUROMACHY for her memorial print room, but gave the money for it & with dry humor sug-jested: "Let's label this: 'Purchased with a fund for prints that Mrs. Rockefeller doesn't like!'"

NB: † Others in at the death:— W.S. and the Lady Macbeth.

AND NOW, following his own brief (but not brief enough) invasion of the Fine Arts field, your scribe returns to the 1st person singular & the cultivation of his own little calligraphic garden in this best of all possible states—ARKANSAS—candidly admitting this is where he feels most at home, either at work or at play, and that he & I, or we, do most heartily approve our State's slogan—THE LAND OF OPPORTUNITY & its symbol—THE RAZORBACK HOG.

With thanks and apologies to Voltaire.

… interested in calligraphy, why I came here to pra… can be started, what paper, pens, and inks I pre… ms, Michigan, wrote a 7 page letter (see excerpts, p… like that. His hand is so free & versatile that I wo… e at all. However, I shall answer a few of the que… his may be helpful to those who wish to improve thei… aphic layout may also find it of value in bringi… en the professional lettering man may* feel the u… itten letters that are needed as foils against the "too p… Let's write letters that permit our imperfections to show, … one to err. In my opinion that is what helps to m… n of error in this way can be disarming—even cha… how to be legible and stylish.

ALAS!—TOO LATE

The Goddess of War and the Liberal Arts*
With her symbol, the owl, wisely imparts
Timely advice for deep meditation—
Thus to dispel our fatal frustration.

Minerva as pendulum swings in the Pit
Lower and lower—soon it will hit!
But to die by the pen instead of the blade
Was the choice your scribe had already made.

* The Greek Athena and the Roman Minerva were goddesses with somewhat similar attributes. The figure your scribe has used as a pendulum is a copy of an Etruscan (sixth century B.C.) bronze sculpture at the British Museum, but it could be taken for a modern piece.

* ie: We could do more of this sort of thing to save time correcting our errors of omission & commission! EASY does it.

MY INTER… n of the work of Edward Johnston and his pupils at the Newberry Library, … fortunate in being able to buy one of the bound manuscripts—THEOCRITUS, IDYL XXIV, transcribed by Margaret Alexander. It has been a source of inspiration & despair ever since for I have found it inimitable! That is as it should be, considering the wisdom of an ancient saying which I have recently subscribed to, BETTER IS ONE'S OWN PATH THOUGH IMPERFECT THAN THE PATH OF ANOTHER WELL MADE, from the BHAVAGAD GITA, a Hindu book of knowledge more than 5,000 years old. Elbert Hubbard said it more concisely some 60 years ago—BE YOURSELF. But with regard to handwriting I would add:—KNOW FIRST THE RULES—lest you make a too hasty departure from your own "scratch line".§

§ Learning to write is like learning to sprint, so don't "jump the gun", you may be penalized for a false start.

IT TOOK the first 15 centuries of the Christian era to develop a good legible style of handwriting. While the rules are not immutable as time marches on, they are still the best we have to write by, even as the 10 commandments, in spite of man-made amendments, are still the best rules we have to live by… Lest I be weighed in the balance and adjudged a hypocritical old

Read the gospel according to ARRIGHI'S La Operina, 1522, also (lest we forget) The Decalogue, Holy Bible, Exodus XX, 2-17.

scribal Pharisee who pays lip-service to the written letter of the law while breaking it as I go about my work, I shall not evangelize further on this phase of my subject. Amen.

Preferred Paper & Mounted Board

FOR handwritten correspondence I use Strathmore Writing, warm off-white, either laid or wove; and for occasional "one-shot" jobs such as plaques and individual certificates, I generally use a heavier weight of Strathmore Parchment which has an excellent surface for the broad pen. At least 90% of my work has always been for advertising and publishing. For this I use a Crescent stock that comes mounted on each side of a 14 ply paste board, intended, I believe, for signs, show cards, and for the mounting of photographs, rather than for pen & ink work. Several years ago while mounting & labeling entries on this board for an exhibition, I found its "matte" surface would take a thin ink line and hold it, even under heavy erasure. It also has a certain "velvety pull" of the ink that I like to feel at the edge of the nib, perhaps due to the slightly chalky surface, but it does not pile up on the nib if the pen is held lightly (as it always should be). It also permits a bit of scraping with a razor blade—or with a scalpel such as the earlier scribes used on vellum in smoothing the rough edges of their letters, or in scraping them off entirely for reuse of the vellum.

Don'ts & Dos

THIS scraping should be held to the minimum, and the same rule applies to retouching with white, for one can quickly pass the point of no return in piling on white and, depending on one's pride in producing a neat job, one may find himself doing it all over—again and again and again. This is true particularly with regard to "one-shot" jobs that MUST be done directly with neither white nor patching permissible. The scribe strives for spontaneity—sometimes he sweats blood to make it look easy—for spontaneity is the life-blood of calligraphy. The rules are not so rigid with regard to the execution of calligraphy & lettering done for reproduction.

SCRAPERS—Ancient & Modern

Scalpel* of the Arrighi Era.

ONE advantage of the Crescent board for commercial work is that it can be stripped easily for quick respacing of letters and words—even the shifting of a whole line. A little practice with a razor blade will show how deep to make the cut around the part to be stripped up, then it can be stuck down again in the desired position with rubber cement which dries quickly and any excess of it around the cuts can be rubbed off, serving as its own eraser, leaving the correction neatly mortised in on the same level.

Please do NOT fold or bend

STILL another advantage is that the whole sheet can be stripped off, rolled & put in a tube for mailing at less cost for postage because of lighter weight and less risk of damage in the mails, which I have found can happen even when the larger flat package has

*The scalpel was used mainly as a knife to cut quill pens.

been stiffened with packing board plainly marked "Please do NOT bend or fold". However, if the mailing tube is to be used, I don't use the razor-blade rubber cement technique because the cemented pieces will not strip with the rest of the sheet. Instead, I lay a piece of transparent layout paper on the part to be corrected, make the necessary alterations, then with the lettering face down on a separate board, fasten the edges of the paper with tape to prevent curling under moisture, and paint it white to make it opaque. When dry, cut the paper from the board & stick the corrections in place with rubber cement. Thin edges of the layout paper covered with a wash of white will make the patch barely noticeable. Either Grumbacher Gouache White or Carter's Tempera White will dry dull for an almost perfect match with the slightly off-white "matte" surface of the Crescent board.

Stripping & Patching

THANKS to the makers of such every-day necessities as paper towels and toilet papers that come wrapped around cylindrical cores, I have never had to buy a single mailing tube; add to this dozens of larger tubes up to 3 feet in length that have come to me gratis.* But such a bonanza in savings on office supplies & postage is not the reason for my incomprehensible (to some folks) hegira to this outpost of culture. I'll tell why a bit later.

*Not to mention hollow reeds plucked from White River cane brakes!

I USE 3 KINDS of pens—fountain, dip, and reed or bamboo. The revival of calligraphy which has won many converts to the cause of legibility in recent years, has made the fountain pen manufacturers pay particular attention to the making of chisel-edged nibs suitable for Italic writing. They are all good for correspondence that is more-or-less informal & speedily written as shown by the examples on the preceding pages. I find the main difficulty with fountain pens for commercial jobs is to get a jet black ink that will flow freely and not clog the nib. Most commercial work requires a tight formal technique as in "built up" lettering, in fact, it is some times difficult to tell where calligraphy stops and lettering begins.

Kinds of pens and ink

PHOTO-ENGRAVING for reproduction calls for an original drawing made with jet black ink that can be touched up with white and not bleed through. So far, only carbon-based inks have met that requirement, so I use mostly dip pens—Grumbachers & Hunts—with reservoirs for waterproof ink. The recently improved A.W. Faber-Castel Higgins is most satisfactory. Steel nibs are available in several widths but it some times happens that only an in-between width can do the job just right. The slightest variation of width can then be made by simply cutting back the gradually tapered nib of reed or bamboo to the exact width desired.

THE THICK & THIN pattern of broad-pen writing cannot be successfully faked—imitations lack spontaneity and the "accidentals" look artificial & contrived—the fakery is obvious, at least to the initiated. Scribes seldom have to make letters much larger than regular writing size,

To help ink flow, file nick in spring—like this.

however, in such an emergency, the resourceful scribe, lacking a nib of sufficient width, can use a kneaded rubber eraser as a wedge between 2 pencils & by squeezing it at either end, the space between the pencil points can be adjusted to the required width. Now, with this improvised nib held at the approved 45° angle, the letter can be drawn in outline according to calligraphic Hoyle. Thus the instrument itself, when guided by the scribe's knowledge of the letter form, becomes, in a sense, the designer of the letter; so, the varying widths of the thick & thin strokes are fundamentally correct throughout their entire length—and the curves, too, as in the finest feminine examples, happen in just the right places! The areas between the outlines can be filled in—solid. The outside obtuse angles where the lines cross could be filled out to make perfect curves—omitted from this chart to give a clearer demonstration of certain points & angles* with regard to the Curvaceous Letter shown here as a prime example.

*NB—points & angles — a · b · c · d

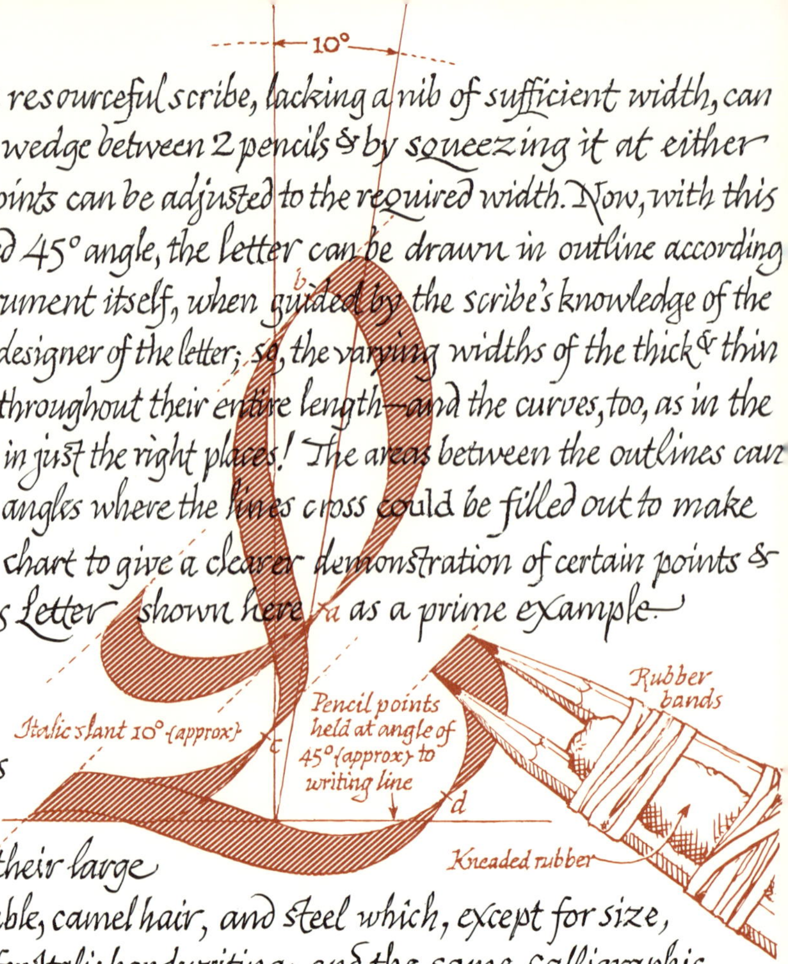

10°

Italic slant 10° (approx)

Pencil points held at angle of 45° (approx) to writing line

Rubber bands

Kneaded rubber

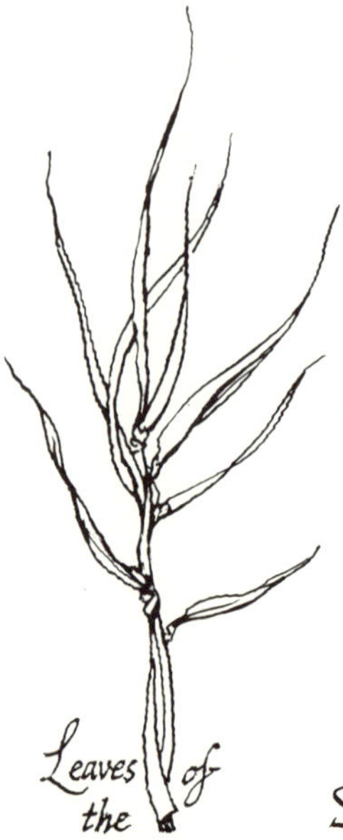

Leaves of the White River Cane

SIGN PAINTERS & Show Card Writers have to be direct and many are indeed exceedingly expert with their large flat square-edged brushes of red sable, camel hair, and steel which, except for size, are quite similar to the pens used for Italic handwriting; and the same calligraphic principles apply to all 3 techniques. For those who work with large letters, this makeshift gadget should prove as helpful as it is inexpensive. It makes Large Letters Look Lively! ... and authentic, no end ...

(Natural History & Local Geography mixed with Calligraphy.

BACK to bamboo & reed pens ... Along the White River near my home, there are cane brakes which less than 100 years ago were inhabited by bears. In a raid on one of those brakes less than 10 years ago I gathered in less than 10 minutes, enough cane for a supply of reed pens to last for a life-time spent in scribal writing at the rate of 10 hours a day. All this at no cost & with no interference by bears. They had been killed off for their "bar grease & erl" which was put in hollowed-out logs, or troughs, roped together as rafts & floated down the White River to the Mississippi & on to New Orleans where it was used as a substitute for butter. That's how Oil Trough, our neighboring village, got its name. In those early days the cane brakes were impenetrable except by following the trails—long stygian tunnels made by the bears. The nervy (and nervous!) hunters had to crouch through those low-ceilinged labyrinths, which also served as the winter quarters for flocks of migratory birds. A protective blanket of snow laid atop this thickly matted jungle never touches the ground.

This sketch was made entirely with a 1/16" reed nib held at various angles to show the range of strokes from thick to thin made possible by manipulation of the broad pen

For very fine lines, only the corner of the nib was used.

GENUINE BAMBOO can be had cheap at most florist shops. It's tougher than the White River reed and will hold its edge longer, but whether it comes from China or from Arkansas, each has its own peculiar quality of line. Artists use both kinds, not necessarily as pens for writing but as sticks dipped in waterproof ink for making sketches on which washes can be laid without "feathering" the lines. Of late years it has been gratifying to find that art critics have noticed the calligraphic quality of line in such drawings, praising it. It is said that the famous Mexican artist, Covarrubias, studied calligraphy to acquire that kind of line. It is also apparent in the pictorial delineations of such artists & graphic designers as Jeanyee Wong, George Salter, Warren Chappell, Franklin McMahon, Oscar Ogg, Fritz Kredel, Everett McNear, Bill Jackson & Philip Grushkin — Americans all.

Portrait of Alfred A. Knopf by Miguel Covarrubias for Typophile Chap Book Number 42

PORTRAIT OF A PUBLISHER

AMONG the many scholarly English scribes, Heather Child is typical of those who have mixed calligraphy & cartography with distinction, extending the graceful scribal line to decorative illustration & heraldic design. Thomas Derrick—another Britisher—is outstanding for his free-flowing renderings of both text & illustration, harmoniously related and modern in style. An inspiring example of his ability as a book designer is apparent in "The Prodigal Son and other Parables" done entirely calligrapically with a brush. His drawing of the repentant sinner and some of the lettering for the title page are shown here with the pubisher's permission—

THE third drawing signed "Saxon" (Anglo, maybe, but probably American) shows a fine flair for the calligraphic brush line. Anyway, it appeared in an American Air Lines ad in Life, Nov. 1968. These 3 drawings have much in common with that of Kanjiro Kuroda on the following page. For centuries oriental artists have written their pictures & pictured their writings with brushes—and the most remarkable facility.

ALMOST any kind of writing instrument in the hand of a calligrapher will make a line that is legible & scribal in style. The ubiquitous ball point pen, however, is definitely not a calligraphic tool—it rolls and slides too easily in all directions, making a monotonous one-weight line. Contrast between thick & thin strokes according to the direction in which the pen is moved without pressure is a prerequisite to calligraphic quality. The one-weight line lacks the æsthetic grace inherent in the broad pen. I find the ball point most useful in making copies that require heavy pressure to register a line through several sheets of paper.

"Flip", the kind of pilot American Air Lines does NOT employ. They say frankly:—
We wouldn't touch him with a 10 foot pole!

THE first ball point was so revolutionary that its inventor made a trip around the world advertising it with this slogan: "Guaranteed, not for a year, not for 10 years, not for a life-time, but for ever!" After 23 years I still have mine—a handsome gold-plated gift from my son just out of the Navy—flushed with victory & flush with money—I know that it was very expensive.

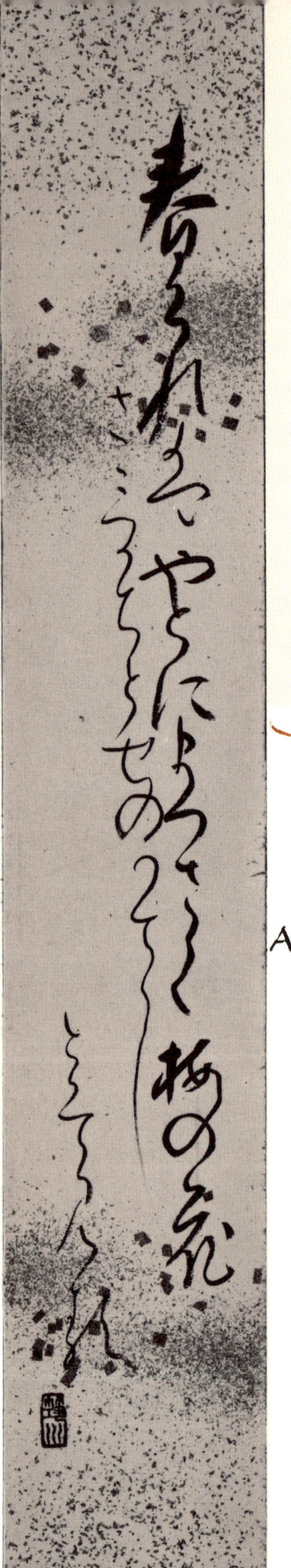

This famous Poem by NARIHIRA ARIHARA is among the best poems chosen to be included in "KOKINSHU" published about 1,000 years ago.

Yo No Na Ka Ni — — — — — — 5 syllables
Ta E Te Sa Ku Ra No — — — — — — — — 7
Na Ka Ri Se Ba — — — — — — 5
Ha Ru No Ko Ko Ra Wa — — — — — — — — 7
No Do Ke Ka Ra Ma Shi — — — — — — — — 7

Yononaka — World
Sakura — Cherry Blossoms
Haru — Spring time
Kokora — Heart, Feelings
Nodokekaramashi — It might be joyous

This is the style of Japanese Poems (Apparently the rules for poetic composition are as strict as for a sonnet)

If the cherry blossoms did not exist in this world, the feelings in the spring might be gay. (Which means, my heart becomes restless and sad, when the cherry blossoms come to bloom in the spring time)

Calligraphy by CHIKUSEN KURODA — Mr. Kanjiro Kuroda a contemporary artist who is also a lecturer of the Ochanomizu University of Tokyo. From an Exhibition of his work held at The University of Arkansas in November 1967

AFTER a short period of use this ball point under pressure began to retreat upward into the barrel and, even when it wrote under light pressure, the line was often broken & indistinct. That was before the day of ball points guaranteed to write under water and through chicken fat. So—on the strength of its fabulous guarantee, my first ball point went back to the factory for repairs, but it soon developed the same unsatisfactory habits. I still cherish it for sentimental reasons & because I can't bear to throw gold away; that precious metal is, at least, not tarnished by time. Meanwhile I remained a "sucker" for every new writing gadget of plastic, felt, nylon, iridium, etc., until finally I saw a Parker Pen ad in Life Magazine for December 22, 1967. It said in part: "Are you still pushing an old-fashioned writing stick with a funny little ball that hops, skips, & jumps?... Throw it away... write with your heart, not with your arm... Ladies and gentlemen, the ball is over!" Coming from Parker, who, by the most conservative estimate, must have made millions of (and on) ball point pens, I figured they must have been through a great conversion. They had seen the light!—I rejoiced greatly, for the more than life-size illustration of the pen showed its nib to be broad—square tipped—not round. I have one now, and it's great!

Illustrating the use of pressure on the flexible pointed pen versus the broad nib used without pressure

4 ADDRESSED ENVELOPES

Louis A. Di Gesare's control of hand under varying pressures is unbelievable and surely unattainable by all but a few who would try to emulate his most skillful penmanship. It is conceivable that a myopic postal clerk in need of corrective glasses & a course in art appreciation might relegate such a work of art to the dead letter office. Item 2, combining the graces of Roundhand, Spencerian & Engraver's Script, is also beautiful; it would cause no delay at the P.O. unless the clerk paused to admire it. Items 3 & 4 by exemplars of the broad pen are legible at a glance—and practical. "Beauty is in the eye of the beholder," says Emerson. That puts all these items at your mercy, dear beholder. If writing is first legible & practical, beauty will follow—

1 Louis A. Di Gesare—Past President, International Association of Master Penmen, Calligrapher; Gen'l. Electric Co.

2 Kenneth O. Johnson, Designer, Calligrapher; Production Manager Kiwanis International.

3 Paul Hayden Duensing—Private Press Printer; Letter-Founder & Calligrapher.

4 Tom Gourdie of Kirkcaldy, Scotland, teaches Italic handwriting & is the author of manuals on that subject.

as the student gains facility and appreciation through broad-pen practice and the study of good Italic models based on chancery cursive hands of the 16th century.

I FIND it convenient to work between 2 desks & a tracing box with a glass lid over fluorescent light. This box is primarily for tracing layouts and checking type proofs, but it often proves to be a calligraphic convenience. One of my desks has a 42" x 30" board mounted on the base of a dentist's chair. The board can be raised or lowered by pumping the foot pedal and is adjustable to any angle up to almost vertical for use as an easel. The other desk is my favorite for writing because the angle of the board can be changed quickly. When writing at the top of a large sheet of paper, the slant of the board needs to be much steeper than when I am writing at the bottom, so I raise the board almost subconsciously as the writing goes downward; thus I am relaxed & comfortable—never cramped—an important point in writing. This desk is so easily made that any one handy with a hammer, saw & screwdriver can put it together.

Tracing box hinged to wall, slants at convenient angle

A Drawing board hinged to wall.
B Prop hinged to drawing board.
C1 Spring hinge (screen door type)*
D Vertical prop holds drawing board in horizontal position when one stands up to work.
E Spur at end of prop B jabs into soft wood to hold board solid.
To raise board, lift lower edge & spring will pull prop up and hold it at the desired angle.
To lower board, pull prop down & adjust at the desired angle.

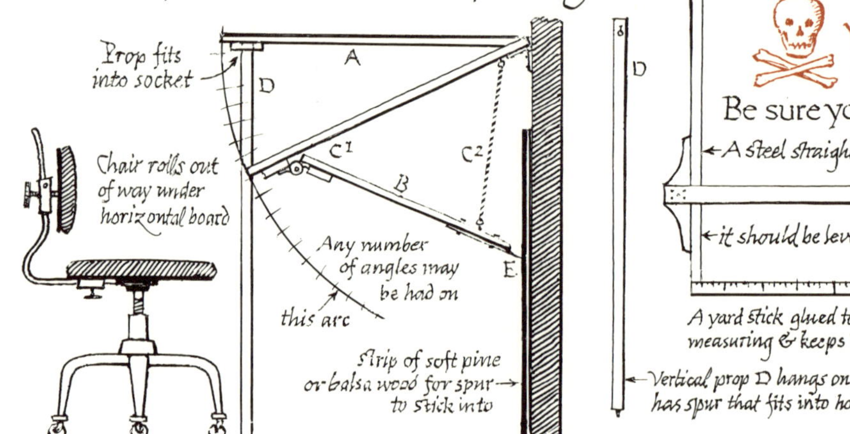

WARNING!
Be sure your board is square
← A steel straight edge is good insurance . . .
← it should be level with the surface of the board.
A yard stick glued to edge of board is useful for measuring & keeps T-square from falling off.
← Vertical prop D hangs on nail when not in use and has spur that fits into hole in floor to hold it in place

* C2 Coiled spring (screen door type) may be used instead of C1 or as an auxiliary to it.

How a study group may be formed.

TWO OR THREE gathered in the cause of legibility can become the nucleus of a group for the improvement of handwriting, and if a competent teacher of Italic is available, so much the better. Even small regional libraries are apt to have historical information and inspirational examples. A list of manuals and the names & addresses of teachers in various parts of the U.S. are shown on page 169. It may also be helpful to know how our Newberry group carried on after the impetus given us there by the British Exhibition in 1939.

OUR great advantage was that the Wing Foundation of The Newberry Library has one of the finest collections of writing books extant, from ancient to modern, & that its custodian, Ernst F. Detterer, made it & other treasures of The Rare Book Room available to us in classes which he conducted gratis one night each week. Himself a pupil of Johnston in 1913, he led the Johnstonian revival in America, first as head of the Art Department of Chicago Normal College, and later as head of the Department of Printing Arts School of the Art Institute of Chicago (1921–31); then on to Newberry. There he led our discussions & studies, not only of the great calligraphers from Arrighi and his illustrious predecessors & contemporary followers, on to Johnston, Koch and Von Larish to Salter and Chappell, but also of the great printers from Gutenberg and Aldus to those of our time: Rogers, Updike, Morison, Goudy, Rudge. This led us to speculate on the ultimate fate of the alphabet. We are sure it will remain essentially as is from A to Z for a long time, even with marvelous technological advances in printing & in spite of such passing fancies as the elimination of capital letters. We studied the Trajan capitals in minute detail from rubbings that Detterer had made above the door of that famous Roman monument, but mainly we concentrated on the less formal chancery hands of such masters as Arrighi, Palatino, Tagliente, Yciar, Ruano, Lucas: to improve our own handwriting,

TRAJAN COLUMN

With Softness strong, with Ornaments not vain;
Loose with Proportion, and with Neatness plain;
Not swell'd, yet full, compleat in ev'ry Part;
And Artful most, when not affecting Art.

JAMES HAYES, *Scripsit* · Quoted from George Bickham's Universal Penman, London, 1743

The fine Italic hand of James F. Hayes is from "Festschrift" honoring Professor Lloyd Reynolds of Reed College. Our Newberry group never numbered over 15. Jim Hayes, Bob Middleton, Rodney Chirpe, Joseph Carter & Ray Da Boll formed its nucleus. When Ernst Detterer died in 1947, Jim Hayes took his place as the monitor of our calligraphy class.

and with sympathetic interest in a project dear to the heart of Detterer, i.e., to publish in facsimile an English translation of Arrighi's La Operina (1522), conceded to be the first & best manual of the Italic hand. Detterer had already translated it and we were anxious for him to go on with the facsimile to be published by Newberry. Then he found that John Howard Benson, America's finest inscriptionist in stone, and an eminent teacher of calligraphy at the Rhode Island School of Design, had "beaten him to the punch" and was well on his way with the same project. Detterer graciously bowed out.

AND THAT is how it came about that Newberry joined Yale University Press to produce "The First Writing Book", duplicated in English by John Howard Benson for Harvard College Library & The Newberry Library in 1954. We have Mrs. Benson's word that her late husband—a perfectionist—rewrote the text at least seven times. We earnestly commend our readers to its study & enjoyment.

WHY?

SINCE moving here in 1952, I have often been asked WHY. Briefly, my boiled-down answer has become because I love country life and after some 40 years of strenuous work as a graphic designer in the Chicago area, I wanted at least semi-retirement. Born & raised on a farm and still at heart a farmer boy at 76, I've had to give up the farming of our "40", but Irene & I enjoy living in the middle of it with our neighbor's cattle, horses and colts that graze our green pastures and drink the still waters of the stock pond we put in shortly after our arrival. Our favorite poet Robert Frost, had reminded us that "good fences make good neighbors" so we fenced our 40, and our good neighbors pay us the added interest of warm friendship. Tho' we kaint see 'em fer th' trees that surround us, its comfortin' to know they're near. Of course the persuasions of our son Briggs & his in-laws who had preceded us on the Circle W Ranch (W=3,500 acres only 20 miles away) were also valid reasons for our move. But we won't go into details about how the beef business "went to pot" in the disastrous 3 years of drought that hit the southern States in the early 1950's causing our son's return to megalopolitan life with its greater assurance of immediate income + polluted air & water. We old folks "stayed put" preferring the simple life with our neighbors & the good friends we've made— many of them members of our local chapter of the National Federation of Music Clubs. Music has always been important to us, in fact, it was what started us all unsuspecting on the long long trail of what is now PART I of this book & its subsequent ramifications which have grown (and groan!) out of it, necessitating painful apologies for delayed publication—

Our "40" is near Newark, Ark. I didn't come here just to practice Calligraphy. Animal Husbandry (beef business) & a pecan Orchard were also on my agenda. —But I have found Farming at 60 to 76 is not like it was at age 6 to 16. I'm still all for Calligraphy & Graphic Design

IT DIDN'T matter much that Arkansas had been represented as an unpromising field for the cultivation of a calligraphy-conscious clientele. Several of my associates advised against this "outlandish move" and predictions were rife at farewell parties that we'd be back in a hurry. On being notified of my new address, the secretary of the American Institute of Graphic Arts took pains to tell me I was the only AIGA member in the state of Arkansas. A recent check indicates that I still hold the same status. These negative reactions only made me feel that I was a pioneer on the new frontier.

by mail with my symbol, a rural box marked with my initials—RFD

MY STUDIO was well equip't but while some work came in as expected from former sources, I needed to hunt up some local trade to fill in the dry spots that were a-gittin' progressively dryer along with the drought that was a-hittin' us farmers. I had heard of the Arkansas Industrial Development Commission and figured it might have some use for my services, so according to the ethical procedures instilled by advertising experience, I contacted the agency that handled the AIDC account in Little Rock. My Chicago friends were right– the agency was not "calligraphy-conscious". I got exactly nowhere. Back home I pondered this seeming impasse and, finally, ignoring ethics, sent my samples directly to Winthrop Rockefeller, chairman of AIDC. His appreciative reply stated regretfully that "currently we are adequately staffed to handle the program as we see it". However, he must have shown my work to William R. Ewald, Jr., his Chief of Development, who wrote saying he wanted to meet me.

& How!

I SHALL never forget his cordial greeting prefaced by the usual question, but with more force–"Why in hell did you come to Arkansas?" From then on, I worked directly with Mr. Ewald.

HE came from Detroit, is of the Campbell Ewald family–pioneers in automotive promotion–knew the graphic arts from A to Z, had studied architecture under Saarinen & written a book on city planning. We worked out the AIDC 6 Point Award of Preparedness given to those communities that have met the necessary requirements before an industrial plant can be installed. I charted river systems for AIDC's Economic Atlas showing the state's water resources, made type layouts & did various odd jobs including ads for auction sales of Santa Gertrudis cattle at Winrock Farm on Petit Jean Mountain.

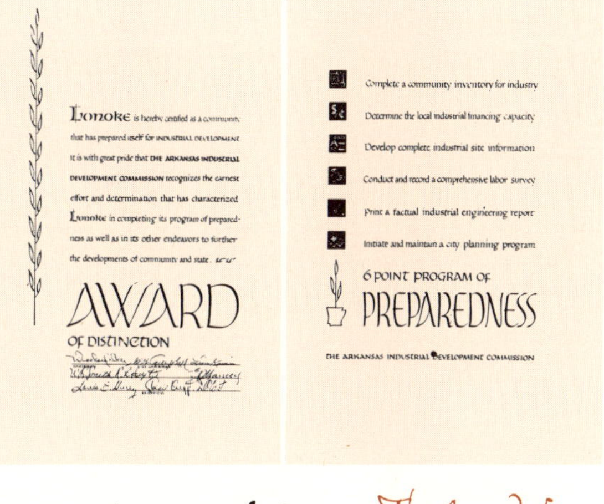

Lonoke is hereby certified as a community that has prepared itself for INDUSTRIAL DEVELOPMENT. It is with great pride that THE ARKANSAS INDUSTRIAL DEVELOPMENT COMMISSION recognizes the earnest effort and determination that has characterized Lonoke in completing its program of preparedness as well as in its other endeavors to further the developments of community and state.

AWARD
OF DISTINCTION

Complete a community inventory for industry
Determine the local industrial financing capacity
Develop complete industrial site information
Conduct and record a comprehensive labor survey
Print a factual industrial engineering report
Initiate and maintain a city planning program

6 POINT PROGRAM OF
PREPAREDNESS

THE ARKANSAS INDUSTRIAL DEVELOPMENT COMMISSION

The Award of Preparedness is presented in 2 parts, each 12 x 19, in silver, deep red and dark gray.

ONE of the most interesting assignments came from Mrs. Rockefeller, who wished to surprise her husband with a birthday gift of TEN ARTICLES OF FAITH IN THE AMERICAN WAY OF LIFE written by his father, John D. Rockefeller, Jr. The text, set in a stiff sans-serif, was sent to me by WR's secretary, Mrs. Jane Bartlett, sans any info as to size, proportion, colors, etc., so I took that as an "all clear" signal to do as I pleased, and submitted an elaborate layout in 4 colors with a fancy initial & the Statue of Liberty to be finished in raised gold. Mrs. Bartlett wrote saying this was not suitable and, since WR's birthday was imminent there was not enough time to do the job over, but they would get in touch with me later. They did.

THIS time it was to be a Christmas gift, and I was given all the time & information necessary to do the job properly. It seems that as a boy of 12, WR had concentrated the sun's rays with a magnifying glass to burn 2 stanzas of Edgar A. Guest's poem–"It couldn't be done" into a large pine board. Guest fans may remember that the first 7 lines of each of the 8 line stanzas pose a problem apparently unsolvable, but the 8th line always ends "...and he did it". This unique rendering of the poem as well as the moral of it

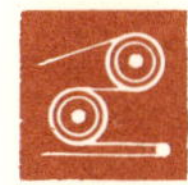

Symbols of the 6 requirements.

thus demonstrated that hard work & persistence have their reward in achievement. This evidence of earnest effort by the boy had understandably become a cherished family treasure and was to be placed as a fitting companion-piece beside the CREED by which his distinguished father had lived.

THE PROBLEM was to make my calligraphy comport with pyrography rather crudely done. It was finally accomplished to MRS. Rockefeller's satisfaction with the aid of a photograph of the "pyro-plaque" and several notes to & fro settling the important details. The result looked like "Ozark folk-art". It was fun to do! I slipped the following jingle into one of my notes to Jane Bartlett:

He sure picked a difficult medium!
Most 12 year-olds would have eschewed it,
But he, undeterred by the Te Deum,
BURNED with ambition and dooed it!

With apologies to Ed Guest & thanks to Red Skelton the inimitable buffoon.

At that time a favorite catch phrase of Red Skelton's was "I dooed it!"

THIS BIT O'BANTER eventually came to MR. Rockefeller's attention and I was relieved to find that it pleased him. He wrote to thank me, saying in part: "I am happy to have it to add to my collection of mementoes". That, however, is not my reason for all this wordy detail.

BRIEFLY, my point is that after "Win" Rockefeller lost in his 1964 campaign for the Governorship of Arkansas, he ran again in 1966 and was elected the first Republican Governor of the State in over 100 years! That was indeed a great achievement. He "dooed it". And we are pulling for him to do it again in 1968.*

*And he dooed it again!

SURELY there is a definite relationship between Calligraphy & Music. The flowing thick & thin Italic line and the undulating melodic line have rhythms in common that run the gamut from spider web delicacy to broad tonal effects — pianissimo to forzando! Adept performers in both professions, sensing these relationships, often find them leading to friendships along with their increased understanding of each other's work. A professional musician may be equally proficient as a calligrapher & vice versa, ie: FR. Palladino (page 108) who is currently broadening his own musicianship by teaching calligraphy to a clarinetist of the Portland (Ore.) Symphony in a friendly exchange of lessons on the clarinet. Fred Eager (page 119) a graduate of The Eastman School of Music, is a teacher of both Italic handwriting & music. His widening circle of influence is in large measure due to correspondence which is

And now — a few music notes

identified by the dual implications of his "Italimuse" symbol of a broad pen nib & the sign of the treble clef. Joe Routon of the Julliard School of Music, plays 2nd violin in the Nashville (Tenn.) Symphony. His colleague, Dr. Lee Zawol, is a violinist in the Kalamazoo Symphony. They are devotees of Italic & correspond regularly to keep in practice—as they say "with 'notes' musical & legible by literate means to ensure mutual comprehension". Refer to the Disciplined Freedom broadside laid into the back cover—Note the method by which the ideal proportions of the upper & lower case letters is determined, and you may be struck, as I was, by the similarity of that graph to the method of notation on the 5 lines of the music staff. Not that this proves

Italimuse · Inc.

Mr. Routon runs a large Florist business.

Dr. Zawol is a Surgeon.

Catharine Fournier, business executive and noted weekend calligrapher, says that she "became a willing victim in the never-ending struggle between the wayward pen and the ideal in the mind's eye."... Disciplined freedom is the Essence of Calligraphy. The equation Precision + Verve = Style applies to many of the arts. In music the composer has one advantage over the performer. If he has an off day, he can discard his failure and start afresh, as can the graphic artist. The performing artist must play his concert when it is scheduled, no matter how travel-weary or indisposed.

This is a bit of editorial writing in the well-disciplined hand of Dorothy Rolph. Mrs. Rolph, an architect by training, studied calligraphy with Paul Standard when she was in her middle fifties. She is an inveterate concert goer, and arrived at the equation she proposes while listening to a brilliant performance by Rudolf Serkin.

anything. However, it does remind us whether we are literate or not, that music is a universal language & if there is any truth in the old saying "Music hath charms to soothe the savage..." we wonder if an international organization, say—Pen Pals for Peace, Inc., might not prove more effective than the occasional notes* that pass between the governing heads of nations. It would be interesting to know what a consensus of the corresponding common people—citizens of the world—would reveal regardless of their personal &/or national isms.

*and talks

I met Hubert Leckie when he was playing 1st violin in a string quintet and was also studying architecture ("frozen music"!). He helped me with an architectural assignment I had fallen heir to. While working on it in my studio, he was severely bitten by the graphic arts bug, never recovered, stayed with me for 7 years. Quick to learn, he was soon teaching a night class in lettering at Moholy Nagy's School of Design. Gaining facility in calligraphy & typography, he set up his own studio in Washington, D.C., became a top designer there and president of the Washington chapter of the AIGA. We helped each other up—and on the way Hubert played, & still plays, his violin on a high professional level.

EDWARD JOHNSTON has called calligraphy "The Dance of the Pen". His statement is borne out by Miss Jenny Hunter, a professional dancer & dance teacher who is also an accomplished performer with the broad-nibbed pen, as shown by the following excerpts from her essay on the art of dancing from the Festschrift book honoring Professor Lloyd Reynolds, her calligraphy teacher. Note the freedom & vigor of her style, written speedily but with the evident discipline which is essential to both dancers & calligraphers — if they are to "go beyond the past into the future ———— ":

The new DANCE-ARTISTS should have knowledge of the past for 3 reasons: (1) So that they may not be foolish (and think they are first; (2) So that they may develop some revealed but unexplored paths of experimentation; and (3) So that they may go beyond the past into the future.

To create in art means not merely to invent, to fill space, but to explore & probe for new ways of seeing, new perception & new expression. It means to risk everything, and to push the boundaries of thought & feeling.

We want to encourage that new generation of Artists who will continue to bring vitality to Dance and Theater expression. We want to see them, with a sense of dedication & craftsmanship become boldly free in their own work.

"Jenny Hunter — professional dancer, choreographer and teacher of Modern Dance; taught 3 summers at Reed College; directs the Jenny Hunter Dance Company and School, Dance West, in San Francisco." (No decline of the West in Terpsichorean Art.)

ALTHOUGH 'Don' Margo & I have never met personally, we have corresponded since 1943. At that time he was with the U.S.A.A.F. in a flying outfit known as "Hell's Angels". Two of my calligraphic pieces, a music chart "From Bach to Gershwin" and "My Rifle—The Creed of a United States Marine" initiated our correspondence. Our common interest in calligraphy & music (war is out!) has kept us in touch for a quarter of a century, and we are agreed that it has been a rewarding experience by which we have come to know each other like the pages of a favorite well-worn old book.

SERENDIPITY — a word coined by Horace Walpole, is derived from the happy faculty of 3 princes of Serendip (Ceylon) who happened to find things of value while searching for something else. Letterwriting is like that; it can turn out to be an adventure in serendipity which may lead to lasting friendships and, incidentally, to rewarding work.

MY ACQUAINTANCE with Howard Paine, Chief of Editorial Layout for National Geographic Magazine, began with a letter in his scribal hand about his interest in writing. He went on to say that, knowing I was a collector of Bountyana, he was sending me a copy of the current issue of the Geographic that featured an article: "I Found the Bones of the Bounty" by Luis Marden who combined photography with writing and was also a calligraphile familiar with my work. He illustrated his story with color photographs, many of them taken under water. So—again, it was mutual interest that eventually brought me 5 solid months of cartographic and calligraphic work on the end-sheet designs for the National Geographic Atlas of the World in 1963, and, subsequently, other work for N.G.S. publications. While doing the end-sheets, I found that my work had first come to Mr. Paine's attention by way of the Balzac classic: "Jesus Christ en Flandre"—the story of a shipwreck & miraculous rescue which I had designed as a Christmas book in 1941. It was indeed gratifying to get such a resounding although long-delayed echo from my previous effort.

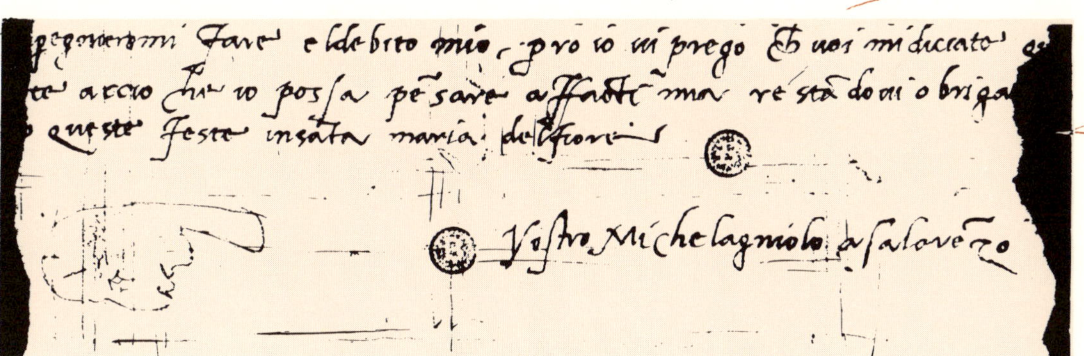
pegonerommi fare el debito mio, pro io vi prego che voi mi diciate
acciò che io possa pensare a' fatti ma' restando obrigato
queste feste in sata maria del fiore
Vostro Michelagniolo

PIETA is the only one of Michelangelo's masterpieces to bear his signature! This part of one of his letters shows that his fluent handwriting is based upon the Chancery Cursive of his day, and yet it expresses perfectly his own vigorous & dynamic personality.

Michelangelo, too, liked to "doodle" in the margins! Shakespeare was right: ". . . What's past is prologue."

THE TEMPEST
Act ii, Scene I

I LIKE TO WORK with people who have wide-ranging ideas. James T. Mangan is such a person; highly successful in advertising and in writing inspirational books on salesmanship, he gave up his agency in 1955 to devote his time entirely to the promotion of Celestia, his Nation of Celestial Space, which he founded in 1948. His claim is legally recorded and the Grand Seal of Celestia was registered by the U.S. Patent Office in 1957. This gave him considerable publicity & a multi-million TV audience, which many of our readers may recall if they witnessed his appearance on a Jack Paar program. Mangan modestly claims only Outer Space—none of the planets or heavenly bodies therein. Of course he was unmercifully kidded: Hermione Gingold kept interrupting with "where are you going to plant your flag?" But Jim is quite an actor—never cracked a smile as he unfurled his Celestial banner—just as he had earlier that day, along with the other flags of the United Nations atop the U.N. Building.

JIM is also an ardent calligraphile—his Charter is written out by Dr. R. Hunter Middleton; his "First Report to The Universe", by David Green in 1956; I did the "Second Report" in 1958—28 pages entirely calligraphed. The following spread shows Jim's famous "Write a Letter".

Write a Letter

It's only a few steps to the nearest mail box —
Write a letter! Take a little chunk of your heart & spread it over some paper
It goes, oh, such a long long way — !

WRITE A LETTER to your mother or father, to your sister, brother, sweetheart, loved ones. Are they dear to you? Prove it with a letter. Are they far from you? Bring them near to you with a letter! Write a letter & give them the same thrill you had when you last received the same kind of letter.

Think of the joy of opening the mailbox and drawing out a warm envelope enriched with the old familiar handwriting! It's a thrill that lasts!

A personal letter – it's good to get one. So send one – write a letter!

Write a letter to the aged relative who hasn't many days to live, the friend of your father, the friend of your family, the one surviving link between your own present and past. Don't wait for that dear soul to die till you act. Act now with a message of love to cheer those last few days on earth. Sit down and start writing!

Write a letter to the author whose story gave you that delightful half hour last night. Write a letter to the cartoonist whose serial strip you avidly devoured this morning; to the teacher who inspired you twenty years ago; to the doctor who saved your baby's life; to your old employer to show him there was something more between you than a pay check. Be a human being — write a letter.

There's a man in public life you admire, believe in, rave about. Write him a letter of praise, encouragement. To be "with him in spirit" is not enough – show your spirit with a letter.

We can't all be pioneers, crusaders,

presidents — But we can help those brave men stay on the track & push through to success if all we ever say is "Attaboy!" Write an "Attaboy" letter!

Write a letter and — give. Give praise, encouragement, interest, consideration, gratitude. You don't HAVE to give these things, but the real letter IS THE ONE YOU DON'T HAVE TO WRITE!

The sweetest, gentlest & most useful of all the arts — letter writing. Great grand characters like Washington, Franklin, Jefferson, Lincoln & the greatest men of all nations, have been regular letter writers. Write a letter!

Write it with a pen, pencil, or typewriter, use any kind of paper, any kind of spelling or grammar. It doesn't matter how you say it, & it doesn't even matter what you say, its beauty, its GOLD lie in the pure fact that it's a LETTER! Each blot is another hand-clasp; every mistake a tear of joy.

Do you see a job? Do you smell an order? Is your mind on business? Write a letter. Then write another letter. No business, no individual, built on the "write-a-letter" rule ever failed. Because you simply can't fail if you write a letter.

Try it, you'll like it. Great joy & many surprises are in store for you. You'll get letters back. You'll get help from unexpected sources. All that you gave in your letters will be returned to you a thousand fold. For a letter is a six*-cent investment in bountiful good fortune.

Write a letter! Whether you say: — "Attaboy!", "Thanks!", "I love you!", always remember: — A LETTER NEEDS NO EXCUSE!

* CURRENTLY, of course (AS OF 1968)

Written out by Ray Da Boll in the Style of — Antonio Sinibaldi, Florence, 1481

The foregoing text by James T. Mangan is generally conceded to be one of the four best advertisements ever written. It first appeared in Advertising Age in 1933 in a copywriters' forum. It received so much acclaim it was afterwards reprinted and millions of copies distributed in Post Offices in the USA and all over the world.

THE MOOD of the writer—his personal feelings with regard to what he is writing, can affect his calligraphic style. The kind of nib—its width & cut—are mechanical factors that also affect pattern & texture. Variations of speed are mood factors that regulate spontaneity & cursive quality. One may even find that a different alphabet is better suited to his text as well as to his mood when writing: take the **uncial** style as an example:—

Now fades the glimmering landscape on the sight,
And all the air a solemn stillness holds,
Save where the beetle wheels his droning flight,
And drowsy tinklings lull the distant folds.

I THINK the above quatrain from Gray's Elegy Written in a Country Churchyard is one of the most beautiful in the English language. Even as an eighth grader I was deeply impressed by the whole poem, and now I wouldn't think of writing it out in any but the most beautiful and dignified hand I am capable of. Yet I have fallen short of the written beauty I would have liked to have given those lines in putting them down on paper.

I BELIEVE that any one who has lived past his allotted 3 score and 10 years—and has loved life and beauty—should be forgiven for musing over Gray's Elegy. That is what I was doing lately with full appreciation and due resignation while listening subconsciously to the sounds of a popular TV program that filtered up to my 2nd floor workshop. And I was constrained—if not inspired—to jot down the following lines in a hand that I felt was appropriate to the occasion:—

Disrupted now is the silence of night,
As the droning Beatles incessant beat
Moves adolescent Girls in heat
To scream with uncontrolled delight!

OF COURSE that's going from the sublime to the ridiculous, but how can one compare 2 opposite poles without mentioning both? A good example of mood that changes from sobriety to inebriety to oblivion is recorded in 'Don' Margo's letter to an "Old Pal". → Such writing is not done with the aid of "hard likker" but with "the art that conceals art".

What else could be expected of a 76 year old "Square", semi-retired on his Arkansas "40", reliving the days of his youth as a farmer boy?

NB: I like the Beatles—it's just those silly gals that get me down!

Egdon H. Margo 4961 MURIETTA AVENUE • SHERMAN OAKS, CALIFORNIA

12 XIV66

Dear Pal

Well; about this time of year, I usually take a few moments out to write a few letters to my good friends. The time, when I remember all of the good things, and I indulge myself to the extent of becoming a little sentimental. It is a blustery evening, and here in my den its cozy and comfortable. I'm sitting before a nice open fire with my pen poised / sort of half listening to the radio and slowly sipping a nice, very dry double Martini. I only wish that you were here... but since you are not, the least that I can do is to toast your health & happiness... so time out, old pal — while I bend my elbow to you!

I just took time out to mix another Martini, and while I was out in the kitchen — I thought of all the time I would waste this evening if I went out to mix another, every once in a while / so I just made up a big pitcher of Martinies and brought it back mw ith me so I'd have it right heere besideeme and wouldnt hav to waste any time making more more of them. So, now I'm all set an here goes!

The greatessts thing in the whole wide world is friendipfriendship. And believe me old pal / you are the greatets anybody evr hadd. Do you rembre all the swell toimes we had togegter ol pal? The wonderdful canpping trisp? I'LL nevre forget the tome you put a dead shunnk in my slepping bag! ha! haha.. Boy! how we laughued didnwee! wee! Nevr did get the sn kout of it! bit it was prity funya anyway. I stil laught aboutit onec in a whole. Not as muhc as I usedto.. But what the heed and aftre all you stil my bestt olpal-pal —— anit a guy cant have a lalnaghg on a treu freidn onee ina whiel / waht the heek!

Dam pitcher is impty / so I jus wantout and ma de annotherone and I sure wisch you weer here olpal to hepl mi drink those Martoni becausthey sure are somply samply delieuccins! Pardn me wheil I lif my flas to you good helahth oneimroe because you are the betst pal I got I got. Of cours why a Pal wolud do a drity thinbg liek ik putting a skrink in a nother Pals sleping bag I'm dambit if I kno!! That was a lousithing for anybodyhdy to do an do an only a frist class hele would dit !!!! Wasna admidaw but funney. Stil stinkks! An you thininkit s funney. Your a drity lous an asifjare as I-m concrened you cn go plum to hell ll and stym ther n styther you drity lous

To hel with ouy Margo

Egdon Margo writes, designs, sets type and prints with only the thumb, fore finger & 1/2 of the next one as the result of a WW II injury to his right hand. While this apparently has not hindered his work as a calligrapher, it stopped his Violin playing but only served to broaden his interest in music.

I have used "opposite poles" as a device to point out the difference between what I've found comforting in poetry & disturbing in music. Mr. Margo's Christmas Eve letter shows that under certain conditions one's personal attitude can range from warm friendship to red hot antipathy. We trust that our illustrations will be taken with a pinch of salt which may help to add savor to our calligraphic efforts.

Now, in the same spirit, let's apply this antipodal device to the world of advertising, citing 2 actual instances with regard to certain attitudes for & against the use of calligraphy for promotional purposes. There must be an "equator" somewhere between the 2 poles.

Let's say one of the poles is the XYZ agency of which Mr. X was the senior partner some 18 years ago when the following situation developed. I had done some commercial art work as a free lance designer for one of XYZ's accounts. We got along very well together and were conveniently located in the same building. But when Dr. P, Librarian of a well-known Chicago Library, recommended me to his friend, Mr. W, president of the G.H. Co., to design an advertising brochure entirely calligraphically, it seems that Mr. X had objected with asperity to Dr. P's meddling with his GH account. Having made the layout, I was summoned to sit in at the conference. I went joyfully unaware of any previous objections to outside help—and to calligraphy in particular. I was happy to find that Mr. W preferred my layout to the one prepared by the XYZ agency; he liked the calligraphy because it was different— and he was looking for something different. But—

Mr. X—verbatim:— "So's a skunk different! Calligraphy stinks! It's for kid's books—it's for the birds—and the monks!"

Being totally unprepared for such an attack, I found myself in the position of the mother skunk who said to her children "Let us spray!" I defended my position as best I could—and left feeling that I was the most despised & rejected of scribes. Next day my spirits rose from the depths on receipt of a letter from Mr. W saying that he "had enjoyed yesterday's polemics*" and that he was going ahead on the basis of my layout. He also enclosed the cc of a letter to the manager of XYZ's Detroit office which serviced the GH account, saying he could see no reason why the agency should not collaborate according to my ideas. There was no collaboration, and never another production order from XYZ from that day to this. But that was OK with me.

* Perhaps intended as a pun about polecats?

My Wife & I received an invitation to spend 2 weeks as guests of Mr. W at GH to absorb the calm but relatively "gay nineties" atmosphere and the unpolluted air & water of the Strait of Mackinac where the GH {Grand Hotel—in case you haven't guest} is located on a lovely wooded island.

PEACE & QUIET reigned. Horse-drawn vehicles, bicycles & rolling chairs provided the only means of transportation unless one walked. Our visit proved both restful and profitable, resulting in my redesigning of 16 pieces of printing, from napkins and cocktail coasters to stationery forms, menu covers & placards. And aside from the 12 page brochure in question, there was also the designing of a mural for the GH snack bar, suggested by DR. P, featuring the first sentence of Raphael Sabatini's novel — Scaramouche — "He was born with the gift of laughter and a sense that the world was mad." A fine painting of Scaramouche by Edgar Miller was part of the composition which required the largest rendering of chancery script I've ever made — lower case 3½ inches high.

The only motorized vehicles were a Fire Engine & an Ambulance. It was delightful!

LATER another letter from MR. W stated that for the first time in GH history, tourists were buying the brochure at 25¢ each, and that R. R. Donnelley & Sons Company had printed an over-run of 5,000 to distribute as an unusual example of their printing.

THE ATTITUDE of Mark Lansburgh, senior partner of Lansburgh & Oldham, the largest advertising agency in Santa Barbara, was exactly 180° opposite that of MR. X. Coming up from calligraphy and hand-press printing at Dartmouth, he had dug deeper & deeper into the motivations and inspiration for Renaissance calligraphy and still earlier manuscript lettering. This got him involved in early manuscript hands, the synthesis of Carolingian scripts and the various writing that evolved during the Romanesque and Gothic times. Of course, tasteful balanced lettering was always a matter of concern to him in his Graphic Arts & Advertising Agency.

X

ML

Opposite Poles

MY INTRODUCTION to MR. Lansburgh came in 1962 by way of a letter asking if I would sell two pieces of my calligraphy which had been shown in an exhibition at the Portland Museum in Oregon. I was happy to give them to him. Shown below are 3 lines of his vigorous & highly individual handwriting: —

the two original pieces you sent are delightful - as are the little folders. I enjoyed it all - especially the Kittredge letter. It is perfect!

FOR THIS I was considerably over-compensated by his gift of the Alfred Fairbank and Berthold Wolpe Book of Renaissance Handwriting in which he had affixed to the fly leaves 3 original specimens of Edward Johnston's handwriting, circa 1908. These were further enhanced by an example of the book hand of his famous pupil, Graily Hewitt, c 1912. This exchange led to a couple of long LD telephone conversations in which he informed me of his intention to take time out from his agency for a year or so of study in Europe with the idea of becoming a professor of palaeography. He did just that in 1962-1963, and it got him so involved in the forms of manuscript lettering & illumination that he became determined to try to understand their relationship to art forms and the mind of men of the Middle Ages. He took over 6,000 colored photographs in libraries & Universities from Ireland to Spain and all European countries in between. And he collected examples of the development of those styles, eventually working his way back to the 15th & 16th centuries and the development of the book hands & book illustration to the great northern Renaissance schools of Albrecht Dürer.

Decoration based on motifs used in the Book of Kells in the Ninth Century

MEANWHILE, his Graphic Arts & Advertising Agency had taken second place to such an extent that he gave it up in 1963. His lecturing at such schools as Dartmouth and Williams, The University of California and Scripps became important to him. In July, 1966 he accepted a position at The Colorado College in Colorado Springs, where he teaches Art History of the Middle Ages, has a laboratory to teach calligraphy and hand press printing, and is also Curator of Rare Books & Graphics.

HE WRITES: "I still use some of the old promotional methods that I learned to impart more drama to my teaching. After all, one can not push people into these studies, but just create a desire to be led. The arts of our civilization are so inter-related with interests in nearly any area (history, linguistics, social sciences) that it is not difficult to catch the fancy of young people today. The grace & beauty of handwritten forms are a perfect springboard into the study of art, history and creativity."

SO—it would seem that comparing MR. Lansburgh's use of lettering and studies of palæography & calligraphy and the success of his graphics in advertising with that of MR. X is not unrelated to comparing a person who is sensitive to the beauty around him and the callous outlook of another who is unable to see these graces of everyday life. Then comes the comparison of consideration & respect for one's fellow man versus the intolerance and unconcern of others.— So much of this is related to the inner person.

AT THE RISK of making too much ado about nothing, I shall append here a calligraphed quotation with poetic application to the differentiation of poles in opposition as a scribal exercise:

One doth not know how much an ill word may empoison liking.

Shakespeare: Much Ado About Nothing. Act iii, Scene I.

THE CALLIGRAPHIC WRITING & DESIGNING of this book have been done sporadically during a period of 3 years under various difficulties, chiefly failing eyesight. I had known for some time that cataracts were limiting my vision, slowing down my production & adversely affecting my penmanship. Early in 1967 the cataracts were removed. Then came the distortions & frustrations incidental to the wearing of cataract glasses & finally, contact lenses, through which I now can see, at least, the promise of vision that will be almost normal after I get used to wearing them. I am truly grateful for the marvelous accomplishments of optical surgery but, even so, must admit there is nothing that can quite take the place of the original "built-in" lenses one is born with.

I AM AWARE of the scribal imperfections that abound in these pages, and have often wished that the idea of adding PARTS II & III had occurred to me immediately after we had given the dialogue which is now PART I. That much of a "headstart" would have beaten the cataracts to the punch and made for a smoother & less delayed scribal performance. I do believe, however, that what others have written is evidence that legibility, individuality & good style can be had in Italic handwriting based on the earlier Chancery Cursive, albeit remotely in a few instances, but the relationship is there & that's what counts. We must not become moribund in copying the excrescences* along with the excellences of Renaissance writing, but we can reflect its humanistic spirit. If what I have added by way of explanatory notes & illustrations to my text throughout proves interesting and helpful to those wanting to improve their handwriting by conversion to modern Italic, I shall feel that my efforts have not been wasted.

What began in fun became something like having a tiger by the tail!

* ie: The clubbed ascenders & descenders of Jo: Geo: Fran: Cresci Scriptor circa 1558

COMMENTING recently on the subject matter of his excerpt (p105) Ralph Douglass writes: "I am neither ashamed nor proud of my past physical impairments. In fact, our testimony might even serve to encourage somebody else in similar difficulty." I heartily concur with him, and would also add that MR. Douglass is a versatile artist — a distinguished painter as well as calligrapher, familiar with the best that the past can offer and interested in extending the boundries of both areas of his art by experimentation. For it is only through intelligent research that one can know what to reject and what to accept for further development. That's a good attitude for students of lettering to take. Jenny Hunter has applied the same reasoning to both choreography & calligraphy.

JAMES H. ILLINGTON, professor of history at Princeton University, states in Life, May 4th, 1968, that "The humanistic heartbeat has failed... the American university has lost its soul amidst unprecedented materialistic growth... public relations with the outside world are more important than human relations within the university. . . marketability—not truth—has become the criterion of intellectual values." The professor makes an eloquent plea for a return to teaching of the humanities which have been neglected since World War II.

The Human Race lives . . . by Art & Reasonings · Aristotle

THE HUMANITIES include calligraphy, which is defined simply as "beautiful handwriting" but, while calligraphy is handwriting, not all handwriting is calligraphy. We leave it up to the reader to make his own judgments—as he will.

IT seems to be characteristic of calligraphers (as well as of other people who write in just too much of a hurry) to end their letters with words to this effect: "I hope you can read this". As a penitent sinner, I confess that I have used such words myself, and with good reason. John Benson and even Arrighi are among the apologists but, I believe, without sufficient reasons. I agree with Gilbert Hanna (p. 120) that Benson's translation of Arrighi's *La Operina* might have been more advantageously placed for quick comparison beside the original—as shown opposite.

READERS will find a note of modest abnegation in Mr. Benson's final word below, but I am charmed by his beautifully rendered translation without which I would not have comprehended what I now find to be the most graceful, gracious & humble apologia I have ever read.

The informal free flowing hand of John Howard Benson was, first of all, legible

J·H·B
1901 - 1956
Rest in Peace

Dear Ray — Thank you for your nice Xmas Card and for the kind words re Arrighi. I am glad that you have a rag edition. The trade will be out soon at $2.50. It means a lot to me that you like it. I just see all of the faults. Yrs.

John.

Reader, if you find aught
to offend
In this little Treatise of Vicenti=
no,
Do not be amazed, Because it is divi=
ne
& not human, to be quite without
fault.

No one can dwell here without
defect
Could any man stand free from
sin,
He would be like to God
who alone is perfect.

Lettor, se truoui cosa che
t'offenda
In questo Trattatel del Vicenti=
no,
Non te marauigliar, Perche diui=
no
& non humano, é quel, ch'é senza
menda.

Qui viuer non si puo senza
defetto
Che chi potesse star senza pec=
cato
Seria simil á Dio
ch'é sol perfetto

The greatest thing in the world is the alphabet
ABC
DEFGHIJ
KLM
as all wisdom is contained therein—except
the understanding
NOPQ
RSTUVW
of putting it together.
From an old German bookplate
XYZ&
RFD
19|69

165

Calligraphy is Handwriting

But not all Handwriting is Calligraphy

[THIS EXERCISE PURPORTS TO BE BOTH]

Broad pens made from bamboo reeds are most satisfactory for the larger sizes of display lettering. The gradual taper of the nib allows for any width of line by simply cutting back the wood—an advantage over steel pens that come only in a few standard sizes. See Edward Johnston's Writing & Illuminating & Lettering for instruction in cutting reed pens—or for any information abt calligraphic writing techniques.

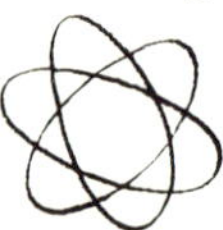

Illustrations may be made by the same method as that used for writing, ie: hold edge of nib at fixed angle while moving pen about in various directions—without manual pressure. Thus picture & text will be properly related.

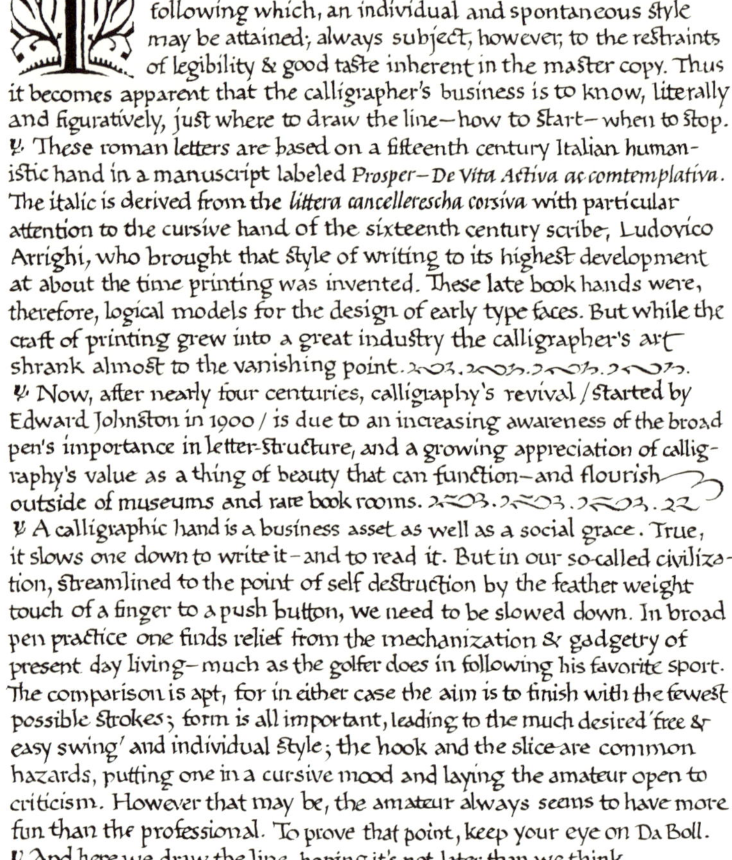

THE DEVELOPMENT of a calligraphic hand requires strict adherence to form and submission to discipline, following which, an individual and spontaneous style may be attained; always subject, however, to the restraints of legibility & good taste inherent in the master copy. Thus it becomes apparent that the calligrapher's business is to know, literally and figuratively, just where to draw the line—how to start—when to stop.

These roman letters are based on a fifteenth century Italian humanistic hand in a manuscript labeled *Prosper—De Vita Activa ac comtemplativa*. The italic is derived from the *littera cancellerescha corsiva* with particular attention to the cursive hand of the sixteenth century scribe, Ludovico Arrighi, who brought that style of writing to its highest development at about the time printing was invented. These late book hands were, therefore, logical models for the design of early type faces. But while the craft of printing grew into a great industry the calligrapher's art shrank almost to the vanishing point.

Now, after nearly four centuries, calligraphy's revival / started by Edward Johnston in 1900 / is due to an increasing awareness of the broad pen's importance in letter-structure, and a growing appreciation of calligraphy's value as a thing of beauty that can function—and flourish outside of museums and rare book rooms.

A calligraphic hand is a business asset as well as a social grace. True, it slows one down to write it—and to read it. But in our so-called civilization, streamlined to the point of self destruction by the feather weight touch of a finger to a push button, we need to be slowed down. In broad pen practice one finds relief from the mechanization & gadgetry of present day living—much as the golfer does in following his favorite sport. The comparison is apt, for in either case the aim is to finish with the fewest possible strokes; form is all important, leading to the much desired 'free & easy swing' and individual style; the hook and the slice are common hazards, putting one in a cursive mood and laying the amateur open to criticism. However that may be, the amateur always seems to have more fun than the professional. To prove that point, keep your eye on Da Boll.

And here we draw the line, hoping it's not later than we think.

RFD

"Birdies" of this sort, done with pointed pen under pressure, fall into category of penmanship known to broad pen devotees as

Spencerian Spinach

Addendum

A SELECTED BIBLIOGRAPHY

ANDERSON, DONALD M. *The Art of Written Forms—The Theory & Practice of Calligraphy.* N.Y.: Holt, Rinehart & Winston, Inc., 1969.

ARRIGHI, LUDOVICO. *Operina* (1522). *The First Writing Book.* English translation in facsimile. Text, introduction and notes by John Howard Benson. New Haven, Conn.: Yale University Press, 1954. (Now available in paperback.)

BENSON, JOHN H. and A.G. CAREY. *Elements of Lettering.* N.Y.: McGraw-Hill, Inc., 1952.

Catalogues of three important Calligraphic Exhibitions held in America

CALLIGRAPHY— *The Golden Age and the Modern Revival.* Portland, Oregon: Portland Art Museum, 1958.

CALLIGRAPHY *& Handwriting in America 1750-1962.* Compiled by P.W. Filby for the Peabody Institute Library, Baltimore, Maryland. Grand Island, N.Y.: Italimuse, Inc., 1962.

2,000 *Years of Calligraphy.* Compiled by P.W. Filby for Walters Art Gallery + Peabody Institute Library + Baltimore Museum of Art. 1965.

CATICH, FR. ED M. *Letters Redrawn from the Trajan Inscription in Rome.* Davenport, Iowa: The Catfish Press, St. Ambrose College, 1966.

CHAPPELL, WARREN. *The Anatomy of Lettering.* N.Y.: Loring & Mussey, 1935.

CHILD, HEATHER. *Calligraphy Today.* N.Y.: Watson-Guptill Publications, 1964.

DOUGLASS, RALPH. *Calligraphic Lettering With Wide Pen & Brush.* N.Y.: Watson-Guptill Publications, 1967.

EAGER, FRED. *Guide to Italic Handwriting.* Grand Island, N.Y.: Italimuse, Inc., 1967.

FAIRBANK & WOLPE. *Renaissance Handwriting.* A monumental work. Cleveland, O.: The World Publishing Company, 1960.

GOURDIE, TOM. *Italic Handwriting—A Simple Modern Style.* London: Studio Books, 1963.

HAYES, JAMES. *The Roman Letter.* Chicago: The Lakeside Press, 1951.

JOHNSTON, EDWARD. *Writing, Illuminating & Lettering.* N.Y.: Isaac Pitman, 1906. Johnston started the scribal revival; his book is known as "the calligraphers' Bible". It has had many reprints.

KOCH, RUDOLF + WOLPE + KREDEL + EICHENAUER. *Das ABC Büchlein.* Alphabets. LEIPZIG, 1934.

NESBITT, ALEXANDER. *The History & Technique of Lettering.* N.Y.: Dover Publications, 1959.

OGG, OSCAR. *The 26 Letters* (7th printing). N.Y.: Thomas Y. Crowell, 1948–1961.

REYNOLDS, LLOYD. *Italic Lettering & Handwriting Exercise Book.* Portland, Ore.: Champoeg Press, 1957.

STANDARD, PAUL. *Calligraphy's Flowering, Decay & Restauration, with Hints for Its Wider Use Today.* Chicago: The Society of Typographic Arts, 1947.

ZAPF, HERMANN. *Feder und Stichel* (*Pen and Graver*). Frankfurt, Germany: Stempel, 1950. A landmark in modern calligraphy and fine press work, with an introduction by Paul Standard in English set in Zapf's Palatino type face.

Teachers of Italic Handwriting

ARKANSAS

Don Marr, Route 4, Conway 72032 – Art Department, Hendrix College

CALIFORNIA

David Green, 176 West Jaxine Drive, Altadena 91001-Otis Art Institute 794-4927

Egdon H. Margo, 4961 Murietta Ave, Sherman Oaks 91403 — 784·3725 · Instructor UCLA

Maury Nemoy, 11558 Kling St., North Hollywood 91602 — 761-3596 UCLA Extension

Theodor Jung, 432 College Avenue, Palo Alto 94306 – 328-1867

Mr. Jean Pratt, 4229 Walnut Avenue, Carmichael 95608 – American River Junior College, Sacramento

CONNECTICUT

H. Edward Oliver, Blue Ridge Road, Wilton 06897. Stamford Museum Private lessons or correspondence 762-3927

ILLINOIS

A. Norman, 1442 North Sedgwick, Chicago 60610

MINNESOTA

Mrs. Roy Justus, 19 South First Street, The Towers B·701, Minneapolis 55401 Minneapolis School of Art

Mrs. Ione Bell, 802 Fourth Avenue, N.W., Austin 55912

MISSOURI

Lawrence Schwarz, 6821 Crest Avenue, University City 63130 (St. Louis)

NEW YORK

Paul Standard, 445 East 65th Street, New York 10021 – 734·8465. Private Instruction

Fred Eager, Grand Island 14072 – 773·2255

OREGON

Lloyd J. Reynolds, Portland Museum of Art School. Call Registrar, Art School

Father John Domin, 3060 S.E. Stark Street, Portland 97214

Jacqueline Svaren, 11182 S.E. Tyler Road, Portland 97266 — Portland Community College and Portland State College. Call Registrar, either college, day or night classes

Shirley G. Orbeck, Portland State College. Register through the college office

Edna M. Boggs, Adult Education, Milwaukie High School, Milwaukie 97222

Sister Loyola Mary, Marylhurst College, Marylhurst 97036 — 636-8141

Sister Grace Taylor, Mt. Angel College, Mt. Angel 97362

PENNSYLVANIA

Arnold Bank, 3 Roselawn Terrace, Pittsburgh 15213 — College of Fine Arts, Carnegie-Mellon University

WISCONSIN

Mrs. June T. Morris, 4128 North Larkin Street, Milwaukee 53211 – 414·962-5623

Note: Most of these teachers are willing to conduct additional classes when there is a demand for such instruction.

INDEX to Those whose Calligraphy appears in this Volume

ENVOI

AFTER calligraphing this book, I understand why Arrighi collaborated with an engraver to produce printing types, and why he later became a printer himself. Surely it was not that he loved the printed more than the written word, but because he wanted a quicker method of multiplying instructions for the benefit of those who wished to write beautifully or, at least, legibly.

FOUR CENTURIES after Arrighi's LA OPERINA, hand compositors could put one letter per second in the stick. A calligrapher might write one in even less time, but the problem of multiplication canceled out his advantages of speed & beauty.

TODAY, electronic typesetting systems capable of painting 10,000 characters per second on the screen of a cathode ray tube are foreseen, promising a revolutionary effect on the printing industry—in a world already plagued with revolution as it gropes for a way out of war.

MODERN INDUSTRY can't get along without business machines, computers & all such marvelous time & labor saving inventions, but from the viewpoint of this simple Ozarkalligrapher, a screaming headline: "It is now vital that American businessmen forget how to write," * which appeared recently in a doublespread ad for IBM, goes a bit too far. Granted, the PR man deliberately overstated his point to help sell his client's merchandise. Granted, I may belabor my point, but I see present-day all-out mechanization as a dehumanizing colossus, whether it affects literacy in the case of IBM or the destruction of millions of human beings via ICBM.

* If they haven't already forgotten, I !

ONE REMEDY, I believe, would be more attention paid to the humanities in education. Even the young school child§ trained to write legibly in modern adaptations of the fine humanistic hands that flowered in the Renaissance, will be started toward a better-balanced life in our speed-ridden age—though I must confess that a qualified handwriting expert once rated me as "mentally disturbed" on the basis of a sample of my own frantically hurried longhand scribbling. Another capable judge—thank heaven!—saved the day for me with "tower of imperturbability" on the basis of my more calmly deliberate calligraphy. Perhaps I may average out as a mean of the 2 extremes—for sometimes I do get perturbed, at least, inside—I do.

§ And that's the time & the place to start.

TYPOPHILE & Calligraphile,—I have worked for over 50 years in the graphic arts as "cub & trained seal", art director & freelancer while learning my ABC's, loving all 26 letters of the alphabet in their infinite variety—and the 10 digits that go with them. I've had this book in mind for several years and am greatly relieved & happy, now that it's finished, regretting only that I didn't start it earlier when my hand was steadier & my eyesight sharper. That so many people have been so patient, generous & helpful during all the vicissitudes of delayed publication this book has suffered goes far beyond my understanding!

FRED

250 copies of this 3-PART memoir,
designed & calligraphed by R.F. Da Boll, Newark, Ark.,
have been reproduced by Casco Print, Portland, Me.,
and specially bound & slipcased at The New Hampshire Bindery, Concord,
New Hampshire.

* * * *

A trade edition of 3,750 copies has been prepared by
the same craftsmen. Both editions printed
on Warren's OldeStyle, substance 70.
Walter R. Prindall, offset photographer
Harleigh P. Clukey, offset flat preparation
John R. Daley, offset presswork
all working under the supervision of
Wendell P. Sargent.

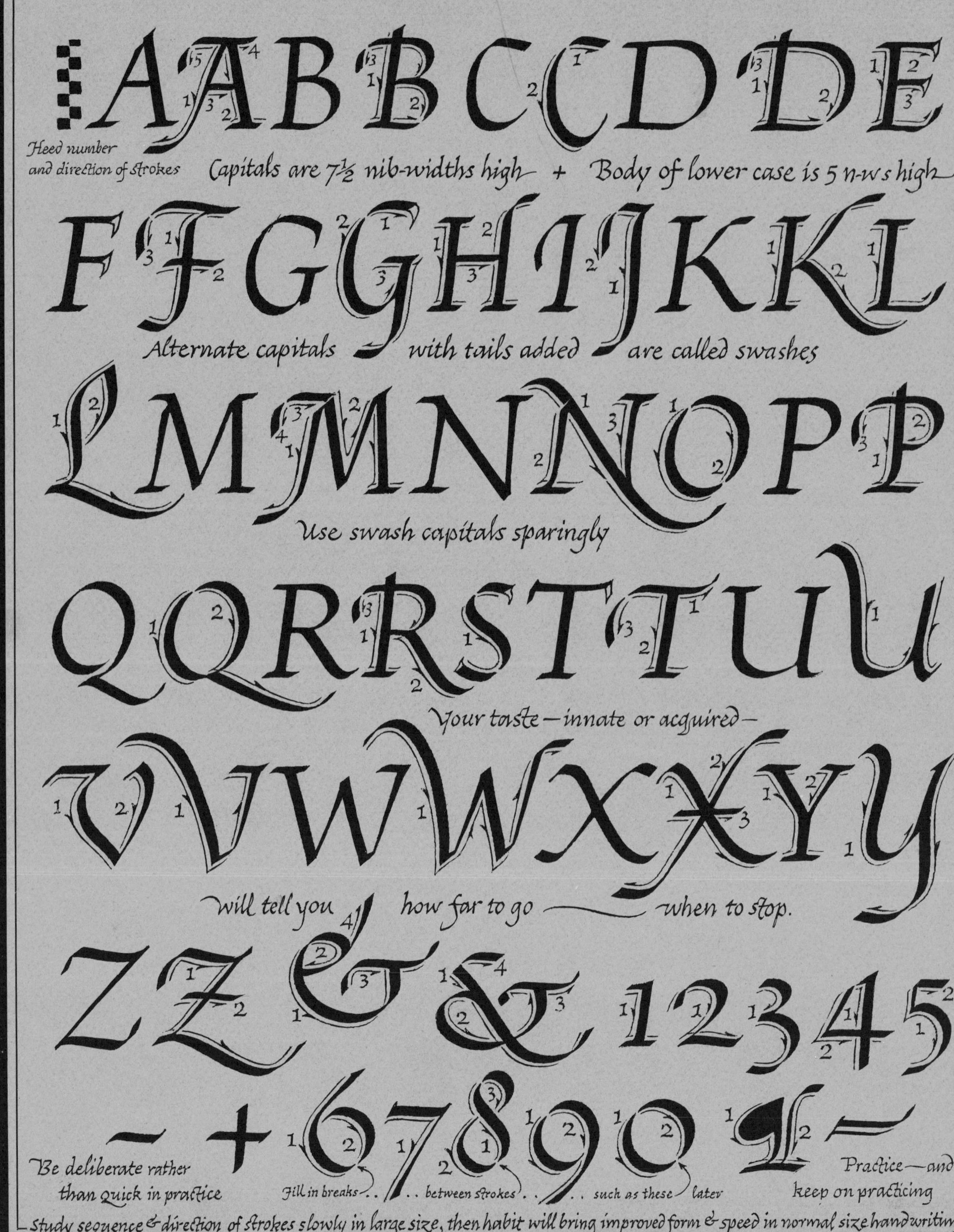
Heed number and direction of strokes
Capitals are 7½ nib-widths high + Body of lower case is 5 n-w s high
Alternate capitals with tails added are called swashes
Use swash capitals sparingly
Your taste—innate or acquired—
will tell you how far to go when to stop.
Be deliberate rather than quick in practice
Fill in breaks . . . between strokes . . . such as these later
Practice—and keep on practicing
Study sequence & direction of strokes slowly in large size, then habit will bring improved form & speed in normal size handwriting